"This book gives back to contemporary psychoanalysis the pleasure of exploring really little-known territories, fascinatingly restoring the connection between the past, present and '*elsewhere*' of communications between human beings, using the Freudian experience as its starting point, in order to reconsider in a reflective way the less visible, sometimes disorienting and mysterious levels of psychoanalytic practice. It offers us an especially valuable reflection on the mysterious communicating paths which put individual and group unconsciouses in contact with each other, often bypassing in an apparently disconcerting manner the border controls."

– **Stefano Bolognini**, *past president of the IPA*
and the Italian Psychoanalytic Society

"Following the thread of thought-transference, Maria Pierri goes through the events of the Freudian endeavour starting from its roots in hypnosis and occultism, through the dialogue with the masters, the pupils and the great female patients, the leading actresses of the cure. In his disquieting curiosity for telepathy, which he shared intimately with Ferenczi, Freud discovers that fortune-tellers, who do not know the future, can read the unconscious of their clients. But the 'golden coin' of occultism, the generative mother-child communication, will be the great discovery of Ferenczi."

– **Luis J. Martin Cabré**, *training analyst, past president,*
Madrid Psychoanalytical Association

"Today we know much about the polyphonic complex of contexts, experiences, relationships and ideas which made psychoanalysis possible and still nourish its current debates. We can be very grateful to Maria Pierri for bringing us up to date with the role and meaning of some little-known aspects of Freud's life and work concerning occultism and the fascinating dialogue of the unconsciouses developed with Ferenczi: what the Author identifies as one of the matrices of the developments of contemporary psychoanalysis."

– **Marco Conci**, *MC, IPA Committee on the History of Psychoanalysis*

Occultism and the Origins of Psychoanalysis

Occultism and the Origins of Psychoanalysis traces the origins of key psychoanalytic ideas back to their roots in hypnosis and the occult.

Maria Pierri follows Freud's early interest in "thought-transmission," now known as telepathy. Freud's private investigations led to discussions with other leading figures like Carl Jung and Sándor Ferenczi, with whom he held a "dialogue of the unconsciouses." Freud's and Ferenczi's work assessed how fortune tellers could read the past from a client, inspiring their investigations into countertransference, the analytic relationship, unconscious communication, and mother-infant relationality. Both Freud and Ferenczi tried in different ways to come close to understanding the infant's occult link with the mother and their secret primal language: their research on thought transference may be identified as a matrix of the developments of current psychoanalysis. Pierri clearly links modern psychoanalytic practice with Freud's interests in the occult using primary sources, some of which have never previously been published in English.

Occultism and the Origins of Psychoanalysis will be of great interest to psychoanalysts in practice and in training, as well as academics and scholars of Freudian ideas, psychoanalytic theory, and the history of psychology, and the occult. It is complemented by *Sigmund Freud and The Forsyth Case: Coincidences and Thought-Transmission in Psychoanalysis*.

Maria Pierri is a psychiatrist and child neuropsychiatrist, formerly researcher and adjunct professor at the Psychiatric Clinic, Medical School, University of Padua. She is a training analyst of the Italian Psychoanalytical Society and International Psychoanalytical Association and member of the editorial board of the *Rivista di Psicoanalisi*.

History of Psychoanalysis
Series Editor: Peter L. Rudnytsky

This series seeks to present outstanding new books that illuminate any aspect of the history of psychoanalysis from its earliest days to the present, and to reintroduce classic texts to contemporary readers.

Other titles in the series:

A Forgotten Freudian
The Passion of Karl Stern
Daniel Burston

The Skin-Ego
A New Translation by Naomi Segal
Didier Anzieu

Karl Abraham
Life and Work, a Biography
Anna Bentinck van Schoonheten

The Freudian Orient
Early Psychoanalysis, Anti-Semitic Challenge, and the Vicissitudes of Orientalist Discourse
Frank F. Scherer

Occultism and the Origins of Psychoanalysis
Freud, Ferenczi and the Challenge of Thought Transference
Maria Pierri, Translated by Adam Elgar

Sigmund Freud and the Forsyth Case
Coincidences and Thought-Transmission in Psychoanalysis
Maria Pierri, Translated by Adam Elgar

For further information about this series please visit https://www.routledge.com/The-History-of-Psychoanalysis-Series/book-series/KARNHIPSY

Occultism and the Origins of Psychoanalysis

Freud, Ferenczi and the Challenge of Thought Transference

Maria Pierri

Translated by Adam Elgar

Routledge
Taylor & Francis Group

LONDON AND NEW YORK

Cover image: © Getty images

First published in English 2023
by Routledge
4 Park Square, Milton Park, Abingdon, Oxon OX14 4RN

and by Routledge
605 Third Avenue, New York, NY 10158

Routledge is an imprint of the Taylor & Francis Group, an informa business

© 2023 Maria Pierri

Translated by Adam Elgar

Published in Italian as Un enigma per il dottor Freud. La sfida della
telepatia by FrancoAngeli 2018

British Library Cataloguing-in-Publication Data
A catalogue record for this book is available from the British Library

ISBN: 978-1-032-15953-9 (hbk)
ISBN: 978-1-032-15955-3 (pbk)
ISBN: 978-1-003-24647-3 (ebk)

DOI: 10.4324/9781003246473

Typeset in Times New Roman
by Apex CoVantage, LLC

Contents

Introduction   xi
STEFANO BOLOGNINI

**Prologue: *a result of character*: the cocaine,
this magical substance**   1

1 Vienna, *Porta Orientis* of the unconscious   5

The force of suggestion: the "wonderful somnambulists" 5
Hypnosis 7
Vienna, laboratory of modernity 10

2 The young Freud   13

A passionate young researcher into nature 13
First love 16
Martha and Bertha: the languages of passion 18

3 The lesson of Jean Martin Charcot   25

At the Salpêtrière 25
The apparatus of language 28
The magic of words 29

4 The lesson of Josef Breuer and the "descent to the mothers"   34

Studies on hysteria 34
A difficult separation: not all debts can be paid 37
A foundation myth: a false pregnancy and a cure with a defect 42

5 Sigmund Freud's lesson   49

The discovery of a false connection 49
Irma's throat and the feminine at the origin of psychoanalysis 51
Dream as desire 56

6 Fliess and the invention of psychoanalysis   61

A secret correspondence 61
My friend in Berlin 64
Freud's heart trouble 67

7 The discovery of infantile sexuality   71

Self-analysis and the writing cure *71*
Cherchez la femme: *the case of Emma Eckstein 76*

8 Original thought requires a rupture   82

The "reader of thoughts" 82
The accusation of plagiarism 85
A future in the image of the past; predestination and superstition 86

9 Occultism made in the USA   92

Spiritualism 92
Medium, *media, and "mental telegraphy" 97*
First hypotheses about the unconscious 103

**10 Jung, spiritualism and countertransference: the world
      of the dead**   108

Jung, poltergeist *phenomena, and séances 108*
The arrival at Burghölzli *111*
First visit to Vienna 113
Easter 1909: Jung's spiritual complex and Sabina 116
The dangerous fascination of the "beautiful Jewess" 119

11 Ferenczi, the unclassifiable   127

The sultan and his "clairvoyant" 127
A psychoanalyst "of a restless mind" 129
Ferenczi and the hidden treasure of Spiritualism 131
The encounter with Freud: a postponed transferential appointment 133

12 A journey to America   137

*Three men and an eventful, mutually analytic crossing:
    the outward journey . . . 137*
. . . and back again 143

13   The Danaan gift   153

The clairvoyant who reads Ferenczi's mind 153
The patient who reads Ferenczi's mind 158
The Palermo incident, or the interpretation of
   paranoia 162
The psychic work of the clairvoyant: two unfulfilled
   prophecies 168

14   An epistolary novel   175

Ferenczi and incestuous countertransferential storms: from
   mother to daughter 175
What is still missing is the fatherly blessing. Fatefulness and
   Oedipal coincidences 179
Elma Pàlos, fragment of the analysis of a seduction 183
The open wound in Ferenczi's heart, a source of
   creativity 185

15   The Saturday goy: getting to know Dr Jones   188

The Welsh liar *188*
Difficult beginnings 190
Freud's first pupil from Britain 194
Dr Jones's stethoscope: rationalisation and censorship
   of excess countertransference 196
A prescribed training analysis in Budapest 199

16   The intergenerational transmission of psychoanalysis   204

Love and death: the three women of the three pupils 204
"If you go to women, don't forget the whip" 210
At school with Freud: the transmission of
   psychoanalysis 212

17   The secret committee   217

The transformations and the desertion of Jung 217
A missed meeting: the "Kreuzlingen gesture" 220
The committee: the Männerbund *and the defence of the*
   *"Cause" (*Die Sache*) 223*
Totem and taboo: unconscious intelligence and intergenerational
   transmission of thought 225

18   1913 – the year before the war   230

The last congress with Jung 230
A black tide of occultism *232*
The question of telepathy 235
The dialogues of the unconscious 238

Epilogue: a fortune-teller visits Freud in Berggasse   244
Correspondence   248
Bibliography   250
Index   265

Introduction

Maria Pierri: *Un Enigma per il dottor Freud. La sfida della telepatia*[1]

Stefano Bolognini[2]

I am quite sure that in these notes, I am introducing a book about psychoanalysis that is profoundly different from all the others that have passed through my hands in the past few years (frequently for the purpose of reviewing or writing a preface).

An Enigma for Dr Freud can be read as an engrossing detective novel of historical reconstruction as it unfolds towards its resolution, elegantly written and full of surprises, and opening up unexpected new fields of thought. Which is not to say that it actually is a novel: rather, it is real History with a capital H, precisely documented (including some fascinating unpublished material discovered by the author herself in the Freud Archives) and with a precise investigative method which revisits one of the most mysterious and controversial areas of Freudian research, that of telepathy.

This book may also be considered a timely addition to historical-scientific knowledge about a highly sensitive and thorny taboo in the analytic institutional sphere: the transmission of the authority to manage and safeguard psychoanalysis (understood in a broad sense as a theoretical, clinical, and organisational corpus) from Sigmund Freud to his daughter Anna – an ordination that was both intimate and foundational: in a sense, as paradoxically *sacred* as an initiation ritual – through personal analysis carried out by her own father.

This book also presents a kaleidoscopic psycho-political portrait of the circle of pioneers who accompanied Freud in those adventurous years when the mill of his genius was turning inexorably and, in his startling writings, bringing about a collective change no less potent than those caused by technological progress and political revolutions across the world, and in the psychoanalytic sphere, through the reconstruction of the personal connections between those protagonists and their conflicting collegial ties, previously unimagined interlinkings emerge between these men and their ideas.

Some blind spots and aporias in our collective knowledge of a part of Freud's work are dramatically highlighted, as is the institutional censorship imposed by some of those authoritative followers who were so fearful of diminishing the prestige and power of the psychoanalytic organisation, wishing to preserve a massive, uncontaminated idealisation of it and going so far as to keep some documents hidden, which they did successfully for many decades. But *An Enigma for Dr Freud* is neither a novel (despite having a novel's passion) nor a historical study, although it is rigorously documented. Rather, it is an original and scientific text in which the author, a passionate and recognised connoisseur of "coincidences," offers us an especially valuable reflection on the mysterious communicating paths which put individual and group unconsciouses in contact with each other, often bypassing in an apparently disconcerting manner the border controls and checking apparatus of individuals' conscious Ego. It's a bit like a scanner revealing deep channels and secret passages which connect different buildings, showing cellars and plumbing that aren't visible from outside. As the polysemic historical reconstruction proceeds, the "coincidences" that are presented, seemingly random at first, reveal the strands of an incredibly rich network of multiple, complex unconscious perceptions which confer meaning and sense, page after page, on cultural events and human experiences of extraordinary profundity.

My personal impression is that this book gives back to contemporary psychoanalysis the pleasure of exploring really little-known territories, fascinatingly restoring the connection between the past, present, and "*elsewhere*" of communications between human beings, using the Freudian experience as its starting point, in order to reconsider in a reflective way the less visible, sometimes disorienting and mysterious levels of psychoanalytic practice.

Notes

1 2018, Franco Angeli, Milan: the original Italian edition of Pierri M. (2022) *Occultism and the Origins of Psychoanalysis* and Pierri M. (2022) *Freud and The Forsyth Case*, Routledge, London.
2 Physician and psychiatrist, past president of the IPA and the Italian Psychoanalytic Society, honorary member of the New York Contemporary Freudian Society (CFS) and the Los Angeles Institute and Society for Psychoanalytic Studies (LAISPS). He is editor of the IPA *Encyclopedic Dictionary of Psychoanalysis* and is the author of many psychoanalytic works which have been translated into various languages to international acclaim, including *Psychoanalytic Empathy* published in English by Free Association Books (2004) and *Secret Passages: the Theory and Technique of Interpsychic Relations* published in English by Routledge (2010).

Prologue

A *result of character*: the cocaine, this magical substance

The discovery of the non-conscious dimension of the psyche was in the air at the end of the nineteenth century, and Freud's particular capacity for receptiveness meant he was ideally placed to accommodate the ideas he was picking up and make them grow in a fertile soil. He had a real nose for catching what was essential and, in his research, even after a limited number of observations, he instinctively identified connections that were likely to find confirmation. This strength, "though sometimes also his weakness," was based on "the quite extraordinary respect he had for the singular fact," for the "single observation" by which he allowed himself to be fascinated.[1]

Freud's initial problem was how to manage a "melting pot" of research projects in a vast range of fields. On more than one occasion, he found himself close to achieving success, but it turned out that others were following the same path and bringing his intuitions to completion. Freud had introduced a new gold chloride method of staining nervous tissue and had studied the course of nerve fibres in the medulla of embryos and children, contributing to the construction of concepts under development such as the neurone theory,[2] which later made Waldeyer (1891) famous, and the theory of evolution.[3] This also happened in the sensational case of the discovery that cocaine could be used as a local anaesthetic, as Freud had been the first to speculate (1884) and which his colleagues L Königstein and K Koller did not hesitate to use in ophthalmology, gaining the credit and the fame, followed in turn by the American W S Halsted in 1885.[4]

That risky and emblematic episode, the subject of repeated working-through in his self-analysis, is well suited to highlighting contrasting elements of Freud's personality that had not yet achieved integration.

In the spring of 1884, when he was still in his 20s, he wrote to his fiancée,

> I am also toying now with a project and a hope which I will tell you about; perhaps nothing will come of this, either. It is a therapeutic experiment. I have been reading about cocaine, the effective ingredient of coca leaves, which some Indian tribes chew in order to make themselves resistant to privation and fatigue. . . . I am certainly going to try it and, as you know, if one tries something often enough and goes on wanting it, one day it may succeed.

DOI: 10.4324/9781003246473-1

We need no more than one stroke of luck of this kind to consider setting up house. But, my little woman, do not be too convinced that it will come off this time. As you know, an explorer's temperament requires two basic qualities: optimism in attempt, criticism in work.

(21/04/1884)

At some expense, Freud was able to obtain the alkaloid from its German producer, Merck, and did not hesitate to experiment with the substance on himself (following a similar test, Halsted would develop an immediate addiction). After experiencing its multiple effects – stimulant, anti-dyspeptic, aphrodisiac, anaesthetic, but also its efficacy against seasickness, vertigo, and loss of appetite – Freud prescribed it to neurotic patients and boldly launched it as a remedy curing addiction to alcohol and morphine, unfortunately recommending it to his colleague E Fleischl, who tragically became a slave to the new substance. Within a few months, on the basis of his experiment, Freud was able to write a monograph on cocaine (1884).

It was one of the first psychopharmacological texts and the most complete treatment of the substance, then unknown to science, though present since 1863 in *Vin Mariani à la Coca du Pérou*, a nerve tonic patented by a Corsican chemist. Only a few years later, thanks to its cocaine content, an extraordinary international success was achieved by the drink invented by the pharmacist J S Pemberton, who, on the introduction of a prohibitionist law, modified the formula of his Pemberton's French Wine Coca, creating the Coca-Cola elixir and syrup.

It was much later that the devastating effects of cocaine were recognised: Coca-Cola, sold as a tonic and antioxidant, contained traces of the alkaloid until 1905.

In the meantime, Freud began to use it without manifesting symptoms of real dependency and, not content with this, sent a quantity to his fiancée as well as recommending it to his sisters. He really didn't hold back, and in June wrote enthusiastically to Martha:

Woe to you, my Princess, when I come. I will kiss you quite red and feed you till you are plump. And if you are forward you shall see who is the stronger, a gentle little girl who doesn't eat enough or a big wild man who has cocaine in his body. In my last severe depression I took coca again and a small dose lifted me to the heights in a wonderful fashion. I am just now busy collecting the literature for a song of praise to this magical substance.

(2/06/1884)[5]

Ernest Jones is not mistaken when, in spite of his veneration for the Master, he claims that Freud was rapidly becoming a public menace.

There are those who believe that cocaine research played an unrecognised role in the birth of psychoanalysis both because these experiments, which placed his person at the centre of observation, anticipated the next method and because the need to free himself from the addiction could have prompted Freud to look for stimulants in the work of research and creative production[6] and to adopt the rigid routine that was imposed around 1896, always getting up before 7 am, working all day with patients and writing late into the night.[7] It certainly accentuated the addiction to cigars.[8]

As the cocaine episode highlights, essential components of Freud's genius were the great freedom and autonomy of his mind together with an opposite tendency to credulity, something similar to what Coleridge describes as "that willing *suspension of disbelief* for the moment, which constitutes poetic faith."[9]

Freud was willing "to believe in the improbable and the unexpected" and such surprising – to Jones – unruliness "enabled him dauntlessly to face the unknown and thus to open up fields of knowledge that had remained closed to more judicial but pedestrian explorers." Jones (with whom we are indirectly starting to become acquainted) reveals how scandalised he is when he has to mention that the Master "had for years absorbed his friend Fliess's amazing numerological phantasies" and even more when he deplores Freud's "credulous acceptance of his patients' stories of paternal seduction," which made him convinced early on of the ubiquitous reality of childhood abuse in the aetiology of the neuroses:[10]

> Most investigators would have simply disbelieved the patients' stories on the ground of their inherent improbability – at least on such a large scale – and have dismissed the matter as one more example of the untrustworthiness of hysterics.
>
> (Jones, *II*, p. 478)

The British disciple confesses how he came to appreciate this ingenuousness of Freud's only after a conversation with James Strachey during their collaboration on the *Standard Edition*.[11] What Jones considered a defect – and Strachey considered a great stroke of luck for psychoanalysis – was Freud's distinctive heuristic tool: putting himself in play, letting himself be influenced like a child and able to risk an error in order to learn from it in a masterly manner.

This was exactly how he came to conceive his revolutionary system of thought: "it was less the result of intellect than of character" (Freud to S Zweig, 7/02/1931).

The scientific circumstances which led to the discovery of psychoanalysis are situated in the encounter between the most advanced and refined histological and neuroanatomical experimental research of the time – which Freud embodied in his training, experience, and talent – and the clinical application of hypnosis to hysteria. This last practice was still crude, rooted in the methods of enchantment and magic, only separated by a century from exorcism, and barely tolerated by the scientific world. Freud succeeded in achieving a creative, indeed personal, synthesis between the most progressive and the most primitive, popular elements of the culture of his time, deeply in accord with Nietzsche's insights:

> Do you believe then that the sciences would have arisen and grown up if the sorcerers, alchemists, astrologers and witches had not been their forerunners; those who, with their promising and foreshadowings, had first to create a thirst, a hunger and a taste for *hidden and forbidden* powers?
>
> (1882, pp. 233–234)

As a scientist, he did not hesitate to engage with and draw on emotional areas which were hard to represent: the primitive and barely manageable affectivity of suggestion, mental contagion, and imitation, which troubled and deeply fascinated him.

Freud later claimed that great decisions in the realm of thought and momentous discoveries and solutions of problems are only possible to an individual working in solitude. But soon after, he would add:

> It remains an open question, moreover, how much the individual thinker or writer owes to the stimulation of the group in which he lives, and whether he does more than perfect a mental work in which the others have had a simultaneous share.
>
> (1921, p. 83)

Notes

1 Jones, I, p. 106.
2 As an essential morphological and physiological unit of the nerve cells and fibrils.
3 Such as persistence of primitive structures in the organism of fully developed appearance.
4 See Jones, *I*.
5 Jones, *I*, p. 93.
6 Lavaggetto, 1985.
7 Markel, 2011.
8 Anzieu, 1959.
9 1817, chap. 14.
10 *II*, p. 478–79.
11 *The Standard Edition of the Complete Psychological Works of Sigmund Freud.*

Chapter 1

Vienna, *Porta Orientis* of the unconscious

The force of suggestion: the "wonderful somnambulists"

> This is that piece of magnet,
> Mesmeric stone
> Originating in Germany
> Which then gained such fame
> In France.
>> *(He touches the heads of the pretend invalids with a piece of the magnet and gently swipes their bodies from head to toe.)*
>
> In a few hours
> You will see
> Through the power of magnetism
> The end of this paroxysm
> And they'll be as they were.
>> [L Da Ponte, W A Mozart, *Così fan tutte* (*la scuola degli amanti*)
>> 26/01/1790, Burgtheater Vienna]

At a time when witch-trials were still not entirely a thing of the past, the confrontation between Father Johann J Gassner (1727–1779), who used exorcism to treat "preternatural" illnesses, the result of demonic possession, and the physician Franz A Mesmer (1734–1815), who, in the presence of the prince elector in Munich, demonstrated identical cures by means of "animal magnetism," opened the way to a method of psychic cure for hysteria that was no longer subject to religion but answered the expectations of an Enlightened era.[1]

Born in the principality of Constanța, Mesmer, a theologian and philosopher as well as a physician, graduated with a thesis on the influence of the planets on diseases and was made a member of the Bavarian Academy of Sciences. He had practised in Vienna (as a friend of the Mozart family, he put on Wolfgang Amadeus's first opera, *Bastien und Bastienne*, in his private theatre) and later, being compelled to move to Paris, at the court of Louis XVI.[2] "His physical appearance

DOI: 10.4324/9781003246473-2

alone made an impression at the first encounter for he was tall and of noble presence," had a "complacent and unshakable patience," and radiated health.[3] He treated his patients with the application of magnets and by means of a fluid which he thought was accumulating in their bodies. Lacking a concept of psychology, he considered this fluid a physical force equal to that exerted by an iron magnet and believed that it enabled him to induce a series of crises in the patient, leading to a cure.

Imported into France, "animal magnetism" also quickly took the form of mass healing events – the flow of the fluid was strengthened in these cases – and, supported by a sort of popular fanaticism, saw a brief period of glory, even establishing its own institute, the *Société de l' Harmonie*, "a strange mixture of business enterprise, private school, and masonic lodge."[4] This treatment, which was fashionable even among the nobility along with the passion for occultism, was poorly regarded by the scientific community since, although they recognised its effects, the cures were neither predictable nor lasting and seemed to be caused more by the power of imagination than by the fluid. In the secret report of the Franklin Commission's inquiry, appointed by the king, reference was made to the dangerous possibility of seduction and of an erotic, even if only "platonic," attachment developing between magnetiser and magnetised.[5]

One of Mesmer's disciples, the Marquis de Puységur, highlighted a primary psychological agent in the treatment – the will of the magnetiser – and discovered that it was possible to induce the phenomenon of magnetic sleep, or *artificial somnambulism*, in his patients: a particular state of consciousness, like a dream, which revealed the presence of a second personality, a much more malleable one endowed with unsuspected imitative abilities but also with heightened awareness. This occult wisdom even made the somnambulist capable of diagnosing their own illness, predicting its progress, and prescribing a cure and also led them to establish a deep and exclusive communion with the magnetiser, an emotional resonance and an intimacy comparable to fusion with him, like being a part of his body. From the beginning, it seemed evident that this extraordinary sensitivity of the magnetised to the magnetiser, which sometimes involved the perception of the other's bodily sensations and thoughts, had characteristics of "reciprocity"[6] and might have erotic-seductive components connected to it. This sexual element was a risk because of the (female) patient's affective subordination, and it was potentially dangerous for the honour of the magnetiser himself: for this reason, it was customary to take the precaution of never magnetising without witnesses. When, as was often the case, a rapport was established between the two, the sexual dynamics could become overt and lead to relations of mutual dependency and to abusive situations in the hands of libertine magnetisers but also to the formation of couplings authorised by marriage.[7]

The craze for somnambulism spread across Europe and was the subject of public demonstrations: from collaborating amateur couples putting on shows in fashionable salons, to the professionals of the philomagnetic societies who strode the stages of theatres commanding large audiences for demonstrations of thought transmission, anaesthesia and hyperaesthesia, tetanic catalepsies, clairvoyance, precognition, and ecstasies.

After the Revolution, public courses in lucid sleep began to be held in France, bringing fame to the Abbé Faria, who, by inducing a state of trance with the simple command, "Sleep!" highlighted the fundamental role played by the patient's suggestibility. In Germany, Romanticism privileged the "sympathetic" relationship between patient and magnetiser, comparing it to the unity of mother and foetus, and emphasised the magnetic phenomenon as the element which Mesmer had called the *sixth sense*: a visionary ability which enabled the perception of distant events, clairvoyance, and the knowledge of other languages. Here there was a flowering of extraordinary subjects to whom researchers devoted impassioned investigations. Friedericke Hauffe (1801–1829), studied by a follower of Mesmer, the poet and physician, Justinus Kerner (1786–1862), in company with philosophers and theologians, became one of the most famous somnambulists, the protagonist of *The Seeress of Prevorst*.[8] This woman had surprising vocal qualities, including a "sonorous and magnificent" form of speech that she called "the original language of mankind, forgotten since the time of Jacob."[9]

The imagination of writers was also fired by these clairvoyants and so-called dual personalities, and there was multitude of novels on the subject in the nineteenth century. The hidden personality offered glimpses of capacities superior to those of waking life but also gave rise to fears of an uncontrolled instinctuality, feeding fantasies of seductions and crimes committed under hypnosis. The most influential and celebrated of these works is R L Stevenson's story *The Strange Case of Dr Jekyll and Mr Hyde* (1886), which is narrated as the product of a feverish dream. The story vividly reflects the fascination then exerted by the double personality in the popular imagination and the anxieties being stirred up by the myths of science, which was becoming the new religion.

In the second half of the century, a renewed interest in the alterations of consciousness was aroused by the arrival in Europe of a strange imported product, spiritualism, or spiritism, a phenomenon that had spread throughout the United States like an epidemic. Spiritualism temporarily eclipsed interest in magnetism, and, by providing the opportunity for studying a particular aspect of the *second condition* – the mediumistic trance – it stimulated a new representation of the human mind and its pathological alterations.

Hypnosis

In 1841, a Manchester physician, James Braid (1795–1860), made magnetism acceptable in medical circles by constructing a physiological explanation of somnambulism as a state of exhaustion of the brain and the medulla. He invented the method of tiring the eyes with bright objects before induction and coined the term *neuro-hypnotism* ("nervous sleep": from the Greek, ὕπνος), which later became simply hypnosis. Only a few years before the introduction of general anaesthetic in 1874, some English surgeons were reporting operations carried out painlessly on patients in a state of hypnotic sleep.[10]

After a period of eclipse, towards the end of the century, hypnosis underwent further development in a neurological setting when two major schools achieved prominence using it as a therapy: one opened in Nancy as a collaboration between a country doctor, Ambroise-Auguste Liébault (1823–1904), and the university professor Hippolyte Bernheim (1840–1919), and one created in Paris, at the *Salpêtrière*, by the most famous neurologist of the time, Jean-Martin Charcot (1825–1893). In 1882, Charcot had succeeded in conferring full scientific respectability on hypnosis when he obtained its recognition by the *Académie des Sciences* (which had in the past refused to give such an acknowledgement to *magnetism*).

Charcot, who used hypnotic suggestion on hysteria, was convinced "that the share of dream-life in our waking state was much more than just *immense*,"[11] and he made clear the very close connection between hysterical states and the hypnotic alteration of consciousness (what he began to call "*double conscience*" or a "second condition"), demonstrating the possibility of causing regression by means of hypnosis but especially of inducing hysterical crises. In this way, elusive, enigmatic, and scandalous hysteria – understood for centuries as a "uterine" disease, the result of witchcraft or possession or, not least, pretence – gained respectable status as a disease of the nervous system. With the recognition of forms of male hysteria and of traumatic hysteria (railway accidents had caused the appearance of *railway-spine*),[12] Charcot contributed to the decline of those theories which connected hysteria to the female genital system and movements of the womb within the body – the very name, hysteria, derived from the Greek ὑστέρα – and continued to advocate the surgical removal of uterus, ovaries, and/or clitoris.[13] In fact, we can still read in Plato's *Timaeus* (360 BCE)

> what is called the matrix and womb in women, which is in them a living nature appetent of childbearing, when it is a long time fruitless beyond the due season, is distressed and sorely disturbed, and straying about in the body and cutting off the passages of the breath it impedes respiration and brings the sufferer into the extremest anguish and provokes all manner of diseases besides.[14]

Though Charcot considered hysteria a hereditary pathology and fundamentally degenerative in nature, he had nevertheless been able to highlight the precipitating psychic role of fixed ideas and suggestions in the production of symptoms.

While the progress of the Paris school justified interest in researching the psychology of patients, the Nancy school of hypnotism went still further and maintained an exclusively psychic interpretation of hysteria and the hypnotic phenomena. According to Bernheim, suggestibility and sensitivity to hypnosis were present to varying degrees in everyone, not only in hysterics, as the "aptitude to transform an idea into an act."[15] On this basis, he demonstrated that miracle cures were also attributable to autosuggestion. Having noted that post-hypnotic amnesia was not always complete and that, once awake, the subject could be helped to recall the "second condition," he took to applying suggestion directly in the waking state without passing through hypnotic sleep and called this mode of treatment "psychotherapeutics": literally, cure of the soul. For all its mystical colouring which revealed its origins in religion and magic, the new therapy began to aspire to the attributes of a science.

French hypnotic therapy spread to many European centres, laying the foundations for a "dynamic" psychiatry in opposition to "biological" psychiatry, which was committed to description and diagnostic classification and more interested in researching pathogenesis in terms of biochemical imbalances, lesions, or degenerations than inclined to believe in the possibility of treatment. In the German world, this second position was represented by the Munich school of Emil Kraepelin (1856–1926) – a disciple of the father of experimental psychology, Wilhelm Wundt (1832–1920) – who introduced the first rational classification of psychiatric nosology, making the first distinction between manic-depressive madness, paranoia, and *dementia praecox*.

At the University of Vienna, the biological outlook prevailed, maintained by Theodor Meynert (1835–1892), the era's greatest anatomist of the brain and one of Freud's teachers, who considered hypnosis a piece of charlatanry suitable for the less educated social classes.[16]

Freud was not of the same opinion. In his autobiography, he reports:

> While I was still a student I had attended a public exhibition given by Hansen the "magnetist", and had noticed that one of the subjects experimented upon had become deathly pale at the onset of cataleptic rigidity and had remained so as long as that condition lasted. This firmly convinced me of the genuineness of the phenomena of hypnosis.
>
> (1925a, p. 16)

It is certainly true that too many elements from occultism and spiritualism, including belief in telepathy, were mixed in with the practice of hypnosis. In France, Liébault was among the first to comment on the connection between telepathy and hypnosis. And Pierre Janet, who came to recognise the distinctively elective nature of the hypnotic *rapport*, after experimenting with the possibility of interrupting the somnambulant influence or transferring it to a third, would also practice "remote" hypnosis by telepathic means.

Credit for the introduction of hypnosis into German psychiatry goes to Auguste Forel (1848–1931), a student of Meynert and professor of psychiatry in Munich, who had initially been involved in anatomical research on the neurone. On becoming director of the *Burghölzli*, the psychiatric hospital of Zurich University, and drawing inspiration from Bernheim's theories, he began to apply hypnotic suggestion there in an entirely original manner: not only on the patients but also on the staff of the clinic, including the nurses (mimicking the selectivity of maternal sleep in relation to the newborn infant's crying so as to ensure that they woke in the night at selected stimuli coming from the inmates).[17] Under him, interest in therapy began to prevail at the *Burghölzli* because of his conviction that the constant personal relationship with the patient, guidance, and moral education were capable of influencing the course and prognosis of even the most severe pathologies.

In 1889 Forel, having left *Burghölzli* to devote himself to the social aspect of psychiatry (he would become president of the *International Order for Ethics and Culture*), published a short text on his experiences of hypnosis, and henceforth the Medical Society of Vienna branded him with the derogatory epithet of "*Forel the Southerner*."[18] Forel later occupied himself with studying the world of ants, which had always fascinated him, and wrote an important treatise on their social life, with acute observations on the neuronal control of instinctive behaviour common to insects and man.

It is Eugen Bleuler (1857–1939), the student who succeeded him as director of the hospital, who gave us one of the rare autobiographical descriptions of hypnotic induction.[19]

Vienna, laboratory of modernity

At the end of the nineteenth century, with the economic wellbeing brought by the second industrial revolution, the appearance of a new middle class, and the crisis of the patriarchal family, individuals were ready for the first time to imagine a personal identity distinct from, and even outside of, the family. The model of happiness constituted a project for individuals as opposed to a path ordained by custom, family, or birth.[20] It was this emerging desire, inseparable from the experience of sexuality, that Sigmund Freud had been able to recognise in the symptoms of hysterical young women at the end of the century, neglected protagonists of the complex emancipatory, individual, and social crisis of modernity.

The dynamic unconscious reflected the aspiration to a new form of privacy, domesticity, and intimacy. Psychoanalysis was a theory and practice of this new experience of a personal life.[21] It was voicing that attention to instinctual reality within the "subject," the bourgeois hero and heir of Romanticism who was manifesting himself in the sciences as in literature, philosophy, and artistic and musical productions. In his 1888 representation of the old Burgtheater, the painting which made him famous, Gustav Klimt chose to revolutionise the view of it by looking not onto the stage and the actors but out to the boxes and the stalls, where people were attending to their own inner thoughts in the private theatre of the mind.[22] Arthur Schnitzler, a Jewish doctor a little younger than Freud, put his characters' stream of consciousness at the centre of his stories about Austrian ladies and soldiers that are so like his colleague's case studies. But it was in *Studies on Hysteria* (1892–95), published by Josef Breuer and Sigmund Freud, that Fräulein Bertha Pappenheim, immortalised as "Anna O.," was able to reveal all the dramaturgies of her internal theatre and be listened to in a way that might interpret it!

At the age of 21, this fascinating young woman of uncommon intelligence, belonging to one of the richest families in the Viennese Jewish business community, had developed a serious hysterical symptomatology with a nervous cough, paralysis and muscular contractions, mutism, inhibitions, and states of psychic confusion. The illness had manifested itself when she was nursing her father, who had been suffering from tuberculosis and died a few months previously. Turning to Dr Breuer for treatment, the young woman had made a formidable contribution to the *talking cure*: speaking at times only in English, this is what she called her doctor's cathartic psychotherapy.

After this debut, Bertha Pappenheim (1859–1936), who was cured but never married, went on to achieve distinction as a journalist and writer and devoted her life to creating a movement for the emancipation of women and the protection of Jewish children.[23] Along with the first feminists, she was taking part in a change that would revolutionise the concept of woman's natural inferiority, reinterpret the image of virility, and overturn the whole universe of the couple, which was no longer tied solely to procreation.

The Freudian concepts of *repressed unconscious*, *infantile sexuality*, *psychic bisexuality*, and *transference* which inaugurate the new century will, for the first time, allow everyone's identity to be understood as the result of a complex, singular, and precarious psychic process. Once they have penetrated deep into the culture, they will contribute to the construction of the twentieth century's new protagonist: the individual who must confront the nature and strength of his instinctuality, with obscure motivations rooted in contingent relational events from early childhood, which come into play in his erotic choices, creativity, and forms of illness.[24]

With his conception of *libido*, accepting and transforming the legacy of the great psychologist that was Friedrich Nietzsche, Freud offered a new form and destiny to the *will to power*: his theorising of the oedipal conflict, locating incest and parricide at the dawn of individual life, offered a way to represent the tragic and vital components of life's very origins.

In the cultural climate of Freud's Vienna, the paintings of Schiele and Kokoschka, building on Klimt's work, will tangibly portray the anxiety about eros and death, and by revealing the violent nakedness of bodies, they will penetrate deep into them with harsh lines and dramatic coloration.

As W Kandinsky observes in "On the Problem of Form" (*The Blue Rider*):

> At the appointed time, necessities become ripe. That is the time when the Creative Spirit . . . causes a yearning, an inner urge. . . . From this moment on, consciously or unconsciously, the human being seeks to find a material form for the new value, which already lives within him in spiritual form.
>
> (1912, incipit)

We may wonder, together with Hugo von Hofmannsthal (1922), if Freudian theories could have come to light anywhere in the European-American cultural world, or if Vienna was their predestined place: Vienna, the capital of an Empire which, without passports or customs controls, had stretched all the way the Near East, the beating heart of a multi-ethnic culture containing more than 15 languages and five religions.

This glorious laboratory of modernity, the arts, sciences, eroticism, extraordinary crossroads for the journeys of Hitler and Trotsky, Tito and Stalin, where at the beginning of the twentieth century the destinies of the West were being played out,[25] "the place where possibilities meet and cross and blend and become one another,"[26] found itself almost inadvertently acting as the cradle of the new thought.

And when a thought, an idea, is historically essential, it does not occur within an epoch but itself makes that epoch (Spengler, 1919).

Notes

1 Ellenberger, 1970.
2 Ibid.
3 Zweig, 1931.
4 Ellenberger, 1970, p. 65.
5 Ibid., p. 152.
6 Ibid.
7 Gallini, 1983.
8 1829. Subtitle: *Being Revelations Concerning the Inner-Life of Man, and the Inter-Diffusion of a World of Spirits in the One We Inhabit.*
9 Ellenberger, 1970, p. 80.
10 Ellenberger, 1970, p. 95.
11 Ibid., p. 92.
12 Traumatic neurosis caused by railway accidents.
13 Bonomi, 2007.
14 Archer-Hind, 1888, p. 341.
15 Ellenberger, 1970, p. 87.
16 His indignation was based on a personal problem, and on his death-bed, he would confess to Freud that he had always suffered from hysteria (Freud, 1900, p. 438).
17 Ellenberger, 1970, p. 88.
18 Freud, 1889, p. 91.
19 Marinelli and Mayer, 2002.
20 Zaretsky, 2004.
21 Ibid.
22 Kandel, 2012.
23 She was a great innovator in childcare and heroically succeeded in rescuing over a hundred Jewish orphans from Russia. The Residential Nursery and Social Work School which she founded in Frankfurt is now a museum (Britton, 2003).
24 Zaretsky, 2004.
25 Illies, 2013.
26 Cacciari, 1980.

Chapter 2

The young Freud

A passionate young researcher into nature

In 1873, at the age of 17, Sigmund Freud graduated *summa cum laude* from the Leopoldstadt Gymnasium in Vienna. Because of his family's poverty, he had from the start set aside any dreams of a poetic or literary career. Thanks to his friendship with his teacher of religion, Samuel Hammerschlag, he was widely known in the city's cultivated Jewish community, and with financial support from Hammerschlag, he decided to enrol in the Faculty of Medicine after hearing Goethe's essay on nature read at a public lecture.[1]

In that essay, attributed to J W Goethe but written by G C Tobler, after repeated conversations with Goethe, we can read these suggestive phrases for our theme of the transference of thought:

> She [Nature] has neither language nor discourse; but she creates tongues and hearts, by which she feels and speaks. . . . Everyone sees her in his own fashion. She hides under a thousand names and phrases, and is always the same.
>
> (1783)[2]

Despite his therapeutic calling, he was in fact "too soft-hearted" for that profession, or at least, that's what his father thought when he tried to dissuade him[3] – and initially he showed a passionate inclination towards natural sciences and research. He felt so little urgency about qualifying that he prolonged his studies by taking courses in zoology, chemistry, botany, and mineralogy and for three years attended the lectures given by Franz Brentano (1838–1917), vacillating about whether to take his doctorate in philosophy. In his third year, he had also spent some months at the centre for experimental zoology in Trieste, carrying out a study of the gonadic structure of eels.

Even after graduating, and without giving much thought to his straitened financial circumstances, Freud had pursued his researches at the Institute of Physiology, where he was directed in his study of histology and neuroanatomy by the Berliner Professor Ernst W Brücke (1819–1892), following the rigorous model of the Helmholtz School. He did earn a little money from translations in which

DOI: 10.4324/9781003246473-3

he could give space to his passion for writing and language: he began with the philosophical and political writings of John Stuart Mill and moved on to translate the works of Bernheim and Charcot.

Meeting Martha Bernays (1861–1951) in April 1882 gave his life a new direction. The "Conquistador"[4] dreaming of a career in the experimental sciences but forced to reckon with the need to earn a living if he was to marry – his fiancée would not bring a substantial dowry with her – did not hesitate to leave Brücke's Institute and undertake a clinical traineeship. So he spent some years at the *Allgemeines Krankenhaus*, Vienna's General Hospital, which was one of the leading medical institutions in Europe at that time, in order to obtain the necessary experience for private practice: he worked in several departments but made a particularly thorough study of neurology, conducting some further anatomical research in Meynert's laboratory.

For a long time, Freud suffered from the clash between his scientific ambitions and his desire to marry Martha: there was no lack of opportunities for regret and ill-concealed rebuke to his fiancée for the sacrifice of his research activity and his lost career. With hindsight it is perfectly clear that Martha was no obstacle for him. The experience of romance bursting tumultuously into his youthful rationalism was the means of realising his destiny: while the longed-for scientific achievement would have to wait – arriving after the age of 40 – it would exceed all his previous, and by comparison very modest, expectations of success. Martha opened up the road that led to the change and brought Sigmund to discover the value of sexuality and "nature" for the first time in his own personal life. Freud had not had any previous amorous relationships. He was no ascetic but a man with powerful desires and even more potent values.[5] Jones writes:

> He had been torn by love and hate before, and was to be again more than once, but this was the only time in his life – when such emotions centred on a woman – that the volcano within was near to erupting with destructive force.
>
> (*I*, p. 152)

And Donald W Winnicott adds:

> Freud was by nature the very opposite of promiscuous. Perhaps this one fact more than any other gave Freud the right to startle and disturb the world's mind, and made possible the launching of such ships as the Dynamic Unconscious, Infantile Sexuality, the Oedipus Complex, Psychic Reality, and many others that have proved to be not so much pleasure steamers as battleships in the Truth war.
>
> (1962, p. 476)

Followed by five sisters and a brother ten years his junior, Sigismund Schlomo Freud had been the first-born child of Amalia Nathanson (1835–1930), the young and beautiful third wife of Jacob Kolloman Freud (1815–1896). Jacob was a wool merchant from Galicia who had moved from Tysmenitz to set up a business in Freiberg in Moravia. A non-observant Jew from the Hasidic tradition, he had been the first teacher of the son to whom he gave his father's name: Rabbi Schlomo had died a few months before Sigismund was born on 6 May 1856. Meghnagi noted that as interpreter of the unconscious, the only language common to humanity, Freud would later compare himself to another Shlomo, Solomon, who was able to speak the languages of animals (2000).

Through bad luck or lack of application, the father Jacob never succeeded in bringing financial ease to his family, who almost always depended on help from relatives and friends. After an early childhood remembered as free and untroubled, Freud's puberty and adolescence were marked by precariousness and the struggle to survive, a fate common to the Jews of central Europe under the emotional, economic, and political pressures of the time.

The move from the village where he was born to the Leopoldstadt district, the ghetto in Vienna where immigrants from the East were crammed together, must have heightened his sense of privation and of not belonging, together with the desire to distinguish himself which his mother had instilled in him. And Amalia Freud's expectations can hardly have been moderate if it is true that, marking her own birthday according to the Gregorian calendar, she had chosen the birthday of Emperor Franz Joseph![6]

A beautiful woman with a proud and indomitable spirit, Amalia Freud easily let herself get carried away by her feelings. She never stopped recalling the beauty of her "*little blackamoor*," her "*goldner Sigi*": from the moment of his birth *en caul*, she had felt that her firstborn was blessed by fortune and destined for great achievements. An old woman's prediction[7] and other subsequent premonitions had confirmed her belief, and Sigismund grew up with the challenging burden of this maternal "prophecy" to fulfil. Once he had achieved success, serenely reflecting on the achievements of great men from the past, he would add a comment on the oedipal dreams of possessing the mother:

> I have found that people who know that they are preferred or favoured by their mother give evidence in their lives of a peculiar self-reliance and an unshakable optimism which often seem like heroic attributes and bring actual success to their possessors.
>
> (1900, footnote added in 1911, p. 398)

However, Sigismund's relationship with his young and inexperienced mother was not quite so rewarding and was soon disturbed by the arrival of an unbroken series of rivals which exposed him to violent jealousies: he was especially dismayed by the birth of another little boy, Julius, born a year and a half after him, who lived only six months.[8] This traumatic event for mother and son came to form part of Freud's precious but tortured relationship with the magic of early childhood and encouraged the "prophetic" expectations which impelled him towards his formidable discovery but also fed his periodic tendency to depression and the conviction that he would die young: every stage attained by the Ego put its own constitution, however solid, at risk. Threatened in this way, Freud placed many obstacles in the way of his own success, which he only achieved after his father's death. While he acknowledged the decisive role played in his own destiny by the working-through of this bereavement, he always seemed to have little awareness of the conflicts in his relationship with his mother. His repeated assertions to the contrary only confirm the intensity of his ambivalence. In the opinion of Sándor Ferenczi, who understood him at a deep level, Freud had had to defend himself against the passionate and extremely excitatory nature of a sexually demanding mother by resorting to idealisation. Moreover, Ferenczi thought that the castration theory of femininity may have been derived from Freud's "personal aversion to the spontaneous female-oriented sexuality in women."[9]

First love

Gisela Fluss, a childhood friend rediscovered by Sigismund during a visit to Freiberg when he was 16, was the first girl capable of replacing his mother, arousing in him a sudden, silent, and almost paralysing infatuation. This transferring of love onto a someone his own age was not pain free: it gave Sigismund an agonising toothache which, after he took pure alcohol as a painkiller, resulted in an outright intoxication, with its attendant vomiting and stupor to complete the initiation. Confined to bed, he was the object of the attentions of Gisela's mother, Eleonora Fluss, who watched over and looked after him as if he were one of her own children. "It would seem that I have transferred my esteem for the mother to friendship for the daughter," Freud wrote to his high school friend.[10] Voluble in singing the praises of Frau Fluss, whom he compared with his own mother, Amalie, he was speechless, like Hamlet, at the fleeting appearance of Gisela, his peer.

In the correspondence with his best friend Eduard Silberstein, the searing subject was addressed in code, in a secret language, Spanish or an approximation of it, inspired by Cervantes, imitating the dialogue between Berganza and Cipion, the two canine protagonists of the story *El casamiento engañoso* (*The Deceitful Marriage*).[11]

It is no surprise that, in those adolescent letters, alongside the exaltation of his love for the angelic young woman, Sigismund evoked images of women as deceitful and disturbing, perhaps because capable of arousing all too conscious carnal desire: "reptiles with venomous fangs, scaly armour, and Mephistophelean ideas" like that particular young woman to whom he gave the name "Ichthyosaur"[12] and who inspired his parody of a wedding song containing verses infused with rancour. About Silberstein, Freud would later write to Martha:

> We became friends at a time when one doesn't look upon friendship as a sport or an asset, but when one needs a friend with whom to share things. We used to be together literally every hour of the day that was not spent on

the school bench. We learned Spanish together, had our own mythology and secret names, which we took from some dialogue of the great Cervantes. . . . He was known as Berganza, I as Cipiòn.

(7/02/1884)[13]

Whereas Eduard-Berganza admitted a new infatuation or boasted about the experience of a kiss, Sigismund-Cipiòn, agonising over whether to follow his friend's example, but above all disheartened by his own shyness, "mischievously" put him on guard:

kisses tend to multiply in proportions that increase with such unusual speed that, soon after the beginning of the series, the facial area no longer suffices and [they] are then forced to migrate.

(13/03/1875)

Ernest Jones, whose puberty was shot through with a much more turbulent sexuality, observes that the young Freud showed little interest in the other sex and says nothing about the adolescent experience of masturbation, even though Freud would later attribute certain aspects of his own neurasthenia to this practice. Sigismund's inhibition was connected to the intensity and ambivalence of his sexual investment in his mother and sisters, especially Anna, of whom he was particularly jealous. We do not know what part jealousy, and the control of it, may have played in the friendship with Eduard Silberstein, who at the start of it had been in love with Anna, or in Freud's bond with his other dear friend, Emil Fluss, the brother of his beloved Gisela.[14] Freud had a vital need for a comrade with whom to share the woman: as we will see, the meeting with his future wife came about within a double sibling relationship. Psychoanalysis would also be born out of encounters with patients in significant and indispensable relationships.

To mark the end of his adolescence, Sigismund decided he wanted to be called Sigmund: he was a good-looking young man, of medium height, well proportioned, with dark hair and eyes, and an intense and penetrating gaze. He could be charming, and he had his work cut out to keep postponing his encounter with women. His belief in female inferiority and his despotic elder-brother behaviour towards his sisters, whose reading he censored, prohibiting Balzac or Dumas,[15] were symptoms of anxiety about his virility in the face of that wild, primitive animal that the female could show herself to be: all instinct, no rationality.[16]

When, shortly before entering adulthood, he travelled to Manchester to visit his English relations, Sigmund Freud speculated about marrying his niece Pauline, a childhood playmate whom his family would have been glad to see at his side, in the hope that he would settle in England and, like them, go profitably into business.

In fact, women had not yet appeared in his student existence or in his *Männerbund*, the group of school and family friends, almost all Jewish, who gathered in their free time to chat in the *Kurzweil* café, playing cards and chess or organising walks in the *Prater* and trips out of Vienna, sometimes joined by his sisters. The close relationships with these friends would last all through Freud's life, and he would spend every Saturday evening, accompanied by Oscar Rie and Ludwig Rosenberg, at the house of Leopold Königstein playing with tarot cards, a passion he had inherited from his mother.

The group was joined by the young Bernays siblings, members of a prosperous, cultivated, and strictly observant German Jewish family (their paternal grandfather had been Chief Rabbi in Hamburg, and an uncle was a famous philologist): they had moved to Vienna after the death of their father, a businessman who, having been declared bankrupt, had left them in a modest, if not precarious, financial situation.[17] As the sisters of his friend Eli, Martha and Minna Bernays took part in the *Bund's* outings. Minna, the younger sister, was already engaged to a member of the group, Ignaz Schönberg, a student of philosophy who would die young of tuberculosis; Martha had become a close friend of Freud's sister Anna, who was already perhaps in love with Eli.

Day by day, Sigmund was discovering the unexpected violence of romantic love. He was 26 years old and was struck by a *coup de foudre* in the tranquillity of his own four walls: coming home one evening in April 1882, the vision of the young woman visiting his family, intently peeling a now legendary apple was, for him, a paradisal manifestation of the feminine. A feminine, it should be added, that was not too threatening in terms of beauty and emancipation, made less wild by the familiar proximity to his friend from the *Bund* and to the sister he was so close to: in some ways, Martha replaced Anna.[18]

Martha and Bertha: the languages of passion

Martha, "the fairy princess from whose lips fell roses and pearls,"[19] introduces Freud to the strange and familiar language of love: she is "my sweet darling girl" in the first secret letter delivered by Eli. The *billets doux* which accompany the red rose he sends her each day will be written in English, Spanish, and Latin. Freud's *courtly love* quickly becomes an ardent and possessive passion, tormented by jealousy. Almost an illness: like the very serious pharyngitis which for many days in August made him unable to swallow or speak and left him with a "gigantic hunger like an animal waking from a winter sleep," as he writes to Martha. "A frightful yearning – frightful is hardly the right word, better would be uncanny, monstrous, ghastly, gigantic; in short, an indescribable longing for you."[20]

Sigmund is afraid that his feelings are not returned with the same intensity or that her love may peter out; he keeps a close eye on her male friendships but also on the manifestations of her dependency on her mother and brother and the link with the family's customs of strict Jewish observance, from rest on the Shabbat to the Kosher rules. He would like her to be autonomous but entirely dependent on him: in fact, he will declare to her that, from the first moment, she has changed his life.[21] In this period of waiting, of growing excitement and sexual frustration – his experience of which will be felt in the first hypotheses about the aetiology of the neuroses – he fears for her health and his own, becoming prey to strange presentiments and somewhat superstitious. As a boy, he had chosen number 17 in a lottery that revealed the character of people, and had drawn the word "constancy," which he now associated with the date of the secret engagement, 17/06/1882.[22]

One day he wrote to Martha:

> Now I have a tragically serious question for you. Answer me on your honour and conscience whether at eleven o'clock last Thursday you happened to be less fond of me, or more than usually annoyed with me, or perhaps even "untrue" to me – as the song has it. Why this tasteless ceremonious conjuration? Because I have a good opportunity to put an end to a superstition. At the moment in question my ring broke where the pearl is set in.
>
> (26/08/1882)

Jones explains that a surgeon had stuck a knife in Freud's throat to relieve an anginal swelling, and in his pain he had slammed his hand on the table. He later learned that at the time Martha was simply eating a piece of cake.[23]

It was a "passionate, conflicted and abstinent love relationship:" the betrothed couple "kissed each other passionately on the streets and in the squares of Vienna, in the woods of Wandsbek, and in Freud's little room at the hospital" – echo of his childhood experience – "for he had been the favourite and despotic child of a despotic mother."[24] Sigmund insists to Martha that he cannot live without her total, complete, and exclusive love and claims possession of her in an almost tyrannical manner: "I have to possess you totally and to be the only one."[25] Martha is not only the source of his desire but becomes a vital point of safety, the "anchor"[26] who supports the storms of his jealousy and impatience with a strength, patience, and compassion surprising in a woman of 20.

It is a mutual sentimental education, and, thanks to her gifts of acute intelligence and intuition, Martha detects Sigmund's fragility in this phase of his life: she protects and calms him and when necessary sets him a limit.[27]

The engagement had to remain a secret until Christmas. The initial plan to celebrate a double wedding between siblings, bringing the two families even closer together, had to be set aside, and instead when Anna married Eli in 1883, Sigmund refused to attend the ceremony. Meanwhile, his future mother-in-law, intuiting the bond between them, and far from considering Freud a good match for her daughter, decided to place a discreet distance between the lovers, returning to live in Wandsbek, near Hamburg. Sigmund and Martha were thinking in terms of a separation lasting a few months, but they had to wait a full four years and needed a substantial gift from one of Martha's uncles before they could marry. It was during this long wait and separation, feeling powerless, brutally confronted with his poverty and the need to seek help from friends and acquaintances, a period made more bearable by a shared secret diary and a correspondence of over 1500 letters[28] – letters in which the nuclei of his later psychoanalytic thought appear in embryonic form – that the would-be bridegroom began both to research cocaine, hoping in this way to change his destiny and hasten his marriage, and also to use it as an anaesthetic.[29]

The intensity of his amorous passion, the contact with his own fragility, mitigated by the epistolary dialogue with Martha, allowed Freud to move gradually closer, in his clinical work as well as his personal life, to the living sexed body of woman, to pass from neuroanatomy to psychology. In fact, without knowing it, or even consciously wanting it, Freud was already on the path that would make him famous.

The neurological work that he had undertaken out of mere expediency required him to accept mentally ill patients in particular. The therapeutic practices had consisted of massages, baths, and rest, and ranged from electrotherapy (a recent addition to the impressive medical armoury) to hypnosis. Freud discussed this last treatment with an older colleague, Josef Breuer (1842–1925), a friend of Hammerschlag, whom he had met at the Institute of Physiology.

Breuer was a wealthy and much-respected Jewish physician who, after gaining experience in experimental research, had devoted himself entirely to private practice, becoming one of the most highly regarded clinicians in Vienna. A brilliant man, famous for his diagnostic abilities – his "golden hands" – he numbered many celebrities among his patients (he treated Brahms during the final months of his life) and also among his colleagues. He had encouraged Freud's research and, in due course, supported his entry into private practice, sending patients to him, lending him substantial amounts of money, and becoming his doctor too for many years. Having been almost adopted into the family, Freud spent a great deal of time with Breuer, staying late into the night for discussions. Once he was married, he named his first child Mathilde after his friend's wife.

Breuer never doubted his young protégé's genius, although a point came where it troubled him, perhaps because of the "extremely daring and fearless human being"[30] he intuited behind Freud's apparent shyness.

It was in November 1882, not many months after Sigmund had met Martha, that Breuer began to talk about the long, recently concluded treatment that had brought him several times a day to the bedside of Bertha Pappenheim, a childhood friend of Martha's and still a visitor to her house: she was only a couple of years older than Martha and her father, Sigmund Pappenheim, who died shortly after Martha's, had been among the legal guardians of the Bernays girls for that brief period.[31]

Like Martha, Bertha had grown up in a rigidly orthodox Jewish family and had received the kind of education that was then permitted to young women destined for marriage: she was knowledgeable about literature, spoke three foreign languages, played the piano, did housework and embroidery (her passion was lacemaking), and also went riding.[32] Her only sibling was Wilhelm, younger than her by a year: her two elder sisters were both dead, Flora before Bertha was born and Henriette at the age of 17 when Bertha was 8.

The young woman's relationship with her mother had always been difficult, whereas she adored her father with an intensity that left no room for other amorous relationships: Breuer had commented on this in the notes he made at the time.[33] Ten years later, he would confine himself to stating that she had never been in love and that her sexuality was entirely undeveloped.[34]

Good-natured and generous, but moody and inclined to exaggerate both her happiness and her sadness, Bertha was endowed with abundant intelligence, intuition, and critical *esprit*. Her parents had not wanted her to continue her studies after she left high school: her wealth of poetic and imaginative talent had led her to take refuge in a world of fantasies, a "private theatre" of daydreaming with which she punctuated the monotonous and automatic rhythm of household chores and let her mind wander even when she was in company.

Her tendency to hypnoid dissociation had been heightened while she tended to her father when he was ill and was interrupted by the appearance of a symptomatology that was both somatic (coughs, paralysis, anaesthesia, contractures) and functional (disorganisation of language, mutism, deafness, visions and hallucinations, anorexia . . .) but above all by the emergence every evening of a second rebellious, impulsive, and disorganised personality.

Called in because of Bertha's cough, Breuer had intuited the association of the symptom with her father and his tuberculosis[35] and had succeeded in helping her to overcome her mutism by inviting her to talk about her father, discovering a secret resentment towards him.[36] In his conversations with the young woman during her *condition seconde* (initially in English since she was no longer able to use, or sometimes even understand, her mother tongue), Breuer reconstructed the affective history of the hysterical symptoms and, from time to time, "swept" them away. For example, this is how he described the appearance of paralysis in her arm:

> Anna was sitting at the bedside with her right arm over the back of her chair. She fell into a waking dream and saw a black snake coming towards the sick man from the wall to bite him. (*It is most likely that there were in fact snakes in the field behind the house and that these had previously given the girl a fright; they would thus have provided the material for her hallucination.*) She tried to keep the snake off, but it was as though she was paralysed. Her right arm, over the back of the chair, had gone to sleep and had become anaesthetic and paretic; and when she looked at it the fingers turned into little snakes with death's heads (the nails). (It seems probable that she had tried to use her paralysed right arm to drive off the snake and that its anaesthesia and paralysis had consequently become associated with the hallucination of the snake.) When the snake vanished, in her terror she tried to pray. But language failed her: she could find no tongue in which to speak, till at last she thought of some children's verses in English and then found herself able to think and pray in that language.
>
> (Breuer and Freud, 1893–95, pp. 38–39, my italics)

As with the princess's wicked sister in the fairy-tale, out of the mouth of hysteria issued toads, lizards, and snakes . . . The *talking cure* gave conscious access to unpleasant, forbidden, or trapped emotions: the symptoms resolved themselves when they were talked about and relived (catharsis) one by one as they presented themselves. This *coughing up the toad* was what Bertha called "chimney-sweeping," a term which her doctor considered entirely innocent.[37]

Freud was much struck by what his colleague confided in him and by his reading of Breuer's notes. Despite the latter's warning to be cautious, Freud had written to Martha about the case, not concealing the fascination which the seductive girl was stirring up in her doctors, Breuer's commitment to the unusual treatment, and Mathilde's jealousy: Freud revealed to Martha that certain unclear aspects of the case's conclusion were in fact connected to Breuer's need to save his marriage.[38] Freud inevitably had to reassure his fiancée that only vanity was behind the fear that her future husband would make his patients fall in love with him: "For that to happen one has to be a Breuer," he wrote.[39]

Whereas, with her good sense, and knowing Bertha, Martha had intuited that the ability to turn doctors' heads was an integral part of hysteria's tragic destiny,[40] it is significant that Freud ended up having to insist *he would never make his patients fall in love with him* and that for his part, Breuer, to whom we owe the first formulation of psychoanalytic treatment, always emphatically denied any sexual component in Bertha Pappenheim's symptoms and in his therapeutic relationship with her.

On his way to Paris two years later, Freud had the model of the talking cure firmly in mind and planned to discuss it with Charcot. Still not yet 30, he was already a respected researcher and *Privatdozent*, having published first-rate contributions to the fields of histology and neuroanatomy as well as the previously mentioned monograph on cocaine. At the end of his clinical traineeship, preparing to start up his private practice and ready to set the date for his wedding, he had surprisingly decided to compete for a scholarship offered in celebration of the University of Vienna's Jubilee. In his own justification, he wrote to Martha, "But a human being's demon is the best part of him, it is himself. One shouldn't embark on anything unless one feels wholehearted about it."[41] And so, in the winter of 1885–86, he went to complete his training at the Salpêtrière Clinic for Nervous Disorders, though not without stopping at his fiancée's home in Wandsbek on the way to and from Paris.

Romantic experience and professional liberation found a synthesis which enabled him to discover his demon. His time in Paris lent depth to his confrontation with hysteria and with psychopathology. More personally, it responded to his curiosity about the female sexual universe, to the question *Was will das Weib?* (What does Woman want?), which would accompany him for a large part of his life and work.[42]

Notes

1 Freud, 1924a, p. 8.
2 Tobler G C *Goethe – Natur*, English translation by T H Huxley, "Nature: Aphorisms by Goethe" *Nature*, Nov. 4, 1869.

3 Bernays Freud, p. 145.
4 Jones I, p. 382.
5 S Bernfeld (1913) in I. Grubrich Simitis, 1981.
6 Kamieniak, 2000.
7 The oracle of the old woman – "You're a lucky mother! Some day the whole world will talk about this little fellow" (Bernays Freud, 1940, p. 140) – had *read the thought* of the young and ambitious mother.
8 Julius was born in October 1857 and was given the name of his mother's brother, who had died shortly before. When he in turn died, at the age of six months on 04/15/1858, Amalie, who was already pregnant with Anna, had a flare-up of tuberculosis (Roazen 1975). Jones, however, records Julius as living from April to December 1857.
9 4/08/1932. It is likely that the Hungarian, when dealing with the unconscious mother-child identification at puberty, had certain feminine traits of the master in mind – for example, the timbre of his voice (1915).
10 09/04/1872.
11 From M de Cervantes *Exemplary Novels* (1613): the conversation of the two dogs, stationed in front of the hospital de la Resurrección in Valladolid, is perhaps imagined in delirium by an inpatient, Ensign Campuzano, who after being deceived, robbed and abandoned by his wife, has also caught a terrible disease from her.
12 17/01/1875. This nickname was also taken from a novel, a contemporary one in this case.
13 In Freud E L, 1961.
14 See Conci 2016, 1996.
15 Freud to Silberstein, 16/06/1873.
16 The later recollection of a youthful bewilderment in a provincial Italian town (perhaps Trieste?), finding oneself involuntarily and inexorably passing through the same quarter, of whose character he could not long remain in doubt ("Nothing but painted women were to be seen at the windows of the small houses"), will be useful to him in portraying the *Heimlich-Unheimlich* (familiar-disturbing) quality of the feminine in the essay "The Uncanny" (1919, p. 237).
17 Behling, 2002.
18 Steiner, 2013.
19 Jones, *I*, p. 115.
20 Ibid., p. 86.
21 15/06/1882, in Grubrich Simitis and Lortholary, 2012.
22 Jones, *I*, p. 170.
23 Ibid., p. 119.
24 In Steiner, 2013, p. 928.
25 29/08/1982.
26 22/06/1883.
27 Grubrich Simitis and Lortholary, 2012.
28 Only partially published (Freud E L, 1961). In 2011, the complete correspondence (five volumes) began to be published, the first four of which are available (Freud and Bernays, 2011, 2013, 2015 and 2019).
29 Grubrich Simitis and Lortholary, 2012.
30 Sigmund to Martha, 2/02/1886.
31 Schwartz, 1999.
32 Breuer and Freud, 1893–95.
33 Hirschmüller, 1978.
34 Breuer and Freud, 1893–95.
35 Rodrigué, *I*, 1996.
36 Hirschmüller, 1978.

37 Freud to Jung, 21/11/1909.
38 13/07/1883, 05/08/1883 and 31/10/1883, in Forrester, 1990.
39 11/11/1883, in Jones, *I*, p. 247.
40 Martha to Sigmund, 2/11/1883, in Forrester, 1990.
41 31/03/1885.
42 Elms, 2001.

Chapter 3

The lesson of Jean Martin Charcot

At the Salpêtrière

The *Salpêtriere* was a kind of citadel in the heart of Paris. Formerly the arsenal, it consisted of a vast set of one-story buildings arranged in a quadrilateral and separated by courtyards and gardens. St Vincent de Paul worked there in the seventeenth century, and it had been used to confine prostitutes, beggars, and the dispossessed. Following the Revolution, Philippe Pinel reformed it into a psychiatric hospital, and by 1813, it had become a hospice for women that could host more than 5000 people.

Jean Martin Charcot had been the first of its directors to stay there long term and establish a systematic study of chronic nervous disorders. Sensitive to the forms of pathology and with an almost artistic passion for nosography, Charcot had devoted himself intensely and exclusively to hysteria: he had identified its distinctive clinical signs, even when they appeared in isolation (the character of the attack, anaesthesia, disturbances of vision, hysterogenic triggers); distinguished the functional paralyses (hysterical, post-traumatic, and hypnotic) from those caused by a lesion; provided a systematic description of "grand hysteria"; and classified and photographed its highly varied iconography. In 1881, with support from Léon Gambetta, the first chair of neuropathology was created for him, followed by a series of buildings devoted to anatomical and physiological research, clinics for electrotherapy and radiology, a photographic laboratory, and even a small ward for treating men. Charcot had transformed the *Salpêtriere* into one of Europe's most renowned centres of research, and large numbers of foreign students and visitors came to attend his magisterial lectures.

Contact with the master was especially stimulating because he used to "think aloud"[1] during the clinical encounter and let his students participate in his investigation of the patients' histories, in the tireless and repeated observation of symptoms ("He used to look again and again at the things he did not understand"),[2] which led him, sometimes unexpectedly, to find a way of interpreting them. Like Breuer, Charcot was a charismatic individual who had lost his mother early in life, but his manner was completely different, and he knew how to keep his head with women without being troubled by their sexuality. His clinical demonstrations

DOI: 10.4324/9781003246473-4

were extraordinary and attracted journalists, writers, theatre people, and elegant *demi-mondaines*: from Guy de Maupassant to August Strindberg to Sarah Bernhardt, a hugely diverse audience from *tout Paris* arrived as if for a gala, curious to witness the surprising phenomenon of hypnotism and the *arc en ciel* of hysteria.[3]

The fate of most of Charcot's young patients was sealed:

> Hypnotized right and left, dozens of times a day, by doctors and students, many of these unfortunate girls spent their days in a state of semi-trance, their brains bewildered by all sorts of absurd suggestion, half conscious and certainly not responsible for their doings, sooner or later doomed to end their days in the *salle des agités* if not in a lunatic asylum.
>
> (Munthe, 1929, p. 303)

Alone and a little lost in a foreign city, during those first days of his French adventure, the young Freud was prey to dark premonitions and quite often had the feeling that Martha was calling his name.[4] Under the full impact of Paris, he compared it to "a vast overdressed Sphinx" who gobbles up every foreigner unable to solve her riddles. "I feel they are all possessed of a thousand demons" – he wrote to Minna, and added,

> I don't think they know the meaning of shame or fear; the women no less than the men crowd round nudities as much as they do round corpses in the Morgue or the ghastly posters in the streets announcing a new novel in this or that newspaper and simultaneously showing a sample of its content. . . . Paris is simply one long confused dream, and I shall be very glad to wake up.
>
> (Sigmund to Minna, 3/12/1885)

As his main attraction and assistant in clinical lessons at the Salpêtrière, Charcot had for some time been making use of the dancer Jane Avril who, after being "cured," went on the stage at the Moulin Rouge, launching the can-can with her famous red knickers, and then became Toulouse-Lautrec's favourite model. But the outstanding protagonist of the master's performances, which passed through all the three stages of hypnosis that he had recognised and described, was the very beautiful Blanche Wittmann, *reine des hystériques,* who was able to throw herself with greater intelligence and mimetic compliance into the role she was given.

Having entered La Salpêtriere at the age of 18, mainly out of poverty, and acting sometimes as a patient and sometimes as a nurse, Wittmann spent most of her life there, accompanying Charcot's career to the end. After his death in 1893, she worked in the photography laboratory and later in the radiology laboratory, where she was one of the first victims of radiation poisoning: this is the name of the new and quite different luminescent fluid, which repeatedly caused the need for amputations of her limbs. In 1913, shortly before her death, the city of Paris awarded her the Medal of Public Assistance.

In all probability, she is the woman immortalised in the famous painting by Pierre André Brouillet, *A Clinical Lesson at the Salpêtrière* (1887), a painting which photographs an era. Many of the students it portrays will become famous names in neurology: Joseph Babinski, Pierre Marie, Gilles de la Tourette. During his brief stay, Sigmund Freud had the opportunity to meet them and to be introduced into Parisian society as it gathered at Charcot's soirées: it is only by chance that he does not also figure in the picture alongside his colleagues. Reproduced in lithograph, Brouillet's painting had hung at Berggasse in front of his chair, a homage to the teacher whose lectures he had translated into German and a memory of an important moment in his educational development. It is from this scene that psychoanalysis begins to emerge.

The Clinical Lesson of Professor Charcot introduced the woman's living and erotic body into what had until then been the main theatre of Freud's scientific research: the dissecting room. The transformation is remarkable if we place Brouillet's painting alongside *The Anatomy Lesson of Dr Nicolaes Tulp,* the painting commissioned from Rembrandt by the Amsterdam Guild of Surgeons in 1632. In this case, it is the dead body of a man, a criminal who had been hanged, which occupies the centre of the anatomical theatre while, all around it, physicians and students observe Dr Nicolaes Tulp as he demonstrates with his own hand the movements of the fingers corresponding to the tendons he has just isolated in the arm of the cadaver.

In Paris, Freud was busy on two fronts: on the one hand, the training provided by clinical observation of the spectacular demonstrations at the Salpêtriere, and on the other, his gloomy apprenticeship at the Morgue where he never missed the lectures on autopsy given by Paul Brouardel (1837–1906).[5] Later on, he was able to make an extraordinary integration of the two scenes, Eros and Thanatos.

In a context of scientific research, the lectures at the Salpêtrière highlighted the woman's living body and the sexuality hidden in the symptom. The patient, induced into a hypnotic sleep *as if* dead, let herself be seduced and mimetically enacted an orgasm, symbolised it in her crisis under the eye of the doctor, and, while submitting to his desire for power, she in fact held him in check.

On the wall of the great hall, before the patient's eyes – as magisterially reproduced by Brouillet, a picture within the picture – there was a painted representation of the hysterical attack. This further suggestion in the background heightened the element of reciprocity but also indicated the possibility of the seducer and seduced exchanging roles, the risk of a *folie à deux,* a *mise en abîme*, an infinite regress. Perhaps Charcot, *le visuel*, "a man who sees,"[6] was also able to issue the challenge and withstand it because of his slight squint which enabled him to resist the mimetic short circuit. The duet took place in the respectable setting offered by the conception of hysteria as a degenerative organic process: the nobility of anatomy allowed the *indecent* role played by sexuality to be kept in the background.

Freud learned Charcot's lesson in the light of his experience of the talking cure and the Anna O-Breuer couple, transforming it in his own way and moving beyond his teachers and hypnosis. The command "look into my eyes" becomes the fundamental rule of psychoanalysis: "talk to me." The asymmetry of gazes becomes the invitation to think aloud; the centrality of listening will transmute the heated sensoriality of the theatrical public performance into a private dialogue, a representational narrative in a private room.

The apparatus of language

Faced with the enormous quantity of material that the Salpêtrière had given him, Freud's first objective was to try to understand hysterical phenomena within the rigorous scientific model he had learned, systematically distinguishing between hysterical and organic paralyses and between the forms they took in adults and children. In doing so, he benefited immensely from his sound neuroanatomical knowledge – he was known among his colleagues for the exactitude of his diagnoses, confirmed in detail by autopsy – and his research into infant anatomy carried out at the Morgue, along with his experience of paediatric neurology gained in Berlin, where he stopped on his way back from Paris and which he continued when he was back in Vienna. In this way he was able to rid the subject of its domination by organic aetiology, as was still being maintained by Charcot, and to isolate the specific psychological characteristics of the hysterical mechanism underlying the variety and mutability of the symptomatological forms.

In the article he started writing on his return, "Some Points for a Comparative Study of Organic and Hysterical Motor Paralyses" (1888), he noted how, in hysteria, paralyses and anaesthesias originated in different parts of the body with no regard for anatomical nerve centres but according to a conceptual pattern, human beings' popular representation of their own body: hysteria behaved "as though anatomy did not exist or as though it had no knowledge of it."[7]

Citing the example of an entirely paralysed arm, he speculated that the hysterical symptom gave expression to an idea – the image "arm" based on groups of tactile and, above all, visual perceptions – for which the patient had no language. Because of the link with an affectively highly charged subconscious association, the idea had to be expelled from the conscious psychic concatenations – "repressed" – through a species of symbolic aphasia operating in hysteria. And he differentiated hysterical symptoms of language from the aphasias caused by organic damage because of the persistence in the latter of older, rudimentary, automatic linguistic capacities.[8] Hysteria did not respect this functional regression: Anna O, for example, was not able to use her mother tongue but could speak English, a complex and more recent acquisition.

Freud's next two papers, his final works of *neurology*, are no longer about neuroanatomy but neuropathology: *On Aphasia* (1891), which he dedicated to Breuer, and the study of *Infantile Cerebral Paralysis* (1897). Though they continue his previous experimental research (so much so that they are considered by many biographers, including Peter Gay, mere expressions of old interests), they are steps on his advance towards the field of psychopathology. Systematically working to construct the bases for the scientific theorisation of a clearly distinct psyche which was, at the same time, anchored in the body,[9] Freud focused his attention on the psychic function of language, on the development and functioning of a specific apparatus (*Sprechapparat*), the first representation of the psychic structure.

Together with language, it will be sexuality and the borderline concept of libido which build the bridge that will enable him to make the "leap"[10] between body and soul, to complete the scientific, not occultist, representation of a psyche rooted in the soma.

The magic of words

On Easter Sunday 1886, Freud began his private practice, supplied with patients generously sent to him by Breuer. They were ill people whom he had to visit in clinics or boarding houses or at home, hiring a *Fiaker* – people like the very old lady he went to see twice a day for at least six years, to give her a morphine injection and drop some collyrium into her eyes – or patients whom he saw in his consulting room during the afternoons. But he had to wait for this latter clientele. Recalling those times, Martha Freud said that Sigmund's sisters would often occupy the waiting room so that if anyone arrived, they would be impressed by the number of supposed patients and would feel confident about having to wait their turn . . .[11]

The first patients who came to see Freud for regular appointments were mostly hysterics whom Breuer no longer wished to care for in person: the two men discussed diagnoses, prognoses, treatment, and possible consequences for theory, as in a supervision. Despite having publicly declared himself a supporter of hypnosis in his translations, reviews, and prefaces to the works of Charcot, Bernheim, and Forel, Freud did not immediately make use of it.

In the scientific sphere, when this practice was not being openly disparaged, it had aroused little interest because of the temporary and unpredictable nature of its outcomes. Apart from the attention given to it by R Krafft-Ebing and H Obersteiner, as well as Breuer, no sustained work had been done to develop a scientific understanding of how it functioned. Arthur Schnitzler was another practitioner who, before giving up medical practice, had tried hypnosis in the treatment of aphonia and in minor procedures on the nose and larynx, but later on he had become enthusiastic about the versatility of one "medium" whom he induced to enact all kinds of situations and sensations which he enjoyed inventing in the presence of a large audience of colleagues from his department and other hospitals in Vienna.

> from one day to the next I arranged for a murder attempt against myself, which I was able to parry successfully because I was prepared for it, and the patient chose to use a dull letter-opener instead of a dagger.
>
> (1968, p. 271)

After word went around that he was "putting on shows" at the Polyclinic, Schnitzler was forced to reduce the size of his audience and to suspend his para-therapeutic experiments, an additional reason being that the most frequently involved mediums had begun suffering from "a loss of will and a certain damage to their health" caused by the repeated experiments.[12]

As for Freud himself, he gradually added hypnosis to his clinical repertoire without being either discouraged or overenthusiastic about it. Despite compromising himself officially – in the opinion of his former professor Meynert, he had turned from "a physician trained in precise physiology" into a vulgar "practitioner"[13] – he was far from satisfied with the therapeutic results. Returning to France in the summer of 1889, he decided to stop at Nancy this time, bringing one of his patients (Fanny Moser, the Emmy von N of the *Studies*) for a consultation with Bernheim. Freud, Bernheim, and Liébault then went to Paris for the First International Conference on Hypnotism which was being held as part of the World's Fair. This second trip, during which he admired the great, newly opened *Tour Eiffel*, enabled him to perfect his hypnotic technique but also to reflect on the possibility of using word suggestion in the waking state, experimenting with "psychotherapy" like Bernheim. But, unlike Bernheim, Freud was looking for the genesis of the symptoms, as he had learned from Breuer, although he remained convinced that hysterical phenomena formed part of a complex picture, both psychological *and* physiological, as manifestations of functional characteristics of the nervous system. For example, he attributed what was then called "*transfert*," the transferring of sensibility – an anaesthesia, a paralysis, a contracture, a tremor, and so on – from one part of the body to the corresponding part on the other side, to a "physiologically intelligible . . . genuine process" based on associations between symmetrical parts of the body.[14] What surprised him, however, was the transfer of the symptom by the contagion of suggestion (rationalised in the magnet) from one hypnotised person to another and from the patient's body to that of the hypnotiser, who would have supposed himself exempt from it. On the subject of these *transferts*, and the theme of thought transmission which interested him more, he claimed:

> The prodigiousness of the fact that a nervous system can influence another nervous system by means other than the sensory perceptions known to us . . . would add something new, not recognized so far, to our worldview, in a certain way moving out the boundaries of the personality.
>
> (1888d, p. 106)

The article which Freud prepared on his return from France – *Psychical (or Mental) Treatment*, (literally, *Soul Treatment*, 1890) – demonstrates how intently he was reflecting on the relationship of the psyche with the soma, from the hitherto unexplored perspective of the mind's influence over the body. He wrote that

> words are the essential tool of mental treatment. A layman will no doubt find it hard to understand how pathological disorders of the body and mind can be

eliminated by "mere" words. He will feel that he is being asked to believe in magic. And he will not be so very wrong, for the words which we use in our everyday speech are nothing other than watered-down magic. But we shall have to follow a roundabout path in order to explain how science sets about restoring to words a part at least of their former magical power.

(p. 283)

And on the subject of "thought-reading," he stressed its somatic-affective component:

The affects in the narrower sense are, it is true, characterized by a quite special connection with somatic processes; but, strictly speaking, all mental states, including those that we usually regard as "processes of thought", are to some degree "affective", and not one of them is without its physical manifestations or is incapable of modifying somatic processes. Even when a person is engaged in quietly thinking in a string of "ideas", there are a constant series of excitations, corresponding to the content of these ideas, which are discharged into the smooth or striated muscles. These excitations can be made apparent if they are appropriately reinforced, and certain striking and, indeed, ostensibly "supernatural" phenomena can be explained by this means. Thus, what is known as "thought-reading" [*Gedankenerraten*] may be explained by small, involuntary muscular movements carried out by the "medium" in the course of an experiment – when, for instance, he has to make someone discover a hidden object [without giving any ostensible prompting]. The whole phenomenon might more suitably be described as "thought-betraying" [*Gedanken-verraten*].

(p. 288)

Emphasising the psychotherapeutic element present since antiquity in any medical action, Freud likened the responsiveness of the patient to every type of "healer" (what he had called *rapport* in hypnosis) to the child's trusting expectancy towards its parents[15] or the devotion of the lover. He claimed that in such psychic states, capable of activating some of the most effective forces for provoking and healing somatic illnesses, speech showed itself to be a privileged tool and regained the "magic power" it had possessed in childhood.

And so, returning to the use of hypnosis, Freud perfected the use of speech in the waking state: this fundamental step revealed the action of a previously hidden psychic force which opposed the emergence of a memory in conscious verbal associations in the service of that very repression that had originally rejected it.

With the aim of challenging this resistance, in his practice he started to favour a particular way of concentrating his attention, out of which he would develop the method of *free association*.

It was in this period that a gift from a patient, Madame Benvenisti, made its appearance in the consulting room in Berggasse, the couch, the most famous piece of furniture in the history of psychological research, a symbol of psychoanalysis all over the world: a sober, typically Biedermeier, buttoned *dormeuse*, which, covered with rugs and cushions, became the silent and maternal foundation of the psychoanalytic setting and the method of free association.

At almost exactly the same time, the philosopher and physician Pierre Janet (1859–1947) was preparing to take up Charcot's position at the Salpêtrière and experimenting with "automatic talking," a technique which consisted in making the patient talk at random, rather like free association.[16] In 1889, drawing on his experience with his famous patient, a young countrywoman called Léonie, he published *L'Automatisme Psychologique*, which would become a classic of research into trauma and dissociation.

In these same years, the Harvard philosopher and psychologist William James (1842–1910), elder brother of the novelist Henry, likened the *"stream of thought"* to the flow of a river, identifying the essence of mental experience as being in the alternation between association and dissociation (1890).

We should remember that in 1883, Sir Francis Galton (1822–1911) was the first to grasp the deep links that words establish in the mind:[17] it seems that one day in London, walking down a long street, he had been struck by the large number of associations provoked by the objects to which he directed his attention, finding that they brought to mind words, images, and representations connected to experience from his childhood and adolescence, which he had not thought about for a long time and seemed to have forgotten.[18] Later discovering that, in a given situation, mental images revealed hierarchies of probability and the tendency to be organised into prevailing configurations of associations, he tried to represent them in the same formats as those used for meteorological maps.

He studied his own automatism and confirmed that responses were not a matter of chance but were statistically modified by emotions and thoughts from autobiographical memory; this led him to claim that "the mind is continually working along familiar paths without the memory retaining any impression of its travels."[19]

As he saw it, associations laid bare the foundations of a person's thoughts with such wonderful precision, revealing his mental anatomy in such a vivid and faithful manner that it was preferable to keep them to oneself.

Notes

1 Freud, 1892, p. 133.
2 Freud 1893, p. 12.
3 Munthe, 1929.
4 1901a, p. 261.
5 "He [Brouardel] used to show us from post-mortem material at the morgue how much there was which deserved to be known by doctors but of which science preferred to take no notice" (Freud, 1913d, p. 335; see Masson, 1984 and Bonomi, 1994 and 2007).
6 Freud, 1893, p. 12.
7 Freud, 1888, p. 169.
8 Stewart, 1967.

9 "The psychical is a process parallel to the physiological – 'a dependent concomitant'" (Freud, 1915, p. 207).
10 "the leap from a mental process to a somatic innervation – hysterical conversion – which can never be fully comprehensible to us" (Freud, 1909b, p. 157).
11 Freud-Marlé, 2006, in Tögel, 2009.
12 Schnitzler, 1968, p. 272.
13 Freud, 1889, p. 95.
14 Freud, 1888, pp. 48–49.
15 Somnambulism can be said to be physiological in infants.
16 Ellenberger, 1970, p. 524.
17 See Cimatti, 1997 and Basile, 2001.
18 Spadolini, 2004.
19 Basile, 2001, p. 52.

Chapter 4

The lesson of Josef Breuer and the "descent to the mothers"

Studies on hysteria

Having gathered a substantial body of clinical material, Freud decided to follow in Breuer's footsteps by making the results of their experiences known in the scientific world. It was not easy for him to overcome his colleague's reluctance to return to the case of Bertha Pappenheim after ten years and agree to a joint publication. As he confided to his new friend in Berlin, Wilhelm Fliess,[1] the *Preliminary Communication* "has cost enough in battles with my esteemed partner," but it was definitely worth the trouble since it made history on its appearance in January 1893. Hysteria now appeared as a malaise derived "mainly from reminiscences,"[2] linked to memories of traumas suffered in silence, left unmodified with all their affective weight in a split-off area of consciousness, the *condition seconde*, and acting like a foreign body in the production of symptoms.

> each individual hysterical symptom immediately and permanently disappeared when we had succeeded in bringing clearly to light the memory of the event by which it was provoked and in arousing its accompanying affect, and when the patient had described that event in the greatest possible detail and had put the affect into words. Recollection without affect almost invariably produces no result. The psychical process which originally took place must be repeated as vividly as possible; it must be brought back to its status nascendi and then given verbal utterance.
>
> (Freud, 1893, p. 255)

With the new method of psychotherapy, the memories could be re-awoken and abreacted: the encapsulated affect came to the surface in conversation and allowed itself to be represented in normal consciousness. The distinction between a *psychically acquired hysteria*, the product of an active defence against the re-emergence of a serious trauma and a repressed sexual emotion, and a *dispositional hysteria*, caused by a spontaneous tendency of consciousness to hypnoid dissociation, was clear in the two authors' parallel and still compatible aetiological interpretations: the first being maintained by Freud, the second by Breuer.

DOI: 10.4324/9781003246473-5

In the complete text published two years later, the *Studies on Hysteria*, five case histories were described in detail: the first illustrative case of hysteria, which appeared under the pseudonym Anna O., was that of Bertha Pappenheim, treated by Breuer; the other four patients, Emmy von N (Fanny Moser), Lucy R, Katharina (Aurelia Kronich), and Elisabeth von R (Ilona Weiss), all Freud's, were chosen from the many cases he had treated, as was the brief example of Rosalia H. A seventh patient, Cäcilie M (Anna Von Lieben, treated from 1889 to 1893, the most serious case, which the two colleagues had treated together and sent for a consultation with Charcot and perhaps also with Bernheim), appears in brief examples in the text of both.[3]

Unlike Anna O, Freud's examples were of mixed neuroses, combinations of hysteria and anxiety neuroses or phobias. In the complex case of Emmy von N, a 40-year-old woman (the so-called "capercaillie" patient,[4] after the peculiar guttural "clacking" and tics which characterised her speech), hospitalisation had become necessary because of her abulia and anxiety attacks accompanied by hallucinations and deliria: hypnotherapy was given in alternation with rest, massages, hot baths, and bran baths. Only the case of Lucy R, a 30-year-old Englishwoman, was one of pure hysteria like Anna O with an unquestionably sexual aetiology (she had started to smell burnt pudding in connection with the fantasy of marrying the father of the children for whom she was employed as governess), whereas the case of Katharina, the 18-year-old whom Freud met briefly during a holiday in the Hohe Tauern mountains, daughter of the landlady at the climbers' refuge where he was staying, belonged to the condition he defined as "virginal anxiety," a combination of neurotic anxiety and hysteria, at the origin of which he had reconstructed a sexual assault suffered at the age of 14. (Perhaps under pressure from Breuer, the text played down the abuse, attributing it to her uncle instead of her father.)[5]

While Breuer expounded in detail the technique used in his cathartic treatment, Freud was already describing the modifications it led to, and, in the case of Elisabeth von R, he demonstrated that he had completely abandoned hypnosis. This patient, aged 45, suffered from strange pains in her legs which prevented her from walking even a few steps. Like Anna O, she had nursed and lost her father: the hysterical symptoms had presented themselves in their full gravity a few years later, at the bedside of her dead sister, whose husband she had fallen in love with without realising it. The treatment, which enabled the events of that pathogenic scene to be reproduced with violent emotion, was the first complete analysis of a hysteria carried out by Freud using a procedure which the two protagonists likened to the technique of excavating a buried city, layer by layer.

In order to overcome the "resistance" to remembering, Freud had used a technique of concentration which involved applying pressure with a hand on the patient's forehead and inviting her to talk, with her eyes closed, about the first *Einfall* – image, idea, or recollection – which came into her mind. Here the fundamental rule of psychoanalysis became explicit for the first time: the method of free association based on the bringing to light, among the individual *Einfälle*, of threads, concatenations, intersections, and nodal points, psychic "associations" which, while seeming "free" of conscious intention, could be recognised as unconsciously determined.

The text concluded with a theoretical chapter by Breuer and one by Freud devoted to the psychotherapeutic technique.

Breuer showed considerable caution:

> A feeling of oppression is bound to accompany any such descent to the "Mothers" [i.e., exploration of the depths]. But any attempt at getting at the roots of a phenomenon inevitably leads in this way to basic problems which cannot be evaded. I hope therefore that the abstruseness of the following discussion may be viewed with indulgence.
>
> (1893, p. 192)

Freud's last technical chapter also contained some theoretical reflections, quite different from his colleague's, and he spent nearly a year completing it: this time, he was the one symptomatically delaying the work since, having drafted the last case study in May 1894, he had only managed to write the chapter on technique in March 1895, which he did in a rush over the space of ten days.[6]

We can suppose that the block was caused by the personal crisis which Freud had to go through, by the difficulties arising in the collaboration with Breuer, and by the clinical impasse with a patient who was particularly resistant to treatment. Freud even gave up smoking at this time. . . . We will come back to these circumstances, which led him to make a decisive turn in the development of his thought.

In the *Studies*, the divergences between the two authors were evident. After they had together discovered the significance of the sexual content in the cases under observation, Freud judged this to be a fundamental fact and was already applying it to all the neuroses in general: thus, he had distinguished the "actual neuroses" (anxiety neurosis, neurasthenia, and hypochondria),[7] which in his opinion involved habitual sexual dissatisfactions of adult life (coitus interruptus, abstinence, masturbation), from the "neuro-psychoses of defence" (hysteria, phobia, obsessional neurosis, and certain hallucinatory conditions), in which he speculated that the repressed sexual material derived from traumatic experiences in childhood or abuses committed by parental figures or their representatives (tutors, teachers, etc.).

Breuer was not inclined to share this view and considered the subjective hypnoid element central: in other words, the laxity of the nexuses in the state of consciousness which, heightened by automatisms or repetitive activities, undermined the individual's cohesion and predisposed him to dissociation.

In his conclusions, he declared himself in favour of an ecumenical vision with room for the various hypotheses which medicine had hitherto formulated about hysteria: from the old "reflex theory" – by stimulation of the genitals, uterus, and/ or ovaries – to innate disposition; to the recent "splitting of the personality," proposed by Janet; to the new "hysteria of defence" maintained by Freud.

A distinctive innovation of the *Studies* was their highlighting of the patient's mental gifts, describing them as "people of the clearest intellect, strongest will, greatest character and highest critical power."[8] Here Breuer and Freud found themselves united against the widespread hypothesis of innate psychological weakness. Breuer tended to think in terms of "an excessive degree of mental mobility" and even in terms of the triggering action of an "*excess* of efficiency,"[9] while Freud, in considering how the "conversion symptom"[10] is produced, was tempted by the hypothesis of "a superior intelligence outside the patient's consciousness," one capable of managing a large quantity of psychic material for determined purposes and arranging "a planned order for its return to consciousness."[11]

He noted *en passant* how the emergence into the patient's mind of "isolated key-words," at first sight disconnected but susceptible of being elaborated into sentences and connected to ideational contents of the pathogenic factor that was being sought, took the form of words emitted in an almost "oracular fashion."[12]

A difficult separation: not all debts can be paid

The seed of psychoanalysis is sown in the *Studies*, and yet during its troubled gestation, Freud and Breuer's scientific collaboration was decisively compromised. The friendship had already ended in the spring of 1894, and by the time the text went to press in May 1895, the relationship was long over. The second edition in 1908, when psychoanalysis was a known and clearly defined discipline, included two separate authorial prefaces.

Breuer's reads laconically,

> So far as I personally am concerned, I have since that time had no active dealings with the subject; I have had no part in its important development and I could add nothing fresh to what was written in 1895.
>
> (Breuer and Freud, 1893–95, p. xxxi)

The republication entailed an exchange of letters since Breuer saw no justification for it and was unhappy about being associated with psychoanalysis:[13] if the book had to appear, he would have preferred his former friend to put out new chapters to replace his own, with the necessary updates. But he had to resign himself to the reprinting of the unchanged 1893–95 text, which, for Freud, had by this point become a historical record as well as a testament of their association.

As it happened, Breuer could not resist the cheque for 2000 crowns with which Freud aimed to repay the old loan – Breuer claimed to have had no interest in keeping track of what he was owed and considered the debt repaid – and he in return gave Freud the proofs of the *Preliminary Communication* which had remained in his possession: on 19/10/1907,[14] he wrote that he considered the matter entirely closed. Those debts had been settled.

Many people subsequently wondered about Breuer's withdrawal, his stopping short on the threshold of psychoanalysis despite being cited by Freud as part of its beginnings. Binswanger recalls the vivid gestures and facial expressions with which Breuer responded to his naive question about his position *vis-à-vis* Freud since the *Studies*:

> His look of downright pity and superiority as well as the wave of his hand, a dismissal in the full sense of the word, left not the slightest doubt that in his opinion Freud had gone scientifically astray to such an extent that he could no longer be taken seriously, and hence it was better not talk about him.
>
> (1956, p. 4)

The official explanation, handed down to the first generation of analysts, emphasised Breuer's emotional difficulties and his resistances in encountering patients' unconscious sexuality.

The two colleagues had never seen eye to eye about this. With these words, Freud explained the rupture of their relationship in 1914:

> [The breach] had deeper causes, but it came about in such a way that at first I did not understand it; it was only later that I learnt from many clear indications how to interpret it. It will be remembered that Breuer said of his famous first patient that the element of sexuality was astonishingly undeveloped in her and had contributed nothing to the very rich clinical picture of the case. . . . Anyone who reads the history of Breuer's case now in the light of the knowledge gained in the last twenty years will at once perceive the symbolism in it – the snakes, the stiffening, the paralysis of the arm – and, on taking into account the situation at the bedside of the young woman's sick father, will easily guess the real interpretation of her symptoms; his opinion of the part played by sexuality in her mental life will therefore be very different from that of her doctor. In his treatment of her case, Breuer was able to make use of a very intense suggestive rapport with the patient, which may serve us as a complete prototype of what we call "transference" today. Now I have strong reasons for suspecting that after all her symptoms had been relieved Breuer must have discovered from further indications the sexual motivation of this transference, but that the universal nature of this unexpected phenomenon escaped him, with the result that, as though confronted by an "untoward event", he broke off all further investigation. He never said this to me in so many words, but he told me enough at different times to justify this reconstruction of what happened.
>
> (1914a, pp. 11–12)

And in 1925:

> After the work of catharsis had seemed to be completed, the girl had suddenly developed a condition of "transference love"; he had not connected this with her illness, and had therefore retired in dismay. It was obviously painful to

him to be reminded of this apparent contretemps. His attitude towards me oscillated for some time between appreciation and sharp criticism; then accidental difficulties arose, as they never fail to do in a strained situation, and we parted.

(1925a, p. 26)

Freud admitted that, as an innocent and ignorant young academic at the time of the *Studies*, he himself had not been ready to acknowledge the importance of sexuality in the neuroses and had been very surprised to learn that this hypothesis was in fact the most frequent subject of his older colleagues' jokes, though not organised or recognised scientifically.[15] After his initial reluctance, he came to take the sexual aetiology of hysteria much more seriously than his authoritative and disenchanted teachers, Breuer included, and made it the lynchpin of his own theory. So he was able to explain the complete sexual ignorance of his hysterical patients even once they had reached maturity, as motivated by a specific presence of repression which exceeded the usual resistances that go by the name of shame, disgust, and morality. This almost instinctive flight from applying the intellect to the sexual problem corresponded to its exact contrary in the unconscious, to an excessive pressure from the sexual drive: and this is how he went on to complete his hypothesis about hysteria in the *Three Essays on the Sexual Theory* (1905b).

Breuer, while acknowledging its importance – "I do not think I am exaggerating when I assert that *the great majority of severe neuroses in women have their origin in the marriage bed*"[16] – rejected Freud's interpretation, which he considered one of various possible hypotheses, and he never tired of repeating that in the case of Anna O, the hysteria had developed and been resolved without the emergence of any sexual material.

We can find his version of events in his letter replying to Auguste Forel, a friend from university days, who asked him in 1907 for an account of his personal contribution to the development of the Freudian discipline that was now having large claims made for it. At that time, Forel, who like many of his colleagues publicised some important reservations about the doctrine of infantile sexuality, seemed to be alarmed at the "psychoanalytic" developments which Bleuler was introducing at the *Burghölzli* (and probably by the uproar being stirred up in Amsterdam – at the First International Congress of Psychiatry, Neurology, and Psychology – by the public clash between Kraepelin's student Gustav Aschaffenburg and the rising Freudian star Carl Jung).[17]

This is what Breuer wrote:

Dear Sir,
My reply to your friendly letter has been delayed for a few days because I had to look through the old book (*Studies on Hysteria*) again, to refresh my memory. I have now done so and I must confess that, contrary to my expectations, I was really quite pleased with it.

As regards the question of the ideas contributed by the two authors to the various elements of the theory involved and of psychotherapeutic analysis, I must begin by explaining that I am unable to discuss the matter with Freud since for several years I have had no personal contact with him. It is, however, very difficult, purely from one's own memory, to separate the contributions in the case of work carried out so very much in common as that which produced this book. The case which I described in the *Studies* as No.1, Anna O., passed through my hands, and my merit lay essentially in my having recognized what an uncommonly instructive and scientifically important case chance had brought me for investigation, in my having persevered in observing it attentively and accurately, and my not having allowed any preconceived opinions to interfere with the simple observation of the important data. Thus at that time I learned a very great deal: much that was of scientific value, but something of practical importance as well – namely, that it was impossible for a "general practitioner" to treat a case of that kind without bringing his activities and mode of life completely to an end. I vowed at the time that I would *not* go through such an ordeal again. When cases came to me, therefore, which I thought would benefit much from analytic treatment, but which I could not treat myself, I referred them to Dr Freud, who had returned from Paris and the Salpêtrière and with whom I had the most intimate friendly and scientific relations. These cases, their course, their treatment, and whatever contributions to theory arose from them, were naturally constantly discussed between us. In this way our theoretical views grew up – not, of course, without divergencies, but nevertheless in work that was so much carried out in common that it is really hard to say what came from the one and what from the other.

This much, however, I believe I can say:

What follows immediately from the case of Anna O. is mine – that is to say, the aetiological significance of affective ideas, deprived of their normal reaction, which operate permanently like psychical foreign bodies; "retention hysteria"; the realization of the importance of hypnoid states in the development of hysteria; analytic therapy.

As regards phobias, we were naturally very much inclined to suspect the same aetiology; Freud was greatly surprised when analyses often produced no result, but sexual abnormalities (coitus interruptus, etc.) were with ever greater frequency and certainty found to be the cause. (I regard this as a discovery of the greatest practical importance.)

We were no less filled with doubt and surprise when the analysis of severe cases of hysteria (e.g. of "Caecilie M." in the book) led us further back into childhood. This too, of course, was entirely a discovery of Freud's.

Freud is wholly responsible for the "conversion of affective excitation", for the theory of the "neuroses of defence" and for the enormous importance of "defence" in the formation of ideational complexes "inadmissible to consciousness" from which the splitting of the psyche (*double conscience*) arises. In comparison with this, the pathological effects of the "hypnoid states" seemed to him negligible – which was not, I think, to the benefit of his theory.

Together with Freud I was also able to observe the prominent place assumed by sexuality [*das Vordrängen des Sexualen*], and I can give an assurance that this arose from no inclination towards the subject but from the findings – to a large extent most unexpected – of our medical experience.

Freud is a man given to absolute and exclusive formulations: this is a psychical need which, in my opinion, leads to excessive generalization. There may in addition be a desire *d'épâter le bourgeois* [*sic*]. In the main, however, his views on the question are derived, as I have said, simply from experience; and anything that goes beyond it is merely fulfilling the law of the swing of the pendulum, which governs all development. In earlier times all hysteria was sexual; afterwards we felt we were insulting our patients if we included any sexual feeling in their aetiology; and now that the true state of things has once more come to light, the pendulum swings to the other side.

The case of Anna O., which was the germ-cell of the whole of psycho-analysis, proves that a fairly severe case of hysteria can develop, flourish, and be resolved without having a sexual basis. I confess that the plunging into sexuality in theory and practice is not to my taste. But what have my taste and my feeling about what is seemly and what is unseemly to do with the question of what is true? I have already said that personally I have now parted from Freud entirely and naturally this was not a wholly painless process. But I still regard Freud's work as magnificent: built up on the most laborious study in his private practice and of the greatest importance – even though no small part of its structure will doubtless crumble away again.

With kindest regards,

J. Breuer (21/11/1907)[18]

Breuer's ambivalence is obvious in the letter, as is his difficulty in thinking back on an experience which had touched him so deeply that he claimed it had put his very existence in danger, but also a certain irritation and scorn at the emancipatory pretensions of his younger colleague:[19] he did not deny the importance and ubiquity of the sexual material encountered in the patients but insisted on the exceptional nature of his exemplar case study.

He had clearly underlined this in the *Studies*:

> The element of sexuality was astonishingly undeveloped in her. The patient, whose life became known to me to an extent to which one person's life is seldom known to another, had never been in love; and in all the enormous number of hallucinations which occurred during her illness that element of mental life never emerged.

(pp. 21–22)

We may suppose that Breuer was entrenching his position behind the absence in Anna O of manifest problems associated with masturbation, an area that was usually investigated with caution since the medical practice of the time regarded it as peculiarly damaging to the health,[20] and that he had a symptomatic resistance to grasping the elements of repressed sexuality hidden behind the young woman's marked inhibition. For example, colluding with the patient's resistances, he had been eager to rationalise the hallucination of the "black snake" as connected to the sight of snakes in the field behind the house,[21] declining to address the unconscious symbolic meaning of the symptom and forgetting the trauma caused by the girl's daily contact with the body of the ill father she was nursing in her parents' bedroom. And even though, from the outset, he had noted the father's affective importance in the appearance of Anna O's symptoms, he had underestimated the influence of his own lively and attentive masculine and paternal presence at her bedside.

It never seemed to have crossed his mind that the young woman's condition worsened during his absences or that she might be using English in order to be understood only by him and not by the nurse; nor did he ever speculate, for example, about the presence of the Englishwoman who was the children's nanny and the father's lover.[22]

The manifesting of unconscious sexuality in the transference onto the physician had caught the older colleague unprepared.

Freud continued to wonder for a long time what had really happened between Breuer and Bertha Pappenheim.[23] In his autobiography, he wrote:

> But over the final stage of this hypnotic treatment there rested a veil of obscurity, which Breuer never raised for me; and I could not understand why he had so long kept secret what seemed to me an invaluable discovery instead of making science the richer by it.
>
> (1925a, pp. 20–21)

The following year, in his colleague's obituary, he was more specific:

> He had come up against something that is never absent – his patient's transference on to her physician, and he had not grasped the impersonal nature of the process. At the time when he submitted to my influence and was preparing the *Studies* for publication, his judgement of their significance seemed to be confirmed. "I believe", he told me, "that this is the most important thing we two have to give the world."
>
> (1925b, p. 280)

A foundation myth: a false pregnancy and a cure with a defect

> Breuer too has a very high opinion of her, and gave up her care because it was threatening his happy marriage. His poor wife could not stand the fact that he was so exclusively devoting himself to a woman about whom he obviously

spoken with great interest. She was certainly only jealous of the demands made on her husband by another woman. Her jealousy did not show itself in a hateful, tormenting fashion, but in a silently recognized one. She fell ill, lost her spirits, until he noticed and discovered the reason why. This naturally was enough for him to completely withdraw his medical attention from B.P.

(Sigmund to Martha, 31/10/1883)

It has often been on the tip of my tongue to ask you why Breuer gave up Bertha. I could well imagine that those somewhat removed from it were wrong to say that he had withdrawn because he realized that he was unable to do anything for her. It curious that no man other than her physician of the moment got close to poor Bertha, that is when she was healthy she already [had the power] to turn the head of the most sensible of men – what a misfortune for the girl. You will laugh at me, dearest, I so vividly put myself in the place of the silent Frau Mathilde that I could scarcely sleep last night.

(Martha to Sigmund, 2/11/1883)[24]

In 1953, Ernest Jones added some details to the mystery of how Anna O.'s treatment ended and constructed the version of the patient's "hysterical pregnancy" which would be handed down as the official one.

Freud has related to me a fuller account than he described in his writings of the peculiar circumstances surrounding the end of this novel treatment. It would seem that Breuer had developed what we should nowadays call a strong counter-transference to his interesting patient. At all events he was so engrossed that his wife became bored at listening to no other topic, and before long jealous. She did not display this openly, but became unhappy and morose. It was long before Breuer, with his thoughts elsewhere, divined the meaning of her state of mind. It provoked a violent reaction in him, perhaps compounded of love and guilt, and he decided to bring the treatment to an end. He announced this to Anna O., who was by now much better, and bade her good-bye. But that evening he was fetched back, to find her in a greatly excited state, apparently as ill as ever. The patient, who according to him had appeared to be an asexual being and had never made any allusion to such a forbidden topic throughout the treatment, was now in the throes of an hysterical childbirth (pseudocyesis), the logical termination of a phantom pregnancy that had been invisibly developing in response to Breuer's ministrations. Though profoundly shocked, he managed to calm her down by hypnotizing her, and then fled the house in a cold sweat. The next day he and his wife left for Venice to spend a second honeymoon, which resulted in the conception of a daughter; the girl born in these curious circumstances was nearly sixty years later to commit suicide in Vienna to escape the Nazis.

(Jones, *I*, p. 246–247)[25]

Freud never referred to this directly in his published writings, however. He only did so in a letter to Stefan Zweig, who in 1932 had sent Freud a copy of the Italian translation of *Mental Healers*. This is how he began:

> I discovered on page 272 an error of representation which cannot be looked upon as unimportant and which, if you don't mind my saying so, actually belittles my merit. It declares that Breuer's patient under hypnosis made the confession of having experienced and suppressed certain "sentimenti illeciti" (i.e., of a sexual nature) while sitting at her father's sickbed. In reality she said nothing of the kind; rather she indicated that she was trying to conceal from her father her agitated condition, above all her tender concern. If things had been as your text maintains, then everything else would have taken a different turn. I would not have been surprised by the discovery of sexual etiology, Breuer would have found it more difficult to refute this theory, and if hypnosis could obtain such candid confessions, I probably would never have abandoned it.

At this point, he added a recollection from the early period of his friendship with Breuer, shortly after his engagement to Martha:

> What really happened with Breuer's patient I was able to guess later on, long after the break in our relations, when I suddenly remembered something Breuer had once told me in another context before we had begun to collaborate and which he never repeated. On the evening of the day when all her symptoms had been disposed of, he was summoned to the patient again, found her confused and writhing in abdominal cramps. Asked what was wrong with her, she replied: "Now Dr. B.'s child is coming!" At this moment he held in his hand the key that would have opened the "doors to the Mothers," but he let it drop. With all his great intellectual gifts there was nothing Faustian in his nature. Seized by conventional horror he took flight and abandoned the patient to a colleague. For months afterwards she struggled to regain her health in a sanatorium.

This was in fact a retrospective intuition which, right to the end, he would have liked his colleague to confirm:

> I was so convinced of this reconstruction of mine that I published it somewhere. Breuer's youngest daughter (born shortly after the above-mentioned treatment, not without significance for the deeper connections!) read my account and asked her father about it (shortly before his death). He confirmed my version, and she informed me about it later.
>
> (2/07/1932)[26]

And yet, even though Freud said he was convinced he had published the reconstructed story somewhere, it does not feature in the *Collected Works*, and so it could not have been read by Breuer's daughter, nor confirmed by her father on his deathbed.

In the same year, he reflected further on the transferential circumstances of Breuer's withdrawal from his patient in the correspondence with Sir Arthur G Tansley, professor of botany at Cambridge and member of the Royal Society who had had an analysis in Vienna in 1922 and later became a "lay" associate of the British Psychoanalytical Society. In the letter of 20/11/1932, Freud remarked that one should speak in terms of *a cure with a defect*.[27] Recalling how Breuer's case study had already pointed out that it took some time for the young Anna O to regain her health after the conclusion of the treatment, Freud noted how her subsequent life as a woman had not been completely fulfilled, since she had never married or had any amorous relationships. The cure seemed to have happened on the condition that she wholly renounce her sexual function, as Freud interpreted it, and the very work of dedication and care for others to which she passionately devoted her entire life – the defence of women's rights – could be connected to this internal sacrifice.

In reality, after the interruption of the treatment with Breuer, Bertha Pappenheim had to be hospitalised for several months in Switzerland at the Bellevue Sanatorium in Kreuzlingen. As Skues points out, initial historical research considered the cathartic cure a failure, questioning the genuineness of the results reported in the *Studies*: it based this on a hasty interpretation of the statements made by Freud, Jung, and Jones. Bertha did not return to the family but had a period of convalescence and a further hospitalisation linked to the emergence of a trigeminal neuralgia that made her dependent on morphine. Furthermore, the underlying psychic pathology on which the hysterical symptoms had developed had not been resolved.[28]

Freud was well aware of this: from Martha's correspondence with her mother, we can learn how in 1887, the young woman still sometimes suffered from evening hallucinatory states.[29] It was at that time that Bertha Pappenheim decided to devote herself to studies as a social worker: she later became the first German woman to obtain such a diploma and was one of the leading figures of the international feminist movement of the time. She died in 1936 in Neu-Isenburg at the age of 77, shortly after being interrogated by the Gestapo.

It has often been noted how the private version passed on to Zweig and Jones – and subsequently taken as actual fact by most historians of psychoanalysis (even by Peter Gay) – contains some incongruities,[30] beginning with the circumstances of Dora Breuer's conception, given that she was born three months before the treatment was broken off and not after.[31]

As Freud himself informs us, he had never been given the details by Breuer but had reconstructed them retrospectively, based on the re-emergence of a memory from 1883, when his colleague, recounting the case for the first time and urging him not to talk about it to his fiancée, had confided in him about Mathilde's jealousy and depression. But the jealousy cannot only have been Mathilde's: we can imagine that the symptomatology which appeared in Anna O during the closing period of the treatment might represent the patient's jealous reaction to her physician's wife being pregnant, imitating it hysterically as a way of keeping him beside her.

Today we can understand how the legend which passed into history – Anna O's hysterical pregnancy, the treatment being broken off by Breuer as a result, the episode of his departure with his wife, and Dora's conception in Venice – may represent the construction of a foundation myth. It was the product of the ongoing re-elaboration by which Freud who, over the years, had become capable of *guessing* the erotic implication of the treatment through clinical experience with patients' transference, had made sense of that first scene narrated by his colleague and had inserted some unconscious personal elements into it, as we shall see.

It is likely that Josef Breuer loved his patient without realising it and that the desire to impregnate her was, above all, his: this "oedipal hell"[32] cannot have crossed the doctor's mind, but Mathilde had spotted the danger, as had Martha when Freud wrote to her about the case. Breuer had responded with an unusual dedication to the seductive desire for exclusivity demonstrated by Bertha Pappenheim, and something very intimate had come into play between them. The treatment of two or more visits a day for such a long period was certainly not a customary therapeutic model: there were times when Bertha would only take food in the presence of her doctor, and when she made a suicide attempt, he left himself especially exposed by taking personal responsibility for not admitting her to hospital.[33]

It has been pointed out that not only oedipal but traumatic elements may have been stirred up in Breuer by his experience with this particular patient.[34] Both had suffered early bereavements. Moreover, Bertha had the same name as Breuer's mother and was the same age as the mother was when she died in the flower of her youth and beauty while her son was too young to fix any image of her in his memory. And Bertha was the name he had given his eldest daughter, in memory of his mother.[35]

The physician had underestimated the affective experiences of the therapy, and it is no surprise that this very denial had made his wife so jealous that it forced her to take the initiative, first with a new conception and then by presenting herself as ill (there had even been a suicide attempt!).[36] After his exceptional paternal investment in the young woman, Breuer solved the problem of the threat posted to his marriage, anxiety, and sense of guilt by giving up the treatment of hysteria. The situation must have been dramatic indeed to stop Breuer taking any more hysterical patients into treatment.

And we can speculate that it had been his own resistances to addressing the counter-transferential disturbances of the case – the incestuous seduction that remained unconscious – which impelled him to invest his daring and adventurous young colleague, Sigmund Freud, and no one else, with the role of confidant and hidden supervisor, passing on to him at the same time a legacy of knowledge and "defects" to be worked through: a burdensome and challenging legacy, but above all an unconscious one.

Breuer's letter to the authoritative Forel still makes clear the weight of these responsibilities and the need to justify himself, to keep his distance from the result of the collaboration with his student and from the development which Freud had given it against Breuer's wishes.

Sexuality was certainly the cause of Breuer's separation from Freud: Anna O was at the same time "his point of light and his blind spot."[37] But the original founding legend of contagion and hysterical pregnancy, with its repercussions which leap from the scene of the treatment to strike at the physician's marital relations, remains impressive because, like all myths, it contains and conveys an important kernel of truth.

Notes

1 18/12/1892.
2 1893, p. 7.
3 The clinical history was not presented in full for reasons of confidentiality (Freud, 1893, p. 69 n.). However, it was the collaboration on this remarkable case (to Fliess, Freud called her his "teacher," 8/02/1897) that convinced them to write the *Preliminary Communication* (Ibid., p. 178) and perhaps the pseudonym of Anna O. is a homage to her. "She was, so to speak, Freud's own Anna O" (Swales, 1986, p. 6).
4 1893, p. 49.
5 As Freud explains in a note added in 1924.
6 Salyard, 1992.
7 Freud, 1894b, Draft H and 1894a.
8 1893, p. 13.
9 Ibid., p. 233.
10 The term *Conversion* was introduced by Freud (1894, p. 352, n. 2): it is the mechanism of transformation of psychic excitement into somatic symptoms, not corresponding to the innervation, which he had identified as typical of hysteria.
11 1893, p. 272.
12 Ibid., p. 276.
13 Hirschmüller, 1994.
14 Jones, *II*; Hirschmüller, 1994.
15 Freud, 1914a.
16 Breuer and Freud, 1893–95, p. 246.
17 Freud had been invited but refused to participate (Falzeder and Burnham, 2007).
18 In Cranefield, 1958: reproduced with kind permission of the Estate of Paul F Cranefield, *Int. J. Psychoanal.*, Taylor and Francis, and the Archives of the Institute of the History of Medicine, University of Zurich.
19 For him, Freud perhaps represented a younger brother (Pollock, 1968).
20 Hirschmüller, 1978; Bonomi, 2007.
21 Breuer, 1893a, Fräulein Anna O, Case Histories from *Studies on Hysteria*. *SE*, 2, p. 38.
22 As Britton proposes (2003).
23 The question should still be left open, taking into account the possibility of new documentation emerging (Skues, 2006).
24 In Forrester (1990).
25 As evidence, Jones reported a confrontational episode between the two colleagues, about ten years after the end of that treatment, in the period in which they still used to discuss clinical cases together: regarding another patient of his, Breuer reacted with extreme annoyance at Freud's interpretation of a phantasy pregnancy (*I*, p. 248).

26 In Freud E., 1970.
27 See Forrester and Cameron, 1999.
28 Skues, 2006.
29 Jones, I.
30 Pollock (1968), Ellenberger (1970), Hirschmüller (1978), Traversa (1984), Rodrigué (1996), Gilhooley (2002), Roudinesco (2014).
31 Dora Breuer was born in March 1882. Jones introduced a final inaccuracy in connection with her death: she did not commit suicide in New York but died in Vienna, a few years after the *Anschluss*, in 1942 (Anzieu, 1959).
32 Rodrigué, 1996, I, p. 240.
33 Hirschmüller, 1978.
34 Pollock, 1968.
35 Hirschmüller, 1978.
36 See Goleman, 1985.
37 Rodrigué, 1996, I, p. 240.

Chapter 5

Sigmund Freud's lesson

The discovery of *a false connection*

With his intuition about the "false" pregnancy, Freud seems to have "wrested from Breuer paternity for the birth of psychoanalysis,"[1] almost creating an illegitimate parent-child relationship. We can detect something in the legend which the two colleagues experienced independently from each other and for which both were completely unprepared. The transferential experience, which Breuer lived through first and kept repressed, formed the seed that was passed on to Freud, a seed which the student accepted and allowed to germinate. When he in turn found himself having to face the awkward setback of the transference, he was in a position not to reject it completely but to begin working through it, first of all by projecting it back onto his mentor. This was the fact that patients could be made to fall in love with the therapist, that aspect of therapy which Freud had grasped and immediately ruled out because of his own potent and hyper-controlled desires, convinced that it was a charismatic and seductive characteristic that he did not possess, one that could only belong to another, "a Breuer": this enabled him not to flee the burden of responsibility and desire as his colleague had done but to address the phenomenon in a different way and set out to observe it.

The first direct experience of transference (and counter-transference) did not catch him unprepared but set him a test: it was probably something to do with those nine months of obstruction which delayed the conclusion of the *Studies*, accompanied the divergences from his older colleague, and supported the investment in a new friendship with his younger colleague, Wilhelm Fliess.

It is likely that during these months, Freud found himself at an impasse in a treatment where not even applying pressure to the patient's forehead had proved capable of making recollections emerge. It is the case briefly referred to in that final section of the *Studies* which had been so delayed, and which would later be read, re-read, and quoted by his students:

> In one of my patients the origin of a particular hysterical symptom lay in a wish, which she had had many years earlier and had at once relegated to the unconscious, that the man she was talking to at the time might boldly take

DOI: 10.4324/9781003246473-6

the initiative and give her a kiss. On one occasion, at the end of a session, a similar wish came up in her about me. She was horrified at it, spent a sleepless night, and at the next session, though she did not refuse to be treated, was quite useless for work. After I had discovered the obstacle and removed it, the work proceeded further; and lo and behold! the wish that had so much frightened the patient made its appearance as the next of her pathogenic recollections and the one which was demanded by the immediate logical context. What had happened therefore was this. The content of the wish had appeared first of all in the patient's consciousness without any memories of the surrounding circumstances which would have assigned it to a past time. The wish which was present was then, owing to the compulsion to associate which was dominant in her consciousness, linked to my person, with which the patient was legitimately concerned; and as the result of this *mésalliance* – which I describe as a "false connection" – the same affect was provoked which had forced the patient long before to repudiate this forbidden wish.

(1893, pp. 302–303)

Here is the first description of the phenomenon of transference onto the doctor through a "false connection": the product of a compulsion, it is presented as a new hysterical symptom, albeit a milder one, which develops during the treatment, replacing the initial symptom and going on to constitute a new and troubling type of resistance, a possible cause of failure. Freud calls it a *mésalliance*, a serious mismatch between individuals whose situations are quite different: this involvement of the doctor in the symptoms, he writes, is "the worst obstacle that we can come across," but, he confidently adds, it necessarily appears "in every comparatively serious analysis."[2]

It is the *defect* that Breuer did not work through. Freud, for whom failure was out of the question, succeeded in addressing his own personal involvement as a datum to be understood scientifically rather than on the personal level, thereby laying the cornerstone of psychoanalysis.

The protagonist of the session in question has been identified by A Salyard as Emma Eckstein,[3] the sister of Freud's friend Fritz Eckstein. Trapped for months in this patient's new and ruinous resistance to treatment, Freud had to seek help from a colleague, and obviously not from Breuer. During Christmas 1894, called from Berlin for a consultation, Wilhelm Fliess decided to intervene personally in the treatment and to subject the patient to an operation on her nose which had disastrous consequences but also unblocked the situation (it should be noted that, during the same period, he also was operating on the nose of Emma's physician, his own colleague Freud!). It was after this *cut*, symbolic as well as literal, that Freud succeeded in getting through the blockage in Emma's analysis and in the progress of his own thinking, conceived the idea of transference, and completed the writing of the *Studies*.

With this still sketchy intuition – the recognition of the transference – the analytic scene becomes the scene of a repetition, an echo and promise of some other, unfinished scene: the patient's question is addressed to someone else, and Freud appends a new address to it, but, as Derrida says in his critique of Lacan, at least part of the letter inevitably reaches its destination.[4]

It is not out of the question that, as Bertha Pappenheim had done to Breuer, Emma Eckstein had unleashed some kind of disturbance in Freud and in his own marital relationship. By making a simple calculation, we discover that in March 1895, at the same time as Fliess's Solomonic intervention, Sigmund and Martha "accidentally" conceived a sixth child, a girl who would be given the name Anna: is this the *true* pregnancy hidden in the myth that has been passed down to us?

So, like Anzieu, we could read the legend of Anna O's treatment as a projection onto Breuer of the ambivalent infantile desire, which was re-emerging transferentially in Freud, to impregnate the mother and conceive the sister.[5] Not consciously desired but unconsciously determined, the pregnancy that will lead to the birth of the last child will not restore vigour to Freud's marriage – in fact, it will mark the start of a silent marital crisis – but it will influence the conception and parallel gestation of psychoanalysis, which will begin to show itself in the dream of Irma's injection.

As fate would have it, both Anna Freud and Robert Wilhelm Fliess, the eldest child of his friend from Berlin, conceived and born a few days apart in December 1895, being the children of such fathers, decided to devote themselves to psychoanalysis when they grew up.

Precious and unconscious erotic, transferential, and counter-transferential intertwinings, present in the treatments, in the amorous conjugal relationships, and in friendships between colleagues were indeed at the origin of psychoanalysis, of its theories as well as of its myths.

Irma's throat and the feminine at the origin of psychoanalysis

The *Studies on Hysteria* had recently been published, and quite a favourable review had appeared in the *Wiener Medizinischer Presse*[6] when, on the night of 23 July 1895, Sigmund Freud, holidaying with his family in the Belle Vue hotel on the hills near Vienna, had a dream which would be the first to be interpreted in depth and would be used in 1899 as an exemplar for *The Interpretation of Dreams*: it is the dream of Irma's injection. Just at the time when he was freeing himself from Breuer and seemed, with the method of free associations, to have given up any interest in induced sleep and hypnosis, Freud made a bold volte-face and returned to the phenomenon to view it from a new position and master it. He set himself the task of exploring that particular and mysterious condition of consciousness represented by the dream, *the mother of all second conditions*, the hypnoid state *par excellence*.

Since childhood Freud had been a good dreamer, interested in dreams, and in the habit of noting them down.[7] During 1889 and 1890, during the treatment of Emmy von N, he had begun to transcribe the dreams he had over the course of several weeks when he had had to sleep on a harder bed: vivid dreams which he was able to recall vividly on waking. He had attributed them to the need to elaborate on as yet incomplete representations of the previous day and to the compulsion to establish nexuses with the circumstances in which these representations had emerged.[8] Applying the new technique, Freud had started to intuit the profound similarity between the structure of the neuroses and that of the dream:

> My patients were pledged to communicate to me every idea or thought that occurred to them in connection with some particular subject; amongst other things they told me their dreams and so taught me that a dream can be inserted into the psychical chain that has to be traced backwards in the memory from a pathological idea.
>
> (1900, pp. 100–101)

Breuer had previously pointed out the significant presence in Anna O. of day-dreams, *habitual reveries*[9] which preceded and prepared her self-hypnosis.

Deciding to follow Breuer's path all the way, Freud also applied analysis to these enigmatic states, trying to decipher their hidden language. And in this way, beginning with his own dreams, he came to explain that, like the symptom, the dream too had its own meaning:

> I have been driven to realize that here once more we have one of those not infrequent cases in which an ancient and jealously held popular belief seems to be nearer the truth than the judgement of the prevalent science of to-day. I must affirm that dreams really have a meaning and that a scientific procedure for interpreting them is possible.
>
> (1900, p. 100)

Freud discovered that the dream is the satisfaction of an unconscious desire and that its meaning is hidden in its symbolic structure. Like Artemidorus of Daldis before him, the decoding procedure that he constructs takes the fullest account of the dreamer's character and circumstances, but it differs from the interpretations of the ancients (which were unitary and symbolic or rather deciphered the dream element by element) because "it imposes the task of interpretation upon the dreamer himself" (p. 98).

In the dream, as in the psychoneurotic symptom, the satisfaction of the desire must undergo a censorship which masks its meaning, and it is the dreamer's associations which allow it to be revealed by indirect means.

The dream about Irma satisfies more than one of Freud's desires and on various levels of conscious awareness. Today we are in a position to identify and follow multiple associative pathways: he explains some of them directly, while others have been highlighted by later interpreters on the basis of the implicit associations he generously left in the text and in his correspondence: traces and clues for every reader to have a go at.

It is surely the most interpreted dream of all time:

A large hall – numerous guests, whom we were receiving. – Among them was Irma. I at once took her on one side, as though to answer her letter and to reproach her for not having accepted my "solution" yet. I said to her: "If you still get pains, it's really only your fault." She replied: "If you only knew what pains I've got now in my throat and stomach and abdomen – it's choking me" – I was alarmed and looked at her. She looked pale and puffy. I thought to myself that after all I must be missing some organic trouble. I took her to the window and looked down her throat, and she showed signs of recalcitrance, like women with artificial dentures. I thought to myself that there was really no need for her to do that. – She then opened her mouth properly and on the right I found a big white patch; at another place I saw extensive whitish grey scabs upon some remarkable curly structures which were evidently modelled on the turbinal bones of the nose. – I at once called in Dr. M., and he repeated the examination and confirmed it. . . . Dr. M. looked quite different from usual; he was very pale, he walked with a limp and his chin was clean-shaven. . . . My friend Otto was now standing beside her as well, and my friend Leopold was percussing her through her bodice and saying: "She has a dull area low down on the left." He also indicated that a portion of the skin on the left shoulder was infiltrated. (I noticed this, just as he did, in spite of her dress.) . . . M. said: "There's no doubt it's an infection, but no matter; dysentery will supervene and the toxin will be eliminated." . . . We were directly aware, too, of the origin of the infection. Not long before, when she was feeling unwell, my friend Otto had given her an injection of a preparation of propyl, propyls . . . propionic acid . . . trimethylamin (and I saw before me the formula for this printed in heavy type) . . . Injections of that sort ought not to be made so thoughtlessly. . . . And probably the syringe had not been clean.

(1900, p. 107)

For the moment, let's limit ourselves to some of the interpretations which Freud permitted himself to show, reading the preamble to the dream and then setting out the associations to its individual elements in the analysis which he had immediately undertaken on waking. He mentions first of all that the previous day, his colleague and close friend Otto (Oskar Rie, paediatrician and the Freuds' family doctor),[10] had brought him news of a young patient named Irma (the widow Anna Hammerschlag Lichtheim),[11] who was part of their circle of friends (she was the daughter of the Hebrew teacher Samuel Hammerschlag). He begins:

It will be readily understood that a mixed relationship such as this may be a source of many disturbed feelings in a physician and particularly in a psychotherapist. While the physician's personal interest is greater, his authority is less; any failure would bring a threat to the old-established friendship with the patient's family. This treatment had ended in a partial success; the patient was relieved of her hysterical anxiety but did not lose all her somatic symptoms. At that time I was not yet quite clear in my mind as to the criteria indicating that a hysterical case history was finally closed, and I proposed a solution to the patient which she seemed unwilling to accept. While we were thus at variance, we had broken off the treatment for the summer vacation.

(p. 106)

Otto had commented about the patient, "She's better, but not quite well."

Irritated by what he had experienced as a veiled rebuke and a slight to his pride just when he had achieved the goal of publishing the *Studies*, Freud had stayed up late that evening to write Irma's case history with Breuer in mind as his reader, as if he were still there to discuss cases as in the past (he is present in the dream as the supervisor, Dr M, "the leading figure in our circle"). He felt the need for confirmation, or perhaps it was to him above all that he wanted to defend his way of working.

There is another actual circumstance represented in the dream: the party that had been anticipated, Freud explains, was the one being organised in one of the grand salons of the Belle Vue for his wife Martha's 34th birthday (she had recently entered the fourth month of pregnancy). Among the guests would be not only Anna Hammerschlag Lichtheim but also Breuer's former patient, Martha's friend Bertha Pappenheim.[12]

In the dream, alarmed by Irma's pale and puffy appearance, Freud fears that he has failed to notice an organic cause for her illness: he examines her throat and calls in Dr M (Breuer), who is then joined by his colleagues Otto and Leopold (Oscar Rie and Ludwig Rosenberg, Freud's tarot partners, who had been his assistants at Max Kassowitz's paediatric institution). After various exchanges, the dream concludes with the hypothesis that the patient has been infected with an injection administered by Otto, in such a way that the blame falls on the one who had made Freud feel he was at fault. Dr M anticipates the solution – dysentery will eliminate the toxin – and the story of the dream concludes with the admonition, *"Injections of that sort ought not to be made so thoughtlessly. . . . And probably the syringe had not been clean."*

As if he had completed the patient's case history during the night, the first explanation of the dream that Freud allowed himself to make explicit concerns the desire to exculpate himself for the unsatisfactory outcome of the treatment. There is the problem of being compared with teachers and colleagues on the question of the therapy and the aetiology of hysteria and the fragility of what has just been gained. It can be intuited that the search for support and the emergence of rivalry also concern the knowledge and possession of the woman in this gynaecological group visit.

The dream-work which starts out, as Freud explains, from a single set of thoughts that had begun to preoccupy him – "concern about my own and other people's health – professional conscientiousness" – in fact widens to include other areas of responsibility and other mistakes, present and past, connects to less conscious desires, and turns back again to the conflicts of childhood, to the search for a new *solution*: for the present patient, for the woman, but also for himself.

Freud has already identified many mechanisms of the dream-work. He comments that dream-thought knows no contradiction since it uses the logic of the *primary process*:

> The whole plea – for the dream was nothing else – reminded one vividly of the defence put forward by the man who was charged by one of his neighbours with having given him back a borrowed kettle in a damaged condition. The defendant asserted first, that he had given it back undamaged; secondly, that the kettle had a hole in it when he borrowed it; and thirdly, that he had never borrowed a kettle from his neighbour at all. So much the better: if only a single one of these three lines of defence were to be accepted as valid, the man would have to be acquitted.
>
> (p. 119–120)

He shows that the characters in the dream may, through condensation, portray more than one person, combining their features into a single composite dream-image whose associations enable their collective nature (which may even have contradictory aspects) to be revealed.[13]

Thus, in Irma, in a kind of *one for all*, he can recognise not only his patient Anna Hammerschlag but another fascinating friend and widow whom he would have liked to have as a patient – perhaps a more docile and cooperative one – Sophie Schwab Paneth (niece of Samuel Hammerschlag and godmother of Freud's daughter Sophie) but also features of his wife, Martha, who was at that time suffering from a very particular *infection*.

In the fantasy of replacing the patient with another one, just as he would perhaps have liked to replace his wife, the intensity of the desire is made manifest: Freud's wish to possess and also his hostility[14] towards Irma, who does not respond to treatment, and towards Martha, who is unavailable because of her pregnancy.

But that is not all: Freud later confides to Abraham that behind the figure of Irma, because of her "sexual megalomania," were hidden the three women: "Mathilde [Breuer] Sophie [Schwab] and Anna [Hammerschlag Lichtheim] are the three godmothers of my daughters, and I have them all!"[15]

In fact, there are even more than these "three women" condensed into Irma: the associations lead Freud to Fliess's wife Ida Bondy, who is at the same stage of pregnancy as Martha, but we can add Bertha Pappenheim with her hysterical pregnancy, which he may have wanted to treat, and also the other current patient, Emma Eckstein, who had given him very good cause for concern.[16]

In Irma, we find the various ages of woman: from his daughter Mathilde to the very old lady whom Freud used to visit at home for her daily injection. Freud writes:

> *And probably the syringe had not been clean.* This was yet another accusation against Otto, but derived from a different source. I had happened the day before to meet the son of an old lady of eighty-two, to whom I had to give an injection of morphia twice a day. At the moment she was in the country and he told me that she was suffering from phlebitis. I had at once thought it must be an infiltration caused by a dirty syringe. I was proud of the fact that in two years I had not caused a single infiltration; I took constant pains to be sure that the syringe was clean. In short, I was conscientious. The phlebitis brought me back once more to my wife, who had suffered from thrombosis during one of her pregnancies; and now three similar situations came to my recollection involving my wife, Irma and the dead Mathilde. The identity of these situations had evidently enabled me to substitute the three figures for one another in the dream.
>
> (p. 118)

And there's more: we can sum it up by saying that the diphtheria of his daughter Mathilde leads us to the young patients in the paediatric department where Freud had worked with Otto and Leopold. With the hostility towards women and to his own mother, fantasies of abortion, the elimination of children and siblings, come to the fore (Dr M's. abortive *solution*).[17]

The *dirty syringe* which Freud regrets is an allusion to his past mistake with cocaine[18] but, more especially, to his recent responsibility for the unwanted conception and imminent birth of a sixth child which, like his wife's birthday, he doesn't feel inclined to celebrate.

Dream as desire

Freud offers many, and only partial, interpretations of this long and complex dream, the first to be analysed with the systematic application of free associations for each of its manifest elements and the first that leads to the emergence of a latent signification. He cites it in several chapters of the book and returns to it in his private correspondence as well, deepening its interpretation after undertaking his self-analysis and having discovered infantile sexual fantasy. The Irma dream is not *completely* analysed: in any case, in the context of *The Interpretation of Dreams*, Freud's aim is not so much to show the most intimate meanings of his own dreams as to highlight the mechanism by which they are constructed and the possible ways of decoding them.

He was unwillingly compelled to reveal many autobiographical and personal elements in an exposition of his private life which runs throughout the book and which he tried to mask by playing down certain indiscretions, omitting or replacing them where necessary.

Lou Andreas Salome (1861–1937) was struck by this admirable generosity of Freud:

> when I read the *Interpretation of Dreams* and realized what a self-exposure Freud had been forced to make at that time in the use of his data, and in the midst of a scornful and antagonistic crowd at that, I gained respect for the simple heroism of the man's life.
>
> ([1912–13] 1987, p. 163)

The essay is built almost entirely on Freud's own dreams. Some of his family's dreams also appear: one dreamed by his mother, Amalia; one by his daughter Sophie; many by little Anna – who had already been a protagonist, so to speak, of the Irma dream while in her mother's womb – and one by her "twin," Robert Fliess. Freud also reports the dreams of friends and acquaintances, but only a few of patients in psychoanalytic treatment, being reluctant to use them in this first context.

At Berchtesgaden in summer 1908 while preparing the second edition of *The Interpretation of Dreams*, Freud no longer had any hesitation about defending the subjective, autobiographical element of the material being used:

> For this book has a further subjective significance for me personally – a significance which I only grasped after I had completed it. It was, I found, a portion of my own self-analysis, my reaction to my father's death – that is to say, to the most important event, the most poignant loss, of a man's life. Having discovered that this was so, I felt unable to obliterate the traces of the experience.
>
> (p. xxiv)

Today, we are aware of the fact that this bereavement already loomed large in the dream of Irma's injection, since in 1926, Sigmund Freud confided to Marie Bonaparte that, among the dream's premises, he had neglected to mention the fact that, the day before, he had been informed of his father's incurable illness:[19] this precursor naturally brings new lines of interpretation into play and the fear of death as a hidden presence alongside the central theme of sexuality and birth portrayed in the pregnant patient.

With the Irma dream, Freud has the opportunity to resolve an enigma that is both personal and scientific: it really is an exemplary dream, both in the fact of having been dreamed in order to reveal the mystery of dreams and in the fact that it contains and represents the neurotic conflict that is blocking the dreamer's self-realisation. The anticipated solution is internal and external because it realises that creative desire which is the dream of his life.

The simplicity and complexity of the discovery still astonish us. "Insight such as this falls to one's lot but once in a lifetime," Freud will later claim.[20] It is the "stroke of luck" he had been expecting from his study of cocaine,[21] but it is also a hard won and costly achievement, the product of an existential crisis that he had to face in person.

Many things have been said and are yet to be said about this dream, which stands as the *original dream* in the psychoanalytic tradition, the scene of the birth of psychoanalysis, the representative nucleus of an unconscious fantasy against which subsequent generations of students have measured themselves: interpreted in depth but, like all dreams, different and new at each re-reading.

From our viewpoint, it may be interesting to note that in this emblematic dream, dreamed in order to comprehend the difficulties of, and the solution to, the treatment of hysteria that could be shared with the patient herself and with Freud's teachers, it is possible to glimpse the scene of Charcot's "clinical lesson" as it had been reworked in the encounter with the "talking cure" that Freud learned from Breuer and the more recent experience of friendship with Wilhelm Fliess that we are starting to learn about.

The dream prefigures psychoanalysis as Freud is constructing it: here is *his* lesson. Setting out from the death-scene of the anatomical theatre and the spellbinding seduction of gazes in the spectacle of hypnotism, Freud accepts the challenge of hysteria, locates the treatment in the scenario of his own preconscious oneiric area, and puts sexuality and the celebration of love at its centre, the generativity of the couple with its dilemmas and unanswered questions.

The dream's infantile desire is also the desire to know where babies come from: for Anzieu, the dream's architectural spaces, "the female sexual cavity of the *large hall*," the examination of the throat rather than a gynaecological examination, the equation with the folds and cavities of the vagina, the example of the damaged kettle in Freud's associations . . . all of these express his search for the nature of the maternal body and of the scene that resulted in his birth.[22] Standing with Freud around the woman in the lecture theatre-salon, teachers and colleagues, all men, compare notes not only on the cause of the illness – organic or psychic? – and on the topic of conception but also on identity: the patient's and the physician's.

"Who seduced or impregnated the woman? Whose child is this?" Freud seems to take responsibility for what Breuer had not been in a position to maintain: that sexuality manifests itself by transferentially involving the therapist and that the treatment can be represented as a psychic fertilisation. Above all, he is able to address the transgressiveness of his curiosity and desire for achievement, his search for success in his scientific career but also in his relationship with women, the ambitious megalomania and the fear of castration that he experiences as connected to them.

All of this echoes Freud's current situation and his creative desire but also the old "paradisiacal wish to possess the mother's body and to merge the child's body into it."[23] On the subject of the "three women" and the three ages portrayed in Irma, Freud, still struggling with the Sphinx's riddles, warns us that he has failed to come upon any concealed meaning and justifies himself by saying, "There is at least one spot in every dream at which it is unplumbable – a navel, as it were, that is its point of contact with the unknown."[24] With these words, he evokes both the original foetal bond with the mother and the limit and prohibition of incest (the unknown – the not known, in the "biblical" sense).[25]

In exploring the secret of the dream, Freud finds himself plumbing the nurturing tangle of the maternal body, seeking and finding that dream-desire which "at some point where this meshwork is particularly close . . . grows up, like a mushroom out of its mycelium."[26]

But this is not all.

While Freud's lesson is about sexuality, the mystery of generativity, and the feminine, the close focus on the mouth, Irma's throat, introduces us to the therapy of knowledge gained through listening to the particular "perspective of the voice," and not only that of language.[27]

It is here that psychoanalysis, as a practice based on the experience of vocalisation, understood as speech issuing from someone's mouth – a point of tension between the uniqueness of the voice and the system of the language – is differentiated not only from medicine, which uses the gaze to reduce everything to corporeality, but also from philosophy and its deadly logocentrism. The act of speaking, the junction between body and word, "a vibration in throat of flesh, which announces the uniqueness of the one who emits it, invoking the other in resonance,"[28] is acoustic relationality, implying and demanding listening. And it is on the foundation of this listening to "feminine" vocalisation, which constitutes the original nucleus of the psychic in both sexes, that psychoanalysis itself originated.[29]

The dream has often been seen as the expression of Freud's ability to identify with hysterical patients and to achieve a close relationship with hysteria and with the real nucleus of the trauma:[30] the throat is Freud's, a hungry, creative, and suffering throat which expresses the eagerness of his genius. Irma is, above all, Freud himself.[31]

Notes

1 Gilhooley, 2002, p. 97.
2 1893, p. 301.
3 1992.
4 1980.
5 1959.
6 Geller, 2000.
7 Jones, *I*.
8 Freud, 1893, p. 67, footnote.
9 Breuer, 1893b, p. 218.
10 He would later become Fliess's brother-in-law, marrying Melanie Bondy.
11 Anzieu (1959), Hartman (1983), Marie Bonaparte (16/11/1925) in Masson (1984).
12 Anzieu, 1959.
13 To explain how condensation creates a *collective and composite figure*, Freud cites the *composite portraitures* constructed by Francis Galton in an attempt to lay the foundations for a science of heredity that would give a way to perfect the best-endowed specimens of human races – it was he who coined the term "eugenics." Galton superimposed the images of several faces on the same photographic plate, trying to identify

the features of the physiognomy of criminals, the mentally ill, Jews, or members of the same family group. As the author also acknowledged, the resulting images make one think of ghosts: they actually look like the faces of real people, but with typically blurred and faint borders since, while common traits stand out clearly, those that do not match overlap indistinctly.

14 Lopez and Zorzi, 1999.
15 Freud to Abraham, 09/01/1908.
16 Schur, 1972; Roazen, 1975; Clark, 1980.
17 Blum, 1996.
18 See Schur, 1972 and Lavaggetto, 1985.
19 Hartman, 1983; Rodrigué, 1996, I.
20 Preface to the third British and American edition, 1931, Freud, 1900, p. XXXII.
21 Sigmund's letter to Martha cited previously, 21/04/1884.
22 1959, p. 145.
23 Ibid., p. 155.
24 Ibid., p. 111, n. 1.
25 Anzieu, 1959, p. 154.
26 1900., p. 525.
27 Cavarero, 2003, p. 14.
28 Ibid., p. 234.
29 About the seductive appeal of the sonority of the mother tongue in Freud, see Kamieniak, 2000.
30 Freud immediately identified with so-called hysterical patients, women whose symptoms represented the radical rejection of any kind of "oceanic feeling" (Phillips, 2014).
31 Rodrigué claims the authorship of this idea (1996, *I*, p. 35).

Chapter 6

Fliess and the invention of psychoanalysis

A secret correspondence

Today we can still hear the voice of Sigmund Freud in the interview he gave to the BBC in December 1938:

> I started my professional activity as a neurologist trying to bring relief to my neurotic patients. Under the influence of an older friend and by my own efforts, I discovered some important new facts about the unconscious in psychic life, the role of instinctual urges, and so on. Out of these findings grew a new science, psychoanalysis, a part of psychology, and a new method of treatment of the neuroses. I had to pay heavily for this bit of good luck. People did not believe in my facts and thought my theories unsavoury. Resistance was strong and unrelenting. In the end I succeeded in acquiring pupils and building up an International Psychoanalytic Association. But the struggle is not yet over.
>
> (S Freud, Maresfield Gardens, 7/12/1938)[1]

We note that, right to the end, he acknowledged his scientific debt – always and only – to Breuer. Wilhelm Fliess, his privileged interlocutor during the gestation of psychoanalysis, remained entirely unknown for years. In the Preface to the second edition of *The Interpretation of Dreams* (1908), where Freud attributed the autobiographical significance of the book to the working-through of his father's death in 1896 – "the most poignant loss of a man's life" – there was no hint of the vital relationship with Fliess: in the intervening years, this friendship had come to an end, and with as much pain as had accompanied the loss of his father (and the previous break with Breuer).

This relationship encompassed Freud's self-analysis and nurtured the writing not only of the book about dreams but also of *The Psychopathology of Everyday Life* and *Jokes and Their Relation to the Unconscious*: works which, as Freud reiterated in one of his last letters to his friend, he considered – quite apart from the lasting value of their content – "a testimonial" to the role Fliess had played in his life.[2]

DOI: 10.4324/9781003246473-7

In *The Interpretation of Dreams*, Fliess appears frequently as "my friend in Berlin," "my Berlin friend," or "Fl." In the associations to the dream of Irma's injection, he is referred to openly as "*Wilhelm*," a friend who shares and supports his work, "who had for many years been familiar with all my writings during the period of their gestation, just as I had been with his," writes Freud (p. 116); and he adds something more:

> Surely this friend who played so large a part in my life must appear again elsewhere in these trains of thought. Yes. For he had a special knowledge of the consequences of affections of the nose and its accessory cavities; and he had drawn scientific attention to some very remarkable connections between the turbinal bones and the female organs of sex. (Cf. the three curly structures in Irma's throat.) I had had Irma examined by him to see whether her gastric pains might be of nasal origin. But he suffered himself from suppurative rhinitis, which caused me anxiety; and no doubt there was an allusion to this in the pyaemia which vaguely came into my mind in connection with the metastases in the dream.
>
> (p. 117)

This man cannot be a secondary character.

Around Wilhelm who, as Freud writes, "did understand me, who would take my side, and to whom I owed so much valuable information, dealing, amongst other things, with the chemistry of the sexual processes," there gathers an extensive group of positive ideas which are stronger than the unpleasant ideas connected with Otto, "who did not understand me, who sided against me, and who made me a present of liqueur with an aroma of amyl" (p. 294).

Besides condensing multiple people and relationships, the figures in the dream may also represent partial and split-off aspects of both the dreamer and the object:[3] in Jones's interpretation, Freud was committed to working through his unconscious hostility towards his father and may have been helped by the temporary solution of deconstructing the father-person into two, one "good" and the other "bad." The dream's Otto-Wilhelm opposition is matched by the opposition experienced in relation to the two friends and colleagues: "So hatred was directed against Breuer and love towards Fliess – both in an excessive degree out of proportion to the merits or demerits of the persons themselves" (*I*, p. 338).

Freud was aware of the neurotic aspect to his demand for intimacy in his friendships and had tried to find its roots in childhood:

> My emotional life has always insisted that I should have an intimate friend and a hated enemy. I have always been able to provide myself afresh with both, and it has not infrequently happened that the ideal situation of childhood has been so completely reproduced that friend and enemy have come together in a single individual – though not, of course, both at once or with constant oscillations, as may have been the case in my early childhood.
>
> (1900, p. 483)

Freud infused his male friendships with that affectivity which had been revealed as so dangerous in relation to the mother and which he had only experienced fully and exclusively with Martha for a short time. The friendship with Wilhelm Fliess was the deepest of his life. In his comment on the dream, Freud cites his friend's knowledge and imaginative hypotheses, wanting to involve him closely in Irma's treatment, just as he had called on him for help with Emma.

In reality, Freud overestimated him and was already establishing those extraordinary nexuses in the psyche which his colleague mistakenly believed he had found in the soma.

At the time, Fliess represented for Freud a very important part of himself that was still being developed, the *madness* of his genius. When they broke up, he destroyed all Fliess's letters, as if he wanted to wipe out that piece of his own history. Martin Freud, who, as a child, had the opportunity to meet the man who had been his father's best friend for five intense years, did not remember him but thought that a clearly visible portrait in the consulting room was of Fliess.[4] Anna Freud, always more reserved, later mentioned to J M Masson that she had never heard of Fliess until the rediscovery of her father's letters to him: "the breakup was still painful to him, even years afterward."[5]

A good number of his pupils from that time never came to know about the Berlin doctor's existence. Sándor Ferenczi (1873–1933) had some idea of how significant that relationship had been, since he was the only one to whom Freud subsequently granted an almost equal closeness. Karl Abraham (1877–1925) had many opportunities to meet Wilhelm Fliess in Berlin and get to know his scientific and clinical work. In 1916, he asked Fliess to treat his wife and became his patient himself, consulting him in the autumn of 1925 about the lung complaint that would lead to his death a few months later. In that final period of his life, despite being put on guard by Freud, Abraham fell completely under the spell of Fliess's personality and the periodic interpretation of illness that Fliess was still maintaining.[6]

We cannot rule out the possibility that Fliess may have tried, through Abraham, to renew contact with his now-famous former friend.[7] He did indeed maintain an investment in the old friendship, if only in his ideas of persecution. And, to our good fortune, he preserved the letters he received.

Their recovery was due to Marie Bonaparte (1882–1962), descendant of Napoleon and wife of Prince George of Greece. She was in analysis with Freud from 1925. In 1936, the princess bought the letters from a merchant to whom Fliess's widow had entrusted them so that they would not suffer the fate of the books of Freud, Mann, Heine, Zweig, Brecht, and Einstein, burned by the Nazis on the Opernplatz in Berlin in May 1933. Disobeying Freud's order to destroy the documents, Bonaparte took them to Vienna, where she was eventually authorised to read and deposit them with the Rothschild Bank. After the Anschluss, when Marie Bonaparte rushed to Vienna to help Freud, she also managed to save the letters from the Gestapo and bring them to Paris. Jones writes that, because of mines in the Channel, to get them to safety in London, they were "wrapped in waterproof and buoyant material to give them a chance of survival in the event of disaster to the ship."[8]

In 1950, an expurgated first selection of the letters was published (168 out of 284), accompanied by the *Drafts* and *Notes* which Freud had attached to them and by the text of *Psychology* (1895),[9] which had remained in the Berliner's possession. Under the title *Aus den Anfängen der Psychoanalyse* (*From the Origins of Psycho-Analysis*), the volume was edited by Marie Bonaparte, Anna Freud, and Ernst Kris[10] in 1950, but the English translation only became available in 1954. In the meantime, the first volume of Freud's biography by Ernest Jones (1953) had appeared. Jones had unrestricted access to those documents and offered an official interpretation of them. That was when Wilhelm Fliess's role in Freud's *self-analysis* began to be understood and analysts had to rethink both the Master's private life and his work as a whole.[11]

In 1978, J M Masson[12] obtained permission from Anna Freud to publish the complete edition of the Freud-Fliess correspondence and, with the support of K R Eissler, finally completed the project in 1985.

Subsequent editions of Freud's writings began to be augmented by editorial comments referring to letters, notes, meetings, or events which depicted Freud's life, states of mind, symptoms, dreams, and intellectual development in relation to the circumstances of his bond with his friend.

In the letters, we discover the concepts of psychoanalysis in *statu nascendi* as Freud was developing, clarifying, and transforming them: though we read them as an interior monologue because we lack the words of his interlocutor, the correspondence testifies to the original creative dialogue, the indispensable relational and affective counterpart of what he was discovering in his patients and in himself.

My friend in Berlin

Wilhelm Fliess (1858–1928) was an acclaimed otorhinolaryngologist, Jewish, and two years younger than Freud. They are portrayed together in a photograph from February 1895 taken when Fliess was staying in Vienna in order to operate on Emma Eckstein. On that occasion, as I mentioned earlier, Fliess also operated on Freud's nose, and Freud commented on the photo saying, "Beautiful we are not (or no longer), but my pleasure in having you close by my side after the operation clearly shows."[13] Although a strong similarity has been noted in the two faces,[14] and they are gazing in the same direction, what is most striking is the convalescent Freud's expression, the intensity of his closeness to his friend, seeming to lean on his firm, positive gaze. We could call the photo *We Two Together*. Today we know that there is a woman in the background to the scene, the patient who has received a complete psychological and surgical treatment and will survive it . . .

The surgeon from Berlin possessed a fascinating personality. In the screenplay that Sartre wrote for John Houston,[15] Fliess is portrayed as tall and Mephistophelian, with flashing eyes. The description of him given in 1925 by Alix Strachey – Freud's English pupil cannot have known *who* she was really writing about to her husband – is of a seductive, old-fashioned man, but short, "almost a dwarf";

it seems that Fliess had suffered from a disease that had restricted his growth, but Freud, in his overestimation of his friend, always saw him as taller.[16]

In 1887 Fliess had taken a year off to travel around Europe: naturally he had visited Charcot in Paris, but he also went to Britain and Italy, and he was in Vienna when Breuer suggested that he attend the lectures in neuroanatomy being given by Freud at the university. It was Breuer himself who brought them together at one of the soirées he organised at his house. Fliess went back to Berlin having acquired a new friend and also a rich fiancée, Ida Bondy, who belonged to Breuer's circle of acquaintances and patients. Ida became Fliess's wife soon afterwards and was always jealous of his friend Sigmund. The two colleagues quickly developed an intense bond, and during the years from 1893 to 1898, not only did they write passionately to each other, as if they wanted to share every thought, but they organised brief periodic two-man "conferences": in Salzburg, Munich, Dresden, Nuremberg, Breslau, Bad Aussee, Innsbruck . . . and finally at Achensee in the Tyrol during the summer of 1900.

Freud and Fliess had much in common: they were researchers trained in the same rigorous programme of the Berlin Society of Physics which Brücke had introduced to Vienna,[17] they passionately admired Darwin, and both had plans for ambitious projects in the still unexplored field of neuropathology; they soon discovered that they shared the same conviction about the role of sexuality.

However, their professional circumstances were quite different. In Berlin, Fliess had rapidly risen to become a fashionable ear, nose, and throat specialist with a large and stable clientele, a situation favoured by the progressive spirit and less severe antisemitic tensions in Germany at that time. In Austria, by contrast, these were difficult years. With the widening of the suffrage in 1895, the growing antisemitism – and hatred of bankers like the Rothschilds and Ephrussis who had built themselves grand houses on the Ring and controlled so much capital – would lead to the election of Karl Lueger as mayor, despite the opposition of the emperor himself.[18] This state of uncertainty, which had played a part in Freud's abandonment of his university career, influenced the recurring anxieties about the future which he confided to his friend who, now further supported by a good marriage, could cultivate his scientific interests at his leisure.

Fliess had already become known for connecting a complex of diverse symptoms, from the hysterical to the clearly somatic, into a new psycho-organic syndrome which he called "reflex nasal neurosis,"[19] maintaining that its origin was sexual. Based on his conviction that there was a functional connection between nasal and genital mucus – with epistaxis and rhinorrhagia in the former and menstruation and other secretions in the latter – he conceived a treatment which involved small operations on the nasal bones and cauterisations of its mucus, after which the area was dabbed with cocaine. Interest in the therapeutic applications of this substance, for which the Berliner was indirectly indebted to Freud's studies, made a further reason for the two colleagues to get on so well.

For Freud, Wilhelm Fliess represented the man of science through whom he could keep psychology, the vague field of research he had chosen, in contact with the soma and its biological foundations. And he was the person to whom Freud allowed himself to express daringly ambitious fantasies without receiving the "detailed and mannered reply" with which the prudent Breuer had started to challenge him.[20] Because of his age and his character, the older colleague was not easily roused to enthusiasm: "Breuer was in his work reserved, cautious, averse to any generalization, realistic, and above all vacillating in his ambivalence."[21] *Vorsicht* was his favourite word, and, in his relations with the man who had once been his young pupil, he oscillated constantly between criticism and admiration. Noting his pupil's growing distance from him, Breuer would write to Fliess: "Freud's intellect is soaring at its highest. I gaze after him as a hen at a hawk" (5/07/1895).

Since the time of the *Preliminary Communication*, Freud had sought solidarity with his peer, especially with regard to his ideas about sexuality. And, as Jones characterises him, Wilhelm Fliess,

> far from baulking at sexual problems, had made them the centre of his whole work . . . was extremely self-confident, outspoken, unhesitatingly gave the most daring sweep to his generalizations, and swam in the empyrean of his ideas with ease, grace and infectious felicity.
>
> (*I*, p. 326)

Fliess enabled Freud to look to the future with an impressive energy to which he could not remain indifferent: he was a cultured and creative individual, not afraid to venture into improbable hypotheses, and he was able to convince Freud that "behind every absurd popular belief," there may lurk "a kernel of truth."[22] It is Fliess who deserves the credit for suggesting the idea of a constitutional bisexuality in men and women, which played a decisive role in the construction of Freudian sexual theory, and he was also the source of concepts such as the "latency period," "sublimation," and the "erogenous zones."[23] And, as if this were not enough, Jones thinks that the theory of periods may have influenced Freud's conception of the "compulsion to repeat."

Taking bisexuality as his starting point, Fliess hypothesised that in every individual and from one generation to the next, biological rhythms (28 days in women and 22 in men)[24] determined all vital phenomena: from conception, to the sex of the unborn child, the dates of birth and death, all somatic and psychic illnesses, and so on. Freud tried to collaborate with his friend, informing him, on the basis of complicated and often slightly manipulated numerical calculations, about events that had occurred and the worsening or improvement of indispositions, both his own and those of other family members.

Fliess came close to identifying the female fertility cycle, a discovery for which Freud waited in vain and which, in the 1920s, offering the first natural method of contraception – a matter of great importance at that time – would bring fame to the Japanese physician Kyusaku Ogino and his Austrian colleague Hermann Knaus.

On the basis of the biological rhythms, and very unscientifically, Fliess would instead go on to construct an outright numerical mysticism, verging on occultism and delusion, in which he maintained the existence of a cosmic principle capable of influencing all living processes, a profound link between astronomical relationships and the creation of organisms.[25] His vision aimed at explaining all the phenomena of existence, almost in terms of predestination.

Jones comments with some embarrassment:

> He was a brilliant and interesting talker on a large variety of subjects. Perhaps his outstanding characteristics were an unrestrained fondness for speculation and a correspondingly self-confident belief in his imaginative ideas with a dogmatic refusal to consider any criticism of them. . . . The mystical features in his writing, and the fantastic arbitrariness with which he juggled with numbers – he was a numerologist *par excellence* – have led later critics to consign most of his work to the realm of psychopathology. . . . Such was the curious personality with whom Freud was to be concerned.
>
> (I, p. 318 and 320)

In the full flow of his transferential cathexis, Freud will pay a generous tribute of admiration to his colleague and his theory of celestial influence, going so far as to write

> You know that I do not laugh at fantasies such as those about historical periods, and I do not because I see no reason for it. There is something to these ideas; it is the symbolic presentiment of unknown realities with which they have something in common. Since not even the organs in that case are the same, one can no longer escape from acknowledging heavenly influences. I bow before you as honorary astrologer.
>
> (9/10/1896)

Freud's heart trouble

Fliess replaced Breuer in all respects: as a colleague who shared his passion for research, as a consultant in his clinical work, and even as his personal physician. In fact, both the friends had a problematic relationship with the body, and both went through episodes of migraine, rhinitis, and chronic sinusitis. Freud already had a propensity to neurasthenia and was suffering from various gastrointestinal disturbances[26] when, in the autumn of 1893, he started to show the symptoms which led him to fear he had a heart disease: for almost two years, he lived with the prospect of an early death hanging over him. The disturbances culminated in the spring of 1894, with almost daily attacks:

> The most violent arrhythmia, constant tension, pressure, burning in the heart region; shooting pains down my left arm; some dyspnoea, all of it essentially in attacks extending continuously over two-thirds of the day; the dyspnoea

is so moderate that one suspects something organic; and with it a feeling of depression, which took the form of visions of death and departure in place of the usual frenzy of activity.

(Freud to Fliess, 19/04/1894)

Whereas Breuer, suspecting rheumatic myocardiopathy, shied away from declaring it, Fliess showed greater optimism and inclined to the benign diagnosis of "reflex nasal neurosis." Freud, who already doubted that he would live beyond the age of 50, sometimes mistrusted them both, fearing that they were keeping something from him. Even so, he more readily believed the younger man, the friend who "did not mix too much compassion with his sympathy" as Breuer tended to do.[27] Fliess recommended applications of cocaine to the nose and advised Freud to give up smoking; he frequently cauterised the nose and even operated on the turbinate bones. He would remain the only person to achieve the extraordinary feat of keeping Freud off the cigars, even if only for a short time. The months of abstinence from Easter 1894[28] to June 1895 cost Sigmund Freud dearly: a heavy smoker like his father, he consumed as many as 20 cigars a day. To make matters worse, his "cardiac" disturbances grew worse. In June 1894, looking forward to the holidays, he wrote to Fliess: "I have given up mountain climbing 'with a heavy heart' – how meaningful language usage is!" (22/06/1894).

Freud's symptomatology – which Jones considered a functional disorder, Felix Deutsch attributed to nicotine poisoning,[29] and Max Schur called a psychosomatic symptom[30] – was destined to be resolved in a highly distinctive denouement during the self-analysis, gradually taking on the forms of hypochondria, anxiety hysteria, and lastly phobia and inhibition, as if expressing progressively more mature ways of psychically representing suffering. An oedipal crisis *ante litteram*, Freud's heart trouble affected him on various practical levels: on the one hand, there were financial difficulties and the delaying of his professional and scientific fulfilment, and on the other the restrained disillusionment of married life after the fiery engagement, with the birth of six children in less than ten years. His ambitions were at risk and he was finding it hard to keep going. At the peak of his castration anxiety came the illness and death of his aged father.

Rodrigué[31] agrees with Jones in suggesting that Freud's "broken heart" may express both his mourning for and his difficult emancipation from the figure of his father and likewise the shift of his affective investment from Breuer, the "good man" who could not let "any opportunity go by when there is a chance of spoiling the most harmless state of contentment,"[32] to Fliess the "curmudgeon" who endorsed the daring of Freud's research.

"He [Breuer] received my book,"[33] complained Freud to Fliess (8/02/1897), "and thereupon paid a visit to my wife to ask her how the publisher may have reacted to the unanticipated size of this work."

But there is more. It was a process of growth, also provoked by the emotional impact of his hysterical patients. The crisis coincided with the previously mentioned difficulties arising from the treatment of Emma Eckstein and was an expression of the trouble which Freud – faced with the unknown phenomena of transference and counter-transference – did not avoid addressing. He did not abandon his Anna O., and so he ended up having to treat himself as well as Emma Eckstein, finally asking his surgeon friend for help in that *mésalliance* which was so hard to disentangle.

In spring 1895, the nasal operations to which Fliess subjected them both, and a surgical oversight which put Emma's life in danger, had astonishing effects in each case: on Freud's health (he would soon start smoking again), on his married life (conceiving Anna), and, not least, the resumption of Emma's analytic treatment and her cure.

A decisive factor in Freud's emergence from this complicated impasse must have been the impulse to modify the idealisation of his comrade. This led to his recovery and to his recognition of the transference phenomenon; he intuited the significance of dreams and began his "self-analysis," a Nietzschean enterprise of self-conception.

Notes

1 http://www.archive.org/details/bbcInterview.
2 7/08/1901.
3 This *all-for-one* function is illustrated by another photographic device frequently used at the beginning of the twentieth century, a game of mirrors where the subject is broken down into five different images. They are the same person, but being caught from five different angles clearly shows its multiple dimensions, dissociated parts of the Ego (front, back, right, left, male, female . . .) not integrated into unity and present as distinct characters in dreams (see Petrella, 1995). The composition recalls the trance atmosphere of a séance (see the works of artists such as Boccioni, Duchamp, and H P Roché).
4 1958a.
5 Masson, 1985, p. 4.
6 Jones, *II*.
7 Fliess had sent a patient to Abraham in 1911 and read his paper on pituitary syndrome at the beginning of 1919. It should be added that in those years, Fliess's son, Robert, had also begun to take an interest in psychoanalysis: he had attended the Berlin congress in 1922 and then had gone to Vienna, where he had had a long meeting with Freud about his prospective career and his own relationship with his father (Fliess E, 1982). He subsequently trained at the Berlin Institute and joined the IPA; he wrote extensively on the subjects of non-verbal communication, counter-transference, and the analyst's "metapsychology" (1942 and 1953).
8 *I*, p. 317.
9 Published in posthumous writings under the title *Project for a Scientific Psychology* (1950 [1885]).
10 Kris was a relative of Fliess, having married the daughter of Oscar Rie and Melanie Bondy.
11 Rodrigué, *I*, 1996.
12 Regarding the "Masson case," the events of his dismissal from the Freud Archives, and his exit from the IPA see Malcolm, 1983; Migone, 1984; and 2002; Pierri, 2001c.
13 4/03/1895.
14 Mannoni, 1968.

15 A monumental script (1958) that was not used in the film *Freud: The Secret Passion*, 1961, with Montgomery Clift and Susannah York.
16 Strachey and Porge in Rodrigué, 1996, *I*, p. 280.
17 The Hermann von Helmholtz Association was based on the principle that no other forces act in the body than the physico-chemical ones (Robert, 1964).
18 Lueger used Austrian anti-Semitism more as an instrument for gaining power, to incite the crowds, than out of conviction: "Jew-baiting is an excellent means of propaganda and getting ahead in politics" he stated, but in 1899 it was possible for a member of Parliament to invoke a bounty for the killing of Jews, an enormity that was passed over in silence among assimilated Jews who were not looking for confrontation (De Waal, 2010).
19 *Der nasale Reflexneurose*, Fliess (1893).
20 Freud to Minna, 17/04/1893.
21 Jones, I, p. 326.
22 Freud to Fliess, 30/06/1896.
23 Rodrigué, 1996, *I*.
24 He had established the male biological rhythm on the basis of the interval between the menstrual cycles.
25 W. Fliess (1897), *Die Beziehungen zwischen Nase und weiblichen Geschlechtsorganen, in ihre biologischen Bedeutung dargestellt* (The Relationships between the Nose and Female Genital Organs Described in their Biological Signficance), in Kris, 1950.
26 See Anzieu, 1959.
27 Freud to Minna, 17/04/1893.
28 "I have in fact not had anything warm between my lips," Freud to Fliess, 19/04/1894.
29 1956.
30 1972.
31 1996, I.
32 Breuer seems to be the emperor who comments on Mozart's music: "Too beautiful for our ears, and far too many notes, my dear Mozart" (Einstein, 1945, p. 458).
33 "Infantile Cerebral Paralyses," in Nothnagel's volume. (Freud, 1897).

Chapter 7

The discovery of infantile sexuality

Self-analysis and the *writing cure*

Since the *Studies*, Freud had been surprised by the way his analytic illustrations could be read as short stories, and he attributed the phenomenon more to the nature of the subject than to his own inclinations. But in 1934, in an interview with Giovanni Papini, he described himself as "by nature an artist" and "a scientist by necessity," concerned with transforming a branch of medicine, psychiatry, into literature.[1] His famous *Case Studies* would go on to be considered true literary texts – it was not by chance that he won the Goethe Prize – and they would combine a dazzling ability to organise narrative[2] with a rigorous logic and an entirely distinctive clarity of argument.[3] Above all, they would speak a "living language,"[4] precisely because of Freud's intention to give a voice to that living narrative that is the psyche.

Freud was deeply invested in spoken and written language. The ability to pass almost directly from the image to the written word and "the gift of a phonographic memory"[5] were distinctive elements of his creative thinking and of his dazzling capacity for conceptualisation based on the succinct, visual immediacy of the German language. Anzieu's impression is that Freud, in a dreamlike and intuitive state, almost more a figurative artist than a novelist, experienced writing as a kind of drawing, playing carefully with the pictorial form of words and with their related transformations.[6]

Alongside his published works, it is calculated that over the course of his life he wrote over 20,000 letters, of which more than half have been preserved.[7] We have learned about the first epistolary relationship with his schoolfriend Eduard Silberstein and the 1500 letters to Martha: the correspondence with Fliess amounts to nearly 300 letters (284). It is not wrong to say that his self-analysis was a *"writing cure"*:[8] Freud proceeded to interpret his own dreams, noting down everything that came into his mind about every element of the dream.

At Ferenczi's prompting, he realised in 1919 that he had applied the instructions of L Börne's *The Art of Becoming an Original Writer in Three Days*, an essay he had appreciated in his youth:

> Take a few sheets of paper and for three days on end write down, without fabrication or hypocrisy, everything that comes into your head. Write down

DOI: 10.4324/9781003246473-8

what you think of yourself, of your wife, of the Turkish War, of Goethe, of Fonk's trial, of the Last Judgement, of your superiors – and when three days have passed you will be quite out of your senses with astonishment at the new and unheard-of thoughts you have had.

(Börne, 1823, in Freud, 1920a)

As the witness to whom Freud communicated the results of the new theory that was being applied to himself, step by step as he was constructing it, Fliess was his absolutely necessary guarantor.

Today's established doctrine is a historical and evolutionary representation of the journey that Freud was the first to make, successfully discovering and overcoming the barrier of his own unconscious resistance.

After the way was opened up by his understanding of the Irma dream in July 1895, following the harmonious concatenation of symptoms, recollections, waking thoughts, dreams, slips, and mistakes, Freud went on to explore the unconscious processes in his patients and at the same time in himself; he was gradually constructing a representation of the psyche and perfecting a method. In September of the same year, after a "conference" with Fliess in Berlin, he made his last attempt to stay anchored to neurology and his previous scientific training, trying to translate into neuroanatomical and neurophysiological terms the psychological notions that he was deriving from his analytic exploration. Having finally given up that ambitious *Project for a Scientific Psychology*[9] – which, with the advances in the neurosciences, psychoanalysis is only today beginning once again to take into consideration – leaving his teachers behind him and also trying to get over the loss of his father, Freud concentrated on constructing his own theory of the psyche.

In 1896, he introduced the term "psychoanalysis" for the first time,[10] and at the same time, he started to collect antiquities. He would learn to base his interpretations for patients on those creations from the infancy of humanity, using the appropriate statuette to illustrate them, giving visual form to memories and fantasies that were emerging layer by layer in the analysis. In a similar manner, alongside the statuettes that were lining up in the glass cases of his consulting room, the re-reading of myths, legends, or fairy tales helped him when, after running into difficulties in concluding the analyses of his patients – and his own – he came to realise in 1897 that the scenes of seduction being presented as memories were to be considered primarily fantasies, and he discovered infantile sexuality.

In the summer of 1897 – Jones reports – Freud undertook his most heroic feat – a psycho-analysis of his own unconscious:

It is hard for us nowadays to imagine how momentous this achievement was, that difficulty being the fate of most pioneering exploits. Yet the uniqueness of the feat remains. Once done it is done for ever. For no one again can be the first to explore those depths.

(I, p. 351)

Freud wrote to Fliess, "The chief patient I am preoccupied with is myself" (14/08/1897).

The stages of these crucial transitions which punctuate his attack on, and reconciliation with, the figure of his father are documented in the letters to Fliess and in the text of *The Interpretation of Dreams*. Here in particular, when Freud addresses the meaning of typical dreams (*Embarrassing Dreams of Being Naked* and *Dreams of the Death of Persons of Whom the Dreamer is Fond*), and couples this with a psychoanalytic reading of Hans Andersen's fairy tale *The Emperor's New Clothes*, Sophocles' drama *Oedipus Rex*, and Shakespeare's *Hamlet*, he makes clear that the relationship between dreams, fairy tales, legends, myths, or other kinds of creative writing are "neither few nor accidental"[11] and demonstrates the ubiquity of the dream-work and the infantile conflict connect to parricide and incest.

The birth date of psychoanalysis is universally considered 21 September 1897, when, in the famous "equinox letter," Freud communicates his first doubts to Fliess about the concrete truth of the seduction scenes presented to him by his patients:

> And now I want to confide in you immediately the great secret that has been slowly dawning on me in the last few months. I no longer believe in my neurotica [theory of the neuroses].

He is also talking about himself and his newfound ability to "no longer believe" his own childhood memories as they present themselves because he has discovered their hidden meaning.

Recovering his infantile naiveté, like the child in the story who is not afraid to reveal the "tailor's lie,"[12] Freud was able to push past his own repression and intuit the naked truth of the unconscious: a heroic, unique, and unrepeatable event. There was a significant readjustment of the previous theorising: he was forced to abandon the seduction theory – which attributed neurotic symptoms to the repression of the memory of infantile abuses by nannies, governesses, tutors, relatives, and parents – but not so much to deny the importance of such traumas as to reconsider them from the perspective of the child's desire and the development of his emerging and immature sexuality.[13] Freud was discovering that unconscious infantile fantasies could have the same effect on the psyche as material reality.

From the new viewpoint, the repressed psychic content was no longer the memory of factual reality: what required repression was the infantile instinctuality which, reawoken in the encounter with the parents, could with good reason be considered universally present in varying degrees of intensity and be found at the origin of psychic development. In this way, pushing himself beyond the simple aetiology of the psychoneuroses, Freud could build a general theory of the genesis, development, and functioning of the psychic apparatus.

In the next phase, the self-analysis made surprising progress: Freud was now venturing onto solid ground, progressing with such speed in his discoveries that no one would have been able to keep up with him, as on his walks through Vienna and his trips to the mountains where he was always out in front or tirelessly guessing where the best mushrooms were hidden.

He now started to think back over fragmentary recollections which remained in his memory of early childhood and had a doubtful or enigmatic meaning. Recalling a journey from Leipzig to Vienna with his mother before he was three years old, "during which we must have spent the night together and there must have been an opportunity of seeing her *nudam*," he reconstructed the perilous desire he felt and later the jealousy and remorse provoked by the death of the little brother born straight after him. He then interpreted the memory of childhood games and misdeeds with his peers John and Pauline in Freiberg in which the little girl came off worst, and he connected them with his neurotic side and the depth of his subsequent friendships (perhaps also becoming aware of the recent misdeed – deflowering – committed on Emma E, with Fliess's complicity) (3/10/1897).

Beginning with a dream in which his old nanny, a Christian, was telling him off for being "clumsy and unable to do anything" (*ibid.*), Freud reconstructed the historical circumstances which gave meaning to a childhood scene that had been coming back into his mind for years with his being able to explain it. As he later wrote to Fliess, he had asked his mother to explain about his *Nannie*:

> "Of course," she said, "an elderly person, very clever, . . . During my confinement with Anna (two and a half years younger), it was discovered that she was a thief, and all the shiny new kreuzers and zehners and all the toys that had been given to you were found in her possession. Your brother Philipp himself fetched the policeman; she then was given ten months in prison."
>
> (15/10/1987)

And then he was able to understand that strange childhood memory:

> I was crying in despair. My brother Philipp (twenty years older than I) unlocked a wardrobe [*Kasten*] for me, and when I did not find my mother inside it either, I cried even more until, slender and beautiful, she came in through the door. What can this mean? Why did my brother unlock the wardrobe for me, knowing that my mother was not in it and that thereby he could not calm me down? Now I suddenly understand it. I had asked him to do it. When I missed my mother, I was afraid she had vanished from me, just as the old woman had a short time before. So I must have heard that the old woman had been locked up and therefore must have believed that my mother had been locked up too – or rather, had been "boxed up" [*eingekastelt*] – for my brother Philipp, who is now sixty-three years old, to this very day is still fond of using such puns. The fact that I turned to him in particular proves that I was well aware of his share in the disappearance of the nurse.
>
> (*ibid.*)[14]

He discovered that the scene which kept coming back into his mind – and which he now connected to the nanny's dismissal and his mother's pregnancy, resulting in the birth of his sister Anna (31 December 1858), eight months after the death of Julius (15 April 1858), events of which he had no direct memory – condensed all his early childhood experiences of birth and death, the discovery of the father and the intuition about the parents' coupling (the so-called "primal scene"). He could acknowledge the anxieties connected with the loss of the mother/old nanny to the curiosity aroused by her belly/cellar and to what might have been "boxed up" inside it by the father and might come out or stay imprisoned there.

> The child of not yet three had understood that the little sister who had recently arrived had grown inside his mother. He was very far from approving of this addition to the family, and was full of mistrust and anxiety that his mother's inside might conceal still more children. The wardrobe or cupboard was a symbol for him of his mother's inside. So he insisted on looking into this cupboard, and turned for this to his big brother, who (as is clear from other material) had taken his father's place as the child's rival. Besides the well-founded suspicion that this brother had had the lost nurse "boxed up", there was a further suspicion against him – namely that he had in some way introduced the recently born baby into his mother's inside. The affect of disappointment when the cupboard was found to be empty derived, therefore, from the superficial motivation for the child's demand. As regards the *deeper* trend of thought, the affect was in the wrong place. On the other hand, his great satisfaction over his mother's slimness on her return can only be fully understood in the light of this deeper layer.
>
> (1901a, p. 51, n.)[15]

The dream and the memory, with the help of the context provided by his mother, illuminate Freud's particular experience in relation to his little sister Anna, who, arriving to bring hope after the death of Julius like a replacement and *revenant*, had catalysed the violent reactions of jealousy and hatred, the anxieties of castration and death, to which Sigmund had not been able to give psychic form.[16]

The self-analytic work had made real progress since the dream of Irma's injection and was leading Freud to formulate the presence of the oedipal conflict:

> I have found, in my own case too, [the phenomenon of] being in love with my mother and jealous of my father, and I now consider it a universal event in early childhood. . . . If this is so, we can understand the gripping power of *Oedipus Rex*. . . . Everyone in the audience was once a budding Oedipus in fantasy and each recoils in horror from the dream fulfillment here transplanted into reality, with the full quantity of repression which separates his infantile state from his present one.
>
> (15/10/1897)

In December, another dream – of his inhibition about visiting Rome – and the memory of his first train journey when the family had left Freiberg for Leipzig enabled him to understand the origin of his anxiety about journeys and about professional achievements as fear of possessing the mother's forbidden body. The association in *après coup* was between this trip in which, passing through Wroclaw station and seeing gaslights for the first time, he thought of "spirits burning in hell" and the exciting intimacy of a subsequent trip, alone with his mother (3/12/1897).

After being treated by Fliess and using him as a double in the service of developing his own thought and psychically surviving his own discovery, Freud was now becoming his own master, and master of his method and his theory.

It is possible, as the Botellas suggest, that Fliess "loved more the man Freud than his ideas." Fliess's listening and his libidinal gaze of recognition, as well as his claim to be guiding and treating him, would have nurtured Freud's narcissism, enabling him to give himself up to the autoerotic impulse connected to every original thought, making him overcome "the fear of vertigo, the fear of autoerotic madness of unbridled thought."[17]

As Freud later explained it to Ferenczi, the relationship with Fliess was the product of an unconscious homosexual transference. Today we would talk in terms of a narcissistic identification with his comrade or the "mimetic desire" for a double, a supporter of his effort to free himself from his father and from Breuer but also from traditional science and learning.[18] We will see how, throughout his creative journey, because of the persistent tendency to innovative transgressiveness in his thought, and still more because of the ingenuous curiosity and childlike wonder that he will revive in himself time and again, Freud will have to face not only the problem of the human creature's extreme solitude and death but also the necessity and desire for intimacy, for a gaze in which to legitimate the sense of one's continuity and the health of one's own Self, and to allay his need for exclusive love.

We do not have Fliess's words and can only imagine the invaluable warmth and intimacy that he was able to offer Freud during his rite of passage, up to the point when he saw Freud breaking free: "original thought requires a rupture."[19]

Cherchez la femme: the case of Emma Eckstein

The search for comradely intimacy with the schoolfriend, the colleague, and then the pupil became necessary and deep rooted in Freud, especially with regard to woman: the first great discovery of infantile curiosity, the transgressive revelation of the mother and her fecundity, and also the upsetting reawakening of libido. While Breuer had represented the ambivalence of Freud's relationship with his father, Fliess was able to bring back to life both the little brother of whom he had been jealous and who had not survived, and John and Eduard, the companions of his childhood and adolescence with whom he had shared the first daring explorations of the female "dark continent."

From this perspective, we can reconsider the accident suffered by Freud's patient in the spring of 1895.

Emma Eckstein (1865–1924), a beautiful unmarried woman in her 30s, belonged to a very well-known Viennese family of socialists with whom the Freuds and the Bernays were on good terms. Sister of his friend Friedrich, the "philosopher of the Ringstrasse," and of one of the first women to enter Parliament, Emma had already played an active part in the feminist movement. She had never fallen in love and suffered from menstrual and gastric pains, difficulties in walking, and problems linked to masturbation and daydreaming. Her case was one of the first successes which brought public recognition to psychoanalysis, even though the treatment had been so painful and the patient had risked death because of the operation on her nose which left her face permanently scarred. Let's see what happened.[20]

We have learned that in the winter of 1894, still in the throes of his heart trouble and the profound, complex analytic involvement with Emma Eckstein, Freud had asked Fliess for help with his patient. His friend had suggested the surgical intervention. Returning to Vienna in the first half of February, Fliess stayed there a week to operate on Freud (cauterising his sinuses) and on Eckstein's turbinate bone. At the same time, Breuer benefited from the presence of the renowned and trusted specialist to have him operate on his daughter Dora's adenoids.[21]

Before Fliess went back home, the two friends sealed their alliance with the famous photo which portrays them together.

Emma suffered greatly after surgery: the post-operative period was complicated by persistent swelling, pain, purulent secretions, haemorrhages, the expulsion of a fragment of bone, the need for further draining, and recourse to morphine, all reported in detail by Freud to his colleague.

About 20 days after the operation, one of the Viennese surgeons called on by Freud, his friend H Rosanes, extracted from the young woman's nose half a metre of iodoform gauze which Fliess had forgotten about, causing a severe haemorrhage that brought her close to death. Freud was so shocked that he almost fainted and had to leave the room for help, provoking an ironic comment from his courageous patient about the supposedly "stronger sex." This was undoubtedly the first time Freud had fainted, more perhaps because of the loss of faith in his friend than because of the danger to Emma.

From the letter in which he found the courage some days later to tell Fliess in detail what had happened, trying to shield him from any accusation of negligence, we can intuit the extent of his distress (8/03/1995).

The event changed his degree of dependency on Fliess (he also started smoking again) and unblocked the state of impotence with Emma who, having been restored to health, could receive the help she had been hoping for from her analysis: moreover, as he announced at the end of the same letter, Freud was able to resume work on the long-awaited final chapter of the *Studies* and bring it to a conclusion: "So now I am writing page after page of 'The Therapy of Hysteria'."

Far from discouraging Freud, adverse circumstances led him once again, though we do not know through what process, to recognise the value of the new resistance, the "false connection," and its meaning as transference.

> To begin with I was greatly annoyed at this increase in my psychological work, till I came to see that the whole process followed a law; and I then noticed, too, that transference of this kind brought about no great addition to what I had to do.
>
> (1893, p. 304)

Today we can speculate that the whole drama of Emma's operation allowed him to relive an emerging sexual fantasy. Some have intuited that the scene re-enacted the sexual curiosity and fantasies of deflowering which had occupied Freud's imagination in relation to woman since his childhood games with John and Pauline in Freiberg, in which the two little boys reacted as allies to the births of their respective sisters Anna and Bertha.[22] Others have pointed out how Freud, in getting Fliess to operate on him and on Emma, may have been trying out his feminine identification and offering himself as an object of his friend's desire. And still others have shown the unconscious identification between the two men, thanks to which both had made their wives pregnant during that same spring. . . . Now we need to understand what a profound role Emma played in the *mésalliance* between Freud and Fliess.

Emma Eckstein was not just a patient but also one of the first students of psychoanalysis: Freud lent her books from his library and sent her patients on at least one occasion.[23] In 1899, Eckstein wrote an essay on sex education, the first one on the subject to acknowledge the influence of psychoanalysis. In the paper, which contained probable autobiographical elements,[24] following the medical opinion of the time, which Freud himself shared during the early years of his work, Eckstein addressed the dangers of childhood masturbation and of girls' fantasies.

In 1904, she published her best-known book, *Die Sexualfrage in der Erziehung des Kindes*,[25] and went on to write articles with a feminist orientation. Eckstein enjoyed good health for over ten years and was also in a position to take a leading role in managing her family's financial collapse. She always kept in touch with Freud and from 1908 to 1912 had a second course of treatment, probably free of charge.[26] The transference problems cannot have been entirely resolved if in 1905, when she requested a return to analysis, Freud, having to ask her to defer it, and not wishing to be misunderstood, reminded her of the importance and difficulty of recognising the transference.[27]

Subsequently, in connection with a gynaecological intervention, the woman fell into a relapse which, on this occasion, Freud considered inaccessible to further attempts at analysis. It seems that Emma spent her final years withdrawn from the world in a room lined with books, confined to her couch because she was unable to walk.[28]

Despite occupying much of Freud's correspondence with Fliess and much of his thought, Emma's case will never be referred to at length, although many anonymous references to this patient will run through his work, from the beginning to the end.[29]

Jones mentions Emma Eckstein briefly, in association with Minna Bernays, and, in chronological order of appearance, before Loe Kann, Lou Andreas-Salomé, Joan Riviere, and Marie Bonaparte in the representative list of that kind of woman, of a more intellectual and perhaps masculine cast, who, in his opinion had always aroused a significant fascination in Freud but – as Jones feels the need to clarify – "had no erotic attraction for him."[30] In *The Life and Work of Sigmund Freud,* as in the first edition of the letters to Fliess, the whole episode of the nasal surgery was censored. This is the reason that Anna Freud gave to Jones:

> Emma Eckstein was an early patient of my father's and there are many letters concerning her in the Fliess correspondence that we left out, since the story would have been incomplete and rather bewildering to the reader.
>
> (19/11/1953)[31]

It will be Max Schur, in his biography of Freud (published posthumously in 1972), who brings the affair to light and reveals this woman's decisive influence on Freud's relationship with Fliess. Emma Eckstein is discovered to be the authentic protagonist and hidden inspiration of the Irma dream. The scene of the surgical intervention on Emma's nose-throat carried out by Fliess in Freud's presence, and the alarming situation suffered throughout that spring, had formed part of the dream's critically intense contents, alongside those already noted. We understand how the negligence from which Freud was trying to exculpate himself might also mask his disappointment and an accusation against his friend for the mistake he had made.[32] "Freud was alternating between deep sympathy for his patient, Emma, whom he had come to like, and guilt for what had been done to her by Fliess who had been called in at his behest."[33]

Even Masson, who took a highly critical view of the incident, did not fail to highlight how attached Freud was to Emma Eckstein and how he spoke about her with a passion and commitment that he had never shown on other occasions.[34]

It is likely that Freud was also looking back on the experience with Emma years later when addressing the question of Sabina Spielrein with Jung. Commenting on the "personal" elements of treatment and the problem of how to master the counter-transference, Freud told Jung that he himself had had to go through this kind of necessary though painful experience, that they were "hard to avoid," and that he had had a narrow escape.[35]

Notes

1 In Ruitenbeek, 1973, p. 99.
2 Lavaggetto, 1985.
3 Rossi, 2006.
4 Gori, 1985.
5 Freud, 1933, p. 5.

6 1959.
7 Falzeder, 2007.
8 Mahony, 2002.
9 1895[1950], cited previously. In it, he mentioned for the first time the dream of Irma's injection and the analysis of Emma Eckstein.
10 1896a and b.
11 1900, p. 246.
12 Lavaggetto, 1985, p. 113.
13 The change of perspective was not so radical: for Freud, it was a question of equating fantasy and reality – making them part of the series of complementary relationships – and concentrating on the former (1915–1917). In 1924, he added a footnote to *Further Remarks on the Neuro-Psychoses of Defence* (1896): "we need not reject everything written in the text above. Seduction retains a certain aetiological importance, and even to-day I think some of these psychological comments are to the point" (p. 168). Faced with the theoretical oscillations that run through his work, causing the psychic experience to gravitate now towards the unconscious infantile fantasy of the subject, now towards the asymmetrical relationship with the parental couple, there are those who have thought Freud was being ambiguous and others who regard him as hypocritical (Masson, 1984). For Laplanche (1997) both are necessary because the two sides of experience would be indissolubly entangled in the theory, each guaranteeing the existence and limits of the other.
14 Freud does not connect this memory to that of the incarceration of his paternal uncle, discovered splitting counterfeit rubles, an event which had so affected his father Jacob that it turned his hair entirely white (Barron et al., 1991).
15 Footnote added 1924 to the Chapter: *Childhood Memories and Screen Memories*.
16 Schur, 1972; see Pierri, 2022 second volume. She was the only one of Freud's sisters not to be killed in the extermination camps.
17 Botella and Botella, 2001, p. 68.
18 A double that acted in part as an "ideal of the Ego, post-adolescent, grandiloquent and narcissistic," in part as an "Ego ideal, transgressive model of Self, revolutionary and regenerative, fundamentally healthy" (Lopez, 2004, pp. 146–147).
19 Botella and Botella, 2001, p. 68.
20 See Schur, 1972 and Masson, 1984.
21 Masson, 1984; Hirschmüller, 1986.
22 Among others: Rodrigué, 1996, *I* p. 335.
23 Freud to Fliess, 12/12/1897.
24 Masson, 1984.
25 *The Question of Sexuality in the Raising of Children*, quoted in Freud, 1907b, p. 137.
26 Masson, 1984.
27 30/11/1905 in Freud, 1916.
28 Masson, 1984.
29 Shortly after the clinical vignette of the *Studies*, Emma's case appears in a more extensive discussion in the *Project for a Scientific Psychology* (September 1895). Here Freud analyzes the patient's compulsion not to be able to enter a shop alone and shows that he has traced the origin of the symptom in the defensive elaboration of two episodes with traumatic sexual content: one dating back to childhood, the second – a reactivation of the first – in the pubertal period. On this basis, he makes clear the two stages of that *a posteriori* transformational process of memory aroused by the traumatic experience when he defines the *proton pseudos* at the base of the hysterical symptom. Emma is perhaps one of the patients of *A Child Is Being Beaten* (1919b), and we again find a reference to her in the third case of *Analysis Terminable and Interminable* (1937a). After mentioning the difficulties of concluding with the *Wolf Man* and those that emerged

after concluding in a patient who reproached him for not having interpreted the negative transference (Ferenczi), Freud rethinks the reliability of Eckstein's recovery and
the reasons for her relapse.

30 Jones, II, p. 469.
31 In Masson, 1984.
32 Likewise, the condemnation of *Otto*, referring to Rie's recent engagement to Fliess's
   sister-in-law, indicated that the "hated enemy," a split-off part of the figure of "close
   friend" Wilhelm, had already formed in Freud's preconscious.
33 Schur, 1972, p. 84.
34 1984.
35 Freud to Jung, 7/06/1909.

Chapter 8

Original thought requires a rupture

The "reader of thoughts"

The two colleagues had many reasons for seeing each other as alike but also for mutual misunderstanding, and Freud's change, which began after Fliess's accident with Eckstein, revealed their rivalry – initially as scientists – and made it acute. Fliess's biological outlook, which Freud had intended to use as an anchor for his own psychological explorations, contained a weak point, an occultist credo which stopped it offering a secure basis for Freud's new conceptions. Moreover, in his continued monitoring of his friend after treating him, Fliess had no hesitation about using his numerical determinism to predict an early death for Freud, which Freud accepted with resignation as confirmation of his own anxieties.

Fliess was not ready for a change in the relationship. Officially their dialogue ended because of their clash over the definition of psychoanalysis as a science. At their final meeting, in summer 1900, Freud refused to accept the thesis maintained by Fliess that neurotic symptoms had a predominantly organic foundation (the famous male and female cycles), and he defended the discovery of psychic dynamism and his own therapeutic method.

In the autumn, there was a considerable reduction in the number of letters from Berlin, the few that came being mostly concerned with details of family life. Freud nevertheless kept Fliess informed about the progress of his own activities: he was finishing *The Psychopathology of Everyday Life*, drafting the brief text *On Dreams* (1901), and beginning the treatment of a new hysterical patient, a girl of 18, "a case that has smoothly opened to the existing collection of picklocks" (14/10/1900) he commented ironically . . . but the closeness between the two had been lost.

Estranged from what his friend was writing and thinking, Freud advanced along his own path, addressing difficulties and making new discoveries. Young Ida Bauer (1882–1945) would help him to confirm his mode of interpreting hysterical symptoms and dreams but would not yield all that easily to analysis.

The "defect" of the cure would once again consist in the failure to acknowledge the transference, which the patient abruptly enacted in its negative aspect by bringing the treatment to a premature end. Taken by surprise, but recognising

DOI: 10.4324/9781003246473-9

the value of the experience, Freud acknowledged its central role, its responsibility for all resistances in the neuroses: transference, he claimed on this occasion, "is the one thing the presence of which has to be detected [*erraten,* guessed] almost without assistance and with only the slightest clues to go upon, while at the same time the risk of making arbitrary inferences has to be avoided" (1905a, p. 116). At this point, Freud was busy trying to locate the meaning of individual symptoms in the highly complex structure of the neurosis and reconstruct its origin and chronology in the various ages of development, refining his comparison of analytic work with that of the archaeologist and his precious, fragmentary finds.

Ida Bauer appeared under the name Dora in what would become the most famous of his case studies: initially entitled *Dreams and Hysteria: Fragment of an Analysis,*[1] the case was only published some years later (1905a). In the meantime, Freud elaborate his conception of sexuality and began to draft the *Three Essays on the Theory of Sexuality* (1905b), as yet uncertain whether to call it *Human Bisexuality*, in homage to his colleague and almost "co-author" in Berlin.[2]

In May 1901, after long silences, taking the opportunity to send good wishes to Freud on his 45th birthday, Fliess reappeared and opened hostilities, accusing Freud of having abandoned the territory of science for "magic." He questions the value of all Freud's efforts and, not mincing his words, writes that "the reader of thoughts merely reads his own thoughts into other people." We learn this from Freud himself,[3] who firmly replies:

> Your letter gave me by no means the least pleasure, except for the part about magic, which I object to as a superfluous plaster to cover your doubt about "thought reading." I remain loyal to thought reading and continue to doubt "magic."
>
> (8/05/1901)

The following month, noticing that the plans for the holidays did not include "conferences," Freud lamented to Fliess about the "beautiful and difficult" times when, thinking himself close to death, he had felt supported by his friend's confidence (9/06/1901), and in the course of the summer he had to state: "There is no concealing the fact that the two of us have drawn apart to some extent. . . . In my life, as you know, woman has never replaced the comrade, the friend" (7/08/1901).

It was only in September, a year after Achensee, that Fliess explained his coldness, justifying it as the result of an interpretation that he had found intolerable. Here too we know about it indirectly, from what Freud tells Ferenczi in 1910:

> The conviction that his father, who died from erysipelas after many years of nasal suppuration, could have been saved made him into a doctor, indeed, even turned his attention to the nose. The sudden death of his only sister two years later, on the second day of a pneumonia, for which he could blame

doctors, instilled in him the fatalistic theory of predetermined dates of death – as a consolation. This piece of analysis, unwanted by him, was the inner cause of our break, which he effected in such a pathological (paranoid) manner.

(10/01/1910)

What Freud could not know, and we know today, was that the death of Fliess's father had been traumatic in a very different way. Jacob Fliess, a cereal merchant, had not died of erysipelas but had killed himself after his business failed, throwing himself under a train when his son Wilhelm was 20 years old.[4] Fliess was much more vulnerable than Freud was then in a position to recognise, and, despite their close affection, he had not been able to confide in his friend about this bereavement, instead concealing such an important part of his life and his thoughts. Hurt by Freud's interpretation, he tried to hit back, taking advantage of that very power of guarantor that he had been granted, and accused him of talking nonsense. But Fliess knew he was lying.

Freud, disconcerted by his unexpected reaction, continued to make protestations of friendship, but Fliess sought to put a decisive distance between them.

Freud was on his own.

New achievements awaited him. In summer 1901, fulfilling a dream he had always denied himself, he went to Rome for the first time, accompanied by his younger brother Alexander; in March 1902, overcoming other inhibitions, he succeeded in taking the necessary steps to gain the title of Professor Extraordinarius, supported by the combined efforts of two patients.[5] Along with the title of professor, there also came students. From the autumn of 1902 onwards, the "Psychological Wednesday Society" began to meet every Wednesday evening at 8.30 in Freud's waiting room in Berggasse. It consisted of his first Viennese acolytes, Wilhelm Stekel, Rudolf Kahane, Max Reitler, and Alfred Adler, and they were joined by the writer and musicologist Max Graf, a friend of the Freud family. Graf's son, born in 1903, would become famous as the protagonist of the case study *Little Hans* (1909).[6] Stekel, a physician and former patient who had suggested the meetings, used to report its discussions every week in the Sunday edition of the *Neues Wiener Tagblatt*. They were soon joined by Freud's future publisher Hugo Heller, Paul Federn, Eduard Hitschmann, Isidor Sadger, and the young Otto Rank, who in 1906, at the age of 22, was charged with drafting the minutes of the meetings as a salaried secretary.[7]

In early 1904, *The Interpretation of Dreams* was reviewed by the illustrious psychiatrist in Zurich, Eugen Bleuler, who had already appreciated *Studies on Hysteria*.[8] In spite of everything, Freud still insisted on letting Fliess know about his successes, accompanying the good news with three crosses for good luck: "Just imagine, a full professor of psychiatry and my ††† studies of hysteria and the dream, which so far have been labeled disgusting!" (26/04/1904).

Shortly afterwards the professor from Zurich got directly in touch with Freud, initiating a first dialogue between psychoanalysis and institutional psychiatry.

The accusation of plagiarism

However intense the friendship had been, the bitterness and resentment, heightened by misunderstandings, about the paternity of the concept of bisexuality were no less intense. Clarifying a previous misapprehension, using the example of a "symptomatic forgetting" on his part, Freud was careful to insert into *The Psychopathology of Everyday Life* the due acknowledgement that the idea had originally come from Fliess (p. 144).

The coda to the story came in 1904 when the breakup occasioned the painful division of the spoils, the ideas that the two men had generated together. The event provides us with the only three letters from the friend in Berlin which Freud spared from destruction and in which we can finally read Fliess's own words. It is a prejudiced and ferociously inquisitorial Fliess who challenges Freud over the appearance of books by authors close to him,[9] books which describe the bisexuality and periodicity of psychic and biological phenomena without the appropriate acknowledgement of Fliess's prior claim. Fliess presents his suspicion that his former friend had divulged his as yet unpublished theory.

Freud made a first clumsy attempt to deny the accusation, claiming that the topic of bisexuality was a usual part of treatment, to which Fliess reacted by upping the ante and blaming Freud indirectly for being a "robber" of his ideas.

> "Until now I did not know what I learned from your letter – that you are using [the idea of] persistent bisexuality in your treatments. We talked about it for the first time in Nuremberg while I was still lying in bed, and you told me the case history of the woman who had dreams of gigantic snakes. At the time you were quite impressed by the idea that undercurrents in a woman might stem from the masculine part of her psyche. . . . In Breslau I also told you about the existence of so many left-handed husbands among my acquaintances, and from the theory of left-handedness I developed for you an explanation which down to *every detail* corresponds to Weininger's (who knows nothing about left-handedness)."
>
> (26/07/1904)[10]

Freud's reply clarified his present position on the free flow of the process of constructing thoughts:

> the thief can just as well claim it was his own idea; nor can ideas be patented. One can withhold them – and does so advisedly if one sets great store by one's right of ownership. Once they have been let loose, they go their own way. . . . For me personally you have always (since 1901) been the author of the idea of bisexuality; I fear that in looking through the literature, you will find that many came at least close to you.
>   . . . The universality of the bisexual disposition is all that is mentioned in treatment, and that I need there.
>
> (27/07/1904)

Reassuring Fliess, Freud wrote, "On the other hand, there is so little of bisexuality or of other things I have borrowed from you in what I say, that I can do justice to your share in a few remarks," adding bitterly (perhaps thinking back to episodes such as the time when he had been taking cocaine):

> Throughout my life I have freely scattered suggestions without asking what will come of them. I can without feeling diminished admit that I have learned this or that from others. But I have never appropriated something belonging to others as my own.
>
> (*ibid.*)

In the case of this separation, the debts were also difficult to settle: each of the two colleagues tried in his own way to deal with the sense of loss and destructiveness that was in play. In 1906, Fliess was still trying to plead his own cause, publishing a further version of the breakup (*In eigener Sache*) which he attributed to Freud having, out of envy, developed a personal animosity towards him: in Achensee, he had even convinced himself that Freud intended to kill him,[11] but perhaps he was feeling left out of his colleague's growth.

Many years later, Marie Bonaparte will write in her notebook:

> As for bisexuality, if Fliess was the first to talk about it to Freud, he could not pretend to priority in this idea of biology. "And if he gave me bisexuality, I gave him sexuality before that." That is what Freud told me.
>
> (in Masson, 1985, p. 4)

In the pain and violence of this detachment, while the friend in Berlin was concentrating on his fantasies of persecution and grandiosity,[12] Freud was developing his superstitious side.

A future in the image of the past: predestination and superstition

Freud had always had a weakness for prophecies and, concluding *The Interpretation of Dreams*, after he had succeeded in gaining a piece of virgin territory for psychoanalysis, one that had been the domain of popular, divinatory, or mystical beliefs, he nevertheless showed that he was uncertain enough, and willing enough, to ask questions about those prophetic elements which had always linked dreams to occultism:

> And the value of dreams for giving us knowledge of the future? There is of course no question of that. [Cf. p. 5 n.] It would be truer to say instead that they give us knowledge of the past. For dreams are derived from the past in every sense. Nevertheless the ancient belief that dreams foretell the future is not wholly devoid of truth. By picturing our wishes as fulfilled, dreams are

after all leading us into the future. But this future, which the dreamer pictures as the present, has been moulded by his indestructible wish into a perfect likeness of the past.

(p. 621)

Rather than being superstitious, he seemed not to believe in the reality of his achievement.

Almost all Freud's biographers stress that his young mother's emotional invest-ment in her eldest child had a decisive influence on his brilliant destiny and his anxieties about death. The excitatory maternal seductions and expectations, and their early disruption by the birth and rapid death of a second child, would have obstructed the gradual manifestation of the boy's Oedipus complex and would have been at the origin of his regularly occurring fantasy of being "wrecked by success"[13] or, more precisely, his "guilt of the survivor,"[14] probably linked to his repressed hostility.

This state of anxiety emerged in the form of a sensation of unreality and alien-ation during a visit to the Acropolis in 1904 with his brother Alexander. In an open letter to the now elderly Romain Rolland, Freud interpreted it as being linked to the overcoming of the father.

It is one of those cases of "too good to be true" that we come across so often. It is an example of the incredulity that arises so often when we are surprised by a piece of good news, when we hear we have won a prize, for instance, or drawn a winner, or when a girl learns that the man whom she has secretly loved has asked her parents for leave to pay his addresses to her.

(1936, p. 242)

He went on to explain:

As a rule people fall ill as a result of frustration, of the non-fulfilment of some vital necessity or desire. But with these people the opposite is the case; they fall ill, or even go entirely to pieces, because an overwhelmingly powerful wish of theirs has been fulfilled. But the contrast between the two situations is not so great as it seems at first.

(*ibid.*)

Excessive joy would endanger the sense of his own identity and give rise to the annihilating terror of being reabsorbed into the maternal ocean. It was in order to defend himself against this kind of anxiety that at the end of 1899, still in his prime, having to proceed alone towards a successful destiny that would divide him from Fliess, Freud himself took to playing with numbers and to feeling per-secuted. As if calling up his friend's menacing calculations and his theory of pre-destination,[15] in a revisiting of the Kabbalah at the age of 43, he convinced himself that he would die at 62, and for a long time he became sensitive to every recurrence of this number in the everyday events of his life:

> I generally come upon speculations about the duration of my own life and the lives of those dear to me; and the fact that my friend in B[erlin] has made the periods of human life the subject of his calculations, which are based on biological units, must have acted as a determinant of this unconscious juggling.
>
> (1901a, p. 250)[16]

A trigger for this superstitious belief was a banal coincidence that had struck his imagination, perhaps because it involved the hated telephone: he was assigned a new number (1-43-62) on the long-awaited day that the *Traumdeutung* was published.

At the time Freud had been concentrating on the collection of material for *The Psychopathology of Everyday Life* (1901a), where he analysed the meaning of those minor symptomatic phenomena – slips, mistakes, forgettings, or lapses – which, like neurotic symptoms and dreams, could be attributed to the constant activity of preconscious dream-work. In the book's final chapter, "Determinism, Belief in Chance and Superstition – Some Points of View," he set out to organise his reflections on the meaning of superstition, trying to explain that childish magical tendency which made him habitually play with reality – the "superstitions" that Anna would remember so well (Pierri, second volume, Prologue) – but which sometimes frightened him so much that he had to make sure that certain presentiments had no foundation or had to convince himself that the future could not be foreseen.

He seemed to want to reassure himself when he asserted:

> Like every human being, I have had presentiments and experienced trouble, but the two failed to coincide with one another, so that nothing followed the presentiments, and the trouble came upon me unannounced.
>
> (1901a, p. 261)

But in the margin of his own copy of the *Psychopathology* he would add a handwritten note:

> My own superstition has its roots in suppressed ambition (immortality) and in my case takes the place of that anxiety about death which springs from the normal uncertainty of life.
>
> (*ibid.*, p. 260, n.3, 1904 edition)

The very day after *The Interpretation of Dreams* appeared, Freud tried to turn the intense emotions that were gripping him into research.

While complaining to Fliess about how slowly he was progressing with the sexual theory and how he still did not know "what to do with the ††† female aspect," he nevertheless announced that he now understood "how premonitory dreams arise" (5/11/1899). He wrote the brief article "A Premonitory Dream Fulfilled" at great speed, the first in a series of papers that he would keep in his bottom drawer for a long time.[17]

Let's see how he interprets this case.

In analysis, Frau B reports having dreamed about meeting a doctor in the street – a certain Dr K, whom she had not seen for a long time and for whom she had no strong feelings – and insisted that she really did meet him the next day in the very place she had dreamed about. Freud interpreted the impression that a dream has come true as the product of "the creation of a dream after the event."[18] It is this which gives her the immediate conviction of having dreamed what is actually happening – in this specific case, meeting Dr K in exactly that place – whereas the real night-dream, a dream of nostalgia, has been forgotten. The authentic content provided by the dream and the sole reason for the creation of the daydream and the surprising subjective experience connected to it could be understood on the basis of the associations to the session itself, when Frau B recalled a scene she had experienced fully 25 years before: at that time she really had seen the fulfilment of a desire in the unexpected arrival of another Dr K she had been thinking about, a lawyer with whom she was then in love.

That first coincidence had nothing remarkable about it, Freud comments – "accidents which seem preconcerted like this are to be found in every love story"[19] – and it was that desire, that memory, which emerged in a censored form associated with the second K in the subjective experience of the dream that comes true.

Though he acknowledged that falling in love involves the ability to predict the future, Freud felt threatened when he found himself exploring the magical area of the romantic illusion (see also the superstitious crosses beside "female" in the previously quoted letter to Fliess). During his engagement to Martha, he had often been prey to presentiments of misfortune. In a section added to the *Psychopathology* in 1907, he recalled some illusory phenomena experienced as telepathic during his stay in Paris: while far away from his fiancée, he often heard her calling his name and warning him of possible troubles ahead.

He explains: "I then noted down the exact moment of the hallucination and made anxious enquiries of those at home about what had happened at that time" (p. 261). Immersed in a foreign language, he found that the unconsciously evoked timbre and urgency of the "unmistakable and beloved voice" speaking his name brought his disorientation to a peak and perhaps rebuked him for his fantasies of betrayal.

Emulating Oedipus in solving the riddles of the unconscious, Freud also felt the burden of the prophecies from the Delphic oracle. In 1906, when his Viennese pupils celebrated his 50th birthday by giving him a medallion bearing his profile on the front and a Greek image of Oedipus answering the Sphinx on the back, he was very struck by the motto they had inserted, ΟΣ ΤΑ ΚΛΕΙΝ' ΑΙΝΙΓΜΑΤ' 'ΗΙΔΗ ΚΑΙ ΚΡΑΤΙΣΤΟΣ 'ΗΝ ΑΝΗΡ ("Who divined the famed riddle and was a man most mighty"),[20] since the incident was the precise fulfilment, in identical words, of a phantasy from his student days at university.[21] He was deeply disturbed by this dream that was coming true, and in his identification with Oedipus, he found himself faced with the "unavoidable task of taking responsibility for his own destiny."[22]

Writing the last chapter of the *Psychopathology*, Freud tried to work through the experiences of the first fusional relationship with his mother, the difficulties encountered in the development of the process of the illusion, and probably that sadistic component connected to the love that he had had to repress.

In the copy of the 1904 edition, annotated by Freud, there is the observation:

> Rage, anger and consequently a murderous impulse is the source of superstition in obsessional neurotics: a sadistic component, which is attached to love and is therefore directed against the loved person and repressed precisely because of this link and because of its intensity.[23]

The subject was inexhaustible and the chapter remained open ended, lending itself over the years to additions and appendices as Freud advanced in his exploration of the prophetic and telepathic properties of the unconscious: after the "wayward" and "humorous" "dream" work, he started turning his attention to the unconscious psychic work of constructing "magic" and the "occult."

This reflection will soon be shared in his dialogue with his followers – Jung, and especially Ferenczi – and will go on to construct that theoretical chapter which Jones will record under the misleading and yet significant title *Occultism*.

Notes

1 Freud to Fliess, 10/01/1901.
2 Freud to Fliess, 07/08/1901.
3 Ibid.
4 Kris, 1950.
5 Elise Gomperz and Baroness Marie von Ferstel (Appignanesi and Forrester, 1992).
6 The father, who consulted Freud about his son's upbringing, under his supervision subjected the boy to the first infantile analytic treatment. It is suggestive that the child, who would develop a phobia about horses, received a gift from Freud on his third birthday: a rocking horse which Freud himself carried up the four flights of steps leading to Graf's house (Graf, 1942).
7 Guido Brecher, Maximilian Steiner, and Fritz Wittels joined in 1907, Hanns Sachs in 1910.
8 Bleuler, 1896.
9 O Weininger, *Geschlecht und Charakter (Sex and Character)* 1903, and H Swoboda *Die Perioden des menschlichen Organismus in ihrer psychologischen und biologischen Bedeutung (Periodicity in the Life of the Human Organism and Its Psychological and Biological Significance)* 1904.
10 In Masson, 1985.
11 Swales, 1982.
12 "Realizing that Freud in fact paid little attention to the details of his communications might have been even more painful to Fliess than a plagiarism" (Schröter, 2003, p. 152).

13 Freud, 1916; Jones, *III*; Casement thinks this is a condition in which "an unconscious link has been formed between an earlier experience of trauma and the prior sense of safety as if that 'safety' had been warning signal for the pending disaster. Perhaps an unconscious set is formed in which feeling safe and subsequent catastrophe are seen as forever linked" (1985, p. 364, n. 2).

14 Schur, 1972, p. 153.

15 Back at the time of his high school diploma, he had amused himself with ideas about numbers, writing about them to his friend and correspondent E Silberstein (6/08/1873).

16 In the 1907 edition, this sentence was deleted.

17 The manuscript of 11/10/1899 will be published posthumously in 1941, but Freud will present a summary of it in a paragraph added to that last chapter of *Psychopathology*, cited previously, 1907 edition.

18 1899a, p. 624.

19 Ibid.

20 Quoted from Sophocles' Oedipus *Tyrannus*.

21 Jones, *II*, p. 15.

22 Balsamo, 2000, p. 82.

23 1901a, p. 260, n. 3.

Chapter 9

Occultism made in the USA

Spiritualism

In the second half of the nineteenth century, interest in the phenomenon of spiritualism had spread like an epidemic from America to the Old World, engaging much of the scientific and cultural world to an extraordinary degree. For an impression of the complexity of the phenomenon, it should be remembered that the United States was then a young, expanding country with a strong aspiration towards democracy and freedom of thought, but, lacking the support of a centuries-old cultural tradition, it found itself making room for new preachers, religious sects, and fantasies which tended to spread rapidly from its small communities into the population as a whole.

Many people "with a passion for faith but not intelligent enough to make a faith for themselves" – wrote Zweig – became "the grateful and obedient disciples of any one, be he genius or swindler, who ministers to the mystical the religious energy which keeps them going":

> Scattered throughout this strange world of ours, in every town, every village, every hamlet, you will find here and there such kindly people, filled with an obscure religious sentiment, and seeking relief from the weary round of their daily task in the attempt to understand the mystery of existence. . . . Pure in spirit, but weaklings as a rule, longing for a mediator who will guide them whither they should go, they form the best possible recruits for the support of new religious sects and novel doctrines of one kind or another. No matter whether they call themselves occultists, anthroposophists, spiritualists, scientists, primitive Christians, fundamentalists, Tolstoyans, or what not, they are all inspired with the same metaphysical will, the same vague yearning to discover a "higher meaning" in life.
>
> (1931, pp. 135–136)

For example, in 1860, an Irish ex-clockmaker, Phineas Quimby, who travelled around performing public experiments in hypnotism and diagnosing diseases using a form of clairvoyance, perfected his own *mind cure* – based essentially

DOI: 10.4324/9781003246473-10

on persuading subjects that their disease did not exist – which was so successful that patients flocked to him in their thousands, forcing him to resort to treatment by letter and in printed bulletins. In Portland in 1862, at his Science of Health consultancy in the Hotel International, Mary Baker, a wretched young hysterical patient from New Hampshire, dragged herself to see Quimby. For many years, a paralysis had made her barely able to sit upright and yet, within a week, she was miraculously cured. From that moment on, the patient immersed herself with a wild and fanatical enthusiasm in her healer's theories as if he were a Messiah and devoted her life to him. After a few years, with her first pupil and financial backer, Richard Kennedy, she founded Christian Science, a religion and business which caused a sensation in America and became one of the most widespread spiritual movements of the nineteenth century.

> In the breadth of her influence, in the swiftness of her success, in the number of her supporters, this old woman, scarcely more than half-witted, always ailing, and of very dubious character, outstripped all the leaders and investigators of our time. Never beneath the eyes of us moderns has spiritual and religious disturbance emanated from any one to the same marvellous extent as from this daughter of an American farmer, this lone being whom her countryman Mark Twain angrily terms "the most daring and masculine and masterful woman that has appeared on earth in centuries".
>
> (*ibid.*, pp. 105–106)

The mystical tendency of the American people, nurtured by the rigid Protestant faith, had also been fertile ground for the growth of spiritualism. In March 1848, in a town with the suggestive name of *Hyde*sville, near Arcadia, mysterious nocturnal noises in the house of the blacksmith John Fox who had lived there for only a short time with this wife and two daughters, aged 15 and 12, were the starting point for a wave of madness on an extraordinary scale. The mysterious bangs showed themselves to be *intelligent* and repeated the knocking noises made by one of the girls; in the presence of neighbours, some questions asked by the mother received an answer in a rudimentary binary code adopted for the occasion (one bang for no, two for yes), revealing that a man would be killed in that very spot and buried in the cellar. Over the next few days, crowds of curious people ran to the Fox house; when Mrs Fox and her daughters went out, the noises followed them everywhere and communicated with them, claiming to be the spirits of dead people.

Only 150 years before, in the town of Salem, a similar epidemic was started by a small group of adolescents who used to meet to "foresee" the future and, taken over by their own fascinating deception, had begun to display bizarre and then openly hysterical behaviour. This became the origin of a horrific witch hunt which spread across the state of Massachusetts, giving rise to dozens of trials and death sentences. In the mid-nineteenth century, the phenomenon would take a different turn and, no longer being connected to intervention by demonic forces, would find a different and more acceptable cultural interpretation, a different mark of approval, and audience: stigmatisation would be transformed into a new kind of entertainment, a phenomenon on an international scale.

The Hydesville incident, an apparently banal invention, showed itself to possess an important element of inner truth and desire, as is always the case in the birth of fake news, legends, or myths. The rise of spiritualism in America could rightly be interpreted as an expression of the New World's urgent mythopoeic need to construct some ancestors to replace its own lost roots. From this viewpoint, the phenomenon seemed to establish itself as a foundation myth around the *lack* of a heroic birth, with its origin in the magical way the Fox family had needed to represent their mourning for their third child, a boy who had died a long time before at the age of only three (the man killed and buried in the cellar).[1]

From the state of New York, already famous as the "burnt-over" district because of its frequent manifestations of mystical-religious fervour,[2] spiritualism spread itself with extreme rapidity, and as it did so, it perfected its instruments for communicating with the souls of the deceased (from the Ouija board to the planchette) and highlighted the ability of certain individuals to act as intermediaries, mediums, between the living and the dead. These mediums entered a state of trance and transmitted the communications of the spirits by means of free association, automatism – such as automatic writing and drawing – or through their ability to cause physical, visible phenomena: they made tables and other pieces of furniture rise and spin around, produced inexplicable bangs and other noises, made objects levitate or ectoplasm appear, and lastly – and most simply – received telepathic messages.

And so, alongside the performances of the great hypnotists and professional so-called magicians and conjurors such as the famous Hungarian illusionist Ehrich Weisz – alias Harry Houdini who, it should be noted, took care to call himself "not a magician but a mystifier"[3] – the Fox sisters' invocations of spirits started to become public shows, attracting paying spectators in enthusiastic and often tumultuous gatherings. During the 1890s in America, spectacular exhibitions of spiritualism, which drew such large audiences that they had to be held in circus tents, represented the opportunistic and profitable response to a widespread need for magic, the product of a colossal collective illusion. They offered not only a confirmation of the religious faith in the afterlife but also a fascinating form of entertainment with sensational special effects.[4] This popular new attraction – which competed for its audience with the three-dimensional effects of the diorama and the wonders of early photography; the exciting installations of the "roller-coaster" and the "loop of death" which were spreading from Coney Island to the new Luna Parks; and the extraordinary inventions exhibited by Barnum & Bailey's Circus, the "Greatest Show on Earth" (the Fiji Mermaid with the shining tail, the skeleton of Christopher Columbus, the Cardiff Giant, the Siamese twins Chang and Eng Bunker) – was encouraged by the growing market for leisure activities and entertainment. Spiritualists organised spectacles on a par with William Cody's Buffalo Bill's Wild West Show.

Doubt about the authenticity of these performances only increased the curiosity of a public ready to be engrossed and seduced by this collective hypnosis: spiritualism thus showed itself to be playing a full part in the construction of modern media culture, and its characteristic political power is apparent in the skill with which an enormous number of people were kept under the constant influence of its methods of suggestion.[5]

This is one reason the international spread of the contagion was unaffected by the revelations of Maggie Fox who, on 21 October 1888 at the New York Academy of Music, in the presence of her sister Kate and an audience of 2000, showed how she could produce rapping noises audible throughout the theatre by clicking the joints of her toes: the learned men in the stalls leapt onto the stage to ascertain the truth of this.

Fox reported how she and her sister, as very young children, enjoyed themselves by terrifying their mother, an excellent woman but very easily frightened. At night, they used to tie an apple to a string and move the string up and down, knocking it against the floor. Later they added the rappings they had learned to produce with the joints of their toes. What was initially a childish trick found an unexpected echo in the neighbours' interpretation, as the house the Fox had just moved into was said to be inhabited by spirits. The two sisters were no longer able to pull back, also because of the fame so suddenly gained to the family and their home village. Maggie explained how surprised she was to find that so many people, influenced by hearing their knocking, immediately imagined they were being touched by spirits.

Her confession about the origin of the Hydesville knocking, signed and well remunerated, appeared in an article in the *New York World* entitled "Fox Sisters Sound the Death-Knell of the Mediums" on 21/10/1888 (p. 17) with this presentation:

> We commend to the thoughtful attention of every reader of The World the exposure of Spiritualism by its two founders, the famous Fox sisters, Margaret and Kate, printed elsewhere in these columns today. It is claimed that there are 8,000,000 believers in Spiritualism in this country alone, all of whom are victims of the delusion and swindles perpetrated by the horde of impostors who followed the Fox sisters as "mediums" through whom the living could hold intercourse with the dead. Records of the insane asylums have shown that thousands of persons have been driven to lunacy by these charlatans and by their own attempts to hold converse with their dear dead. Unwholesome speculations, the disinheriting of rightful heirs and transfer of large estates to the medium themselves, by direction of "the spirits," have been frequent and unhappy fruits of this nefarious business.
>
> (*New York World*, "The Jugglery of Spiritualism," p. 4)

The *New York World* concluded that many perplexing phenomena may prove surprisingly easy to explain. But the ambiguity of the phenomenon was an integral

part of the attraction, and the growing mass media and entertainment industry continued to exploit its fascination, as still happens today in the organised
fortune-telling that is the daily horoscope. "Here, the occult appears rather insti-
tutionalized, objectified and, to a large extent, socialized," wrote Adorno, and
so the commercialised occult would become part of a "secondary superstition"
unmediated by the encounter with the magician.[6]

It is estimated that in 1870, spiritualism had 100 million followers worldwide.[7]
Some sociologists have tried to explain the rise of spiritualism in America from
the second half of the nineteenth century to the First World War by pointing to the
gradual erosion of authority undergone by major American Christian sects and the
persistent need for mass proofs of immortality. But the most important investiga-
tors into the paranormal were cosmopolitan intellectuals.[8]

When the wave of occultism crossed the Atlantic, Europe witnessed a similarly
unbridled enthusiasm for "table-turning séances" and an unbounded curiosity about
all so-called occult phenomena. In the Old World, these manifestations were not
only held for pleasure, in private gatherings or public performances, but became
the object of study and dramatised the transition from faith in the supernatural
to confidence in scientific rationality. There was always the risk of veering into
superstition, and, as an example, the definition of the subject under investigation –
that which was to be understood by "spiritualism" – symptomatically read, "the
practice of having regular convocations with spirits without a *self-confessed witch*
present" (*sic.* My italics).[9]

In reality, the firm hold that presentations of spiritualism had over public opinion provided a similar opportunity for the proponents of the emerging sciences
and technologies which were gaining the status of new religions and were caught
up in the same mania for performance.[10] It is significant that in 1869, London saw
the appearance of two periodicals devoted to Spiritualism shortly after the first
issue of *Nature*.

In step with a great increase in the number of individuals endowed with mediumistic abilities, especially women, who started performing like professional
actors or used the trance as an opportunity for holding public events, organisa-
tions were being set up with the aim of verifying the authenticity of the phenomena and unmasking fraudsters.

As well as the Fox sisters, who, on arriving in Europe became objects of long-
term study, Daniel D Home (1833–1886) also enjoyed enormous fame in London
despite the fiery condemnations of Charles Dickens in the pages of his *Household
Words*. Home convinced not only Sir Arthur Conan Doyle[11] but also a man of
science like William Crookes. He specialised in telekinesis and levitation and
played musical instruments from a distance. Eusapia Palladino (1854–1918) in
Italy and Teofil Modrzejewski (1873–1943), under the name Franek Kluski, in
Poland achieved stardom especially through their ability to produce ectoplasm –
fibrous material like gauze which emerged from the medium's body, mouth, and
even feet – and for materialising people's limbs or faces.

The most eminent scientists, philosophers, and writers were not above addressing the problem, trying to set up reliable experiments and instruments for

measuring "psychic strength," even putting chains on the mediums in order to evaluate their powers or using the new photographic technology to capture the materialisations.

Victor Hugo, the astronomer Camille Flammarion, and the philosopher Henri Bergson took part in séances and published accounts of them.

However, the spectacle of sparks, blowing winds, rapping noises, and moving furniture in a darkened room left Charles Darwin unmoved. When, in 1874, he engaged the American medium Charles Williams (who had previously charmed Crookes with the intense feeling of genuineness he was able to arouse) and organised a couple of séances in the presence of his family, Francis Galton, the novelist George Eliot and one of his young protégés, Frederic Myers (a professor of literature at Cambridge who went on to dedicate his life to research into occultism), Darwin was in no doubt that it was all trickery and imposture, and made sure his whole family was aware of the fact.[12]

Bitter conflicts arose between supporters of opposed positions: in Italy, Cesare Lombroso, a wholehearted convert to occultism in his later years thanks to the skill and fascination of Eusapia Palladino, entered, with the support of Hereward Carrington (1909), into a fierce debate with Enrico Morselli, who had a decidedly more critical approach (1908). Amid the flowering of scientific discoveries during the nineteenth century, growing interest in spiritualism, which even included the Curies and the physiologist Charles Richet (who won the Nobel Prize for the discovery of anaphylaxis in 1913), was not a mere resurgence of primitive magical thinking or a way to face bereavements that had not been worked through, especially after the losses inflicted by the First World War, but the opening of up a new sphere of research, offering the opportunity to study still unknown aspects of the human mind and to develop a renewed model of the psyche.

The chemist M E Chevreul, who had discovered electromagnetic induction in 1831, demonstrated how the oscillations of the dowser's pendulum or the water diviner's rod were connected to the expectations and suggestibility of the subject's muscular movements (1833, 1854). The physiologist W Carpenter called this the "ideomotor phenomenon" and engaged in a lengthy campaign of articles, pamphlets, and conferences to refute the claims of the spiritualists (1854). Michael Faraday carried out laboratory tests which revealed how the participants at séances transmitted their own movements and tremors to the table without realising it, which paradoxically caused them such surprise because they were expecting them, and provided clear evidence that it was their hands that moved first (1854). The American psychologist Joseph Jastrow devoted much of his career to studying such "ideomotor" movements, creating a strange apparatus for bringing them to light, which he called an *automatographone* (1900).

Medium, media, and "mental telegraphy"

In France, towards the end of the century, the *Revue Philosophique* began to contain papers on thought transmission and long-range suggestion: by Pierre Janet who, in 1884, had succeeded in hypnotising his patient Léonie from a distance of

500 metres; by Charles Richet (1850–1935) who, together with the philosopher Henry Sidgwick and Arthur and Frederic Myers, had replicated Janet's experiments in telepathic hypnotism; and by the young Henri Bergson. who, experimenting with "mind reading" under conditions of hypnosis, discovered the ability of the hypnotised subject to detect images on the hypnotiser's cornea through the activation of a hyperaesthesia of the sense organs. Moreover, Bergson noted the serious state of anxiety caused in the hypnotised by the experience of ever-deeper trances, induced in succession by different hypnotists.[13]

Still in the sphere of research on thought transmission in hypnosis, Frederic Myers coined the happy term *telepathy*, constructing it from the Greek along the same lines as the new words *telescope, telegraph*, and *telephone*, and highlighted similarities between the hypnotic state, genius, and hysteria (1882).

William James, who had become his friend, was passionate about experiments in extrasensory perception and intuited that automatic writing could be employed to explore experiences beyond the mind's perception and still little-known creative faculties (1889). This interest, which remained central to his scientific and philosophical activities, began with observations about a medium of great power in terms of her crypto-aesthetic and telepathic abilities, Leonora E Piper of Boston, who was also studied at length by Myers and the physician Oliver Lodge.

In 1882, a group of learned men led by Myers, Lodge, and Sidgwick set themselves the goal of using the instruments of scientific research to study the marginal area of phenomena denoted by terms such as "*mesmeric, psychic*, and *spiritual*" with the aim of bringing to light new frontiers and realms of knowledge, and founded the Society for Psychical Research in London. Over the years, its presidents included A J Balfour, William James, William Crookes, Charles Richet, Henri Bergson, Gilbert Murray, W McDougall, Camille Flammarion, and many Nobel laureates of the period. Celebrated figures from science, the humanities, and politics became associated with it, and among its members were Marie Curie, Nicholas Murray Butler, Stanley Hall, Pierre Janet, William Butler Yeats, Alfred Tennyson, Lewis Carrol, Mark Twain, and Arthur Conan Doyle.

"Telepathy," which had initially been considered a characteristic feature of hypnosis, began to be distinguished and evaluated in its own right: during the final years of the century, the most varied fields of learning converged around this hybrid subject and enriched each other.

As an illustration of this intrinsic peculiarity and the frustrations endured by the single researcher trying to understand it in its entirety, Francis Galton portrayed it as a chiasmus.[14]

Three founder members of the Society for Psychical Research (SPR), E Gurney, F Myers, and F Podmore, supported the adoption of the new oxymoronic term in place of "thought transmission" in order to indicate "the ability of one mind to impress or to be impressed by another mind otherwise than through the recognised channels of sense."[15] In dealing with phenomena that they defined as "paranormal," the researchers drew on the emerging scientific terminology and hypothesised the presence of waves, radiation, electromagnetic fields, and so on.

In the second volume of *Phantasms of the Living* (1886), an essay in which the three authors collected and classified a wide range of inexplicable phenomena, they compared the results of observations of 149 cases of presumed telepathic dreams.

As well as collecting any manifestation of spontaneous telepathy they came across, the experiments conducted by the members of the SPR continued to calculate the probabilities as being in favour of telepathy and to build models for studying telepathic communication through the transmission of thoughts about playing cards, numbers, and drawings. The Oxford classicist Sir Gilbert Murray preferred to study the induction of more significant contents, for example, a historical episode, and it seems that the evidence of his receptive ability when his daughter was the sender and he attempted to guess the thought in question, in a process similar to free association, exceeds the likelihood predicted by mathematical chance alone.[16]

Some years later, Richet applied statistical tests to observations of telepathy, verifying the independence and distinctiveness of the hypnotic state. In 1894, Henry Drummond, physician and poet, predicted that telepathy would be the next stage in the evolution of language.

And Théodore Flournoy wrote in 1899:

> One may almost say that if telepathy did not exist one would have to invent it. I mean by this that a direct action between living beings, independent of the organs of senses, is a matter of such conformity to all that we know of nature that it would be hard not to suppose it à priori, even if we had no perceptible indication of it.
>
> (p. 387)

There was every kind of experiment in telepathy, but, as the tests multiplied and diversified, no method obtained a monopoly of success, and no particular type of individual seemed to excel, since "the simple countryman was on the same level as the university professor."[17]

And "pseudo" mind-readers started to appear. The American Washington Irving Bishop, an illusionist and music hall entertainer, after touring the theatres revealing the tricks of mediums, introduced himself in 1877 as the "world's foremost expert in the reading of thoughts" and began to put on prodigious displays of telepathy. Examined by a team comprising the editor of the British Medical Journal, the Queen's personal physician, and Sir Francis Galton, many of his performances were connected to his extraordinary capacity for revealing even the smallest "ideomotor" movements by touch.[18] The mentalist Stuart Cumberland, famous for reading thoughts through the muscles, an expert at the "willing game," was entirely hostile to the spiritualists and did not wish to be confused with them.

In the meantime, some surprising inventions had made possible a new and more reliable means of long-distance communication: after Antonio Meucci's "telectrophone" of 1871, at the end of the century, the telephone began to enter people's homes. From the moment of its appearance, the object which today accompanies our existence almost like a prosthesis introduced a source of alarming and importunate anxiety into daily life, and it was not easy to escape its impressive power. Walter Benjamin recalled it like this:

> Not many of those who use the apparatus know what devastation it once wreaked in family circles. . . . At that time, the telephone still hung – an outcast settled carelessly between the dirty-linen hamper and the gasometer – in a corner of the back hallway, where its ringing served to multiply the terrors of the Berlin household. When, having mastered my senses with great effort, I arrived to quell the uproar after prolonged fumbling through the gloomy corridor, I tore off the two receivers, which were heavy as dumbbells, thrust my head between them, and was inexorably delivered over to the voice that now sounded. There was nothing to allay the violence with which it pierced me. Powerless, I suffered, seeing that it obliterated my consciousness of time, my firm resolve, my sense of duty. And just as the medium obeys the voice that takes possession of him from beyond the grave, I submitted to the first proposal that came my way through the telephone.
>
> (1938, pp. 77–78)

Sigmund Freud did not like the telephone at all, or even the radio. His son Martin remembers:

> My father hated the telephone and avoided its use whenever possible. . . . My own theory is that when my father communicated by speech with another human being, the conversation had to be a very personal thing. He looked one straight in the eye, and could read one's thoughts. Then it was an absolute impossibility to attempt to say anything that was not precise truth: not that I had ever any occasion to try to tell him anything but the truth. Father, aware of this power when looking at a person, felt he had lost it when looking at a dead telephone mouthpiece. . . . When, a good many years later, a radio had its place in many homes, none appeared in his own part of the flat nor in the family living-room. . . . He did listen to Schuschnigg making his abdication speech to the Austrian nation; but this is the only occasion I can recall his tolerating the radio.
>
> (1958a, pp. 39 and 121)

In the immediacy of telephonic communication, distance became more evident, and absence took on more definite and definitive perceptual qualities. Hearing his grandmother's voice on the telephone for the first time, Proust already anticipated the pain of her disappearance:

> It is she, it is her voice that is speaking, that is there. But how far away it is! How often have I been unable to listen without anguish, as though, confronted by impossibility of seeing, except after long hours of travel, the

woman whose voice was so close to my ear, I felt more clearly the illusoriness in the appearance of the most tender proximity, and at what a distance we may be from the persons we love at the moment when it seems that we have only to stretch out our hands to seize and hold them. A real presence, perhaps, that voice that seemed so near – in actual separation! But a premonition also of an eternal separation!

(1920, p. 178)

Poetic intuition, between dreaming and waking, sometimes knows how to grasp and give shape to emotions and desires of this genre:

I dream that I'm woken
by the telephone.
I dream the certainty
that someone dead is calling. . .
(W Szymborska, "Receiver,"
Moment, 2002)[19]

The media have been called "collective desires dreamed for a long time until someone takes hold of their vagueness and transforms them into devices,"[20] and it should not astonish us that the emergence of the electrical media was characterised by a close metaphorical and symbolic relationship with the world of the supernatural and a renewed turn towards the occult.[21] Even A G Bell, who competed with Meucci to invent the telephone, was prompted not only by the desire to find a way of communicating with his mother and his wife, who were both hearing impaired, but by the idea that there might be "a way of communicating with the dead."[22] Many researchers who played a role in the development of electrical telecommunications – such as Edison, Crookes, and Lodge – were interested in paranormal phenomena and, on the basis of an analogy between the human brain and the electrical means of communication, came to speculate about communication with the departed. It is no coincidence that the first spiritualist journal in English was called the Yorkshire Spiritual Telegraph.[23]

Samuel L Clemens, alias Mark Twain, the bitterest enemy of occultism but a convinced supporter of telepathy – it seems he often had premonitions of meetings or messages – mocking the technological innovations made from 1878 onwards, gave thought transmission the name "Mental Telegraphy"[24] and declared that it was quicker and more reliable than the postal service, and less intrusive and wordy than the telephone. . .:

Certainly mental telegraphy is an industry which is always silently at work – oftener than otherwise, perhaps, when we are not suspecting that it is affecting our thought. . . . I imagine that we get most of our thoughts out of somebody else's head, by mental telegraphy – and not always out of heads of acquaintances, but, in the majority of cases, out of the heads of strangers.

(1906, vol. 2, Wednesday, March 21)

The emerging concept of "cosmic ether," an updated heir of "animal magnetism," came to designate a territory shared by the fields of science and occultism:[25] so much so that Lodge considered it the basis of both the wireless telegraph and telepathy.[26] At the start of 1900, the Italian *Rivista di Studi Psichici* published an article on "thought-readers" in which it speculated that every idea was propagated in the brain and thence throughout the body, and could reach its recipient with its vibrations by physical contact or, at a distance, through the "ether."[27]

Scientists were revealing themselves to be genuine "wizards," responsible for a new kind of magic. Nikola Tesla (1856–1943) was an imaginative Croatian electrical engineer and physicist whose theories and patents laid the basis for the modern system of alternating current, initiating the second industrial revolution (in the United States, he was called "the Man Who Invented the Twentieth Century").[28] Alongside extraordinary scientific and commercial successes, he accumulated such disasters at the same time that he was regarded as deluded and remained largely unknown. He never made or preserved notes or drawings of what he had intuited (from the induction motor to quantum theory): he literally saw his inventions and, being the son of a clairvoyant, sometimes put on shows like a "magician."

Marconi, too, was motivated by almost visionary intuitions,[29] and when in 1896 he asked the Ministry of Posts and Telegraphs to finance his "wireless telegraph" – which used radio waves – he was thought crazy and had to go to London to obtain the patent which won him the Nobel Prize in 1909.

Radio broadcasting would only be developed in the 1920s, but as early as 1912, a sensation was caused by the news that many of the *Titanic*'s passengers had been rescued because of the SOS sent via radio by the radiotelegraph operator who had stayed at his post even once the water had reached the bridge. Radiotelegraphy and Morse code, the first form of digital communication to translate verbal information into electrical signals, led the way to a cultural transformation comparable to the one recently brought about by the Internet.

From the beginning of the century, Max Planck had begun to revolutionise physics and current hypotheses about the nature of matter. And in 1915, when Albert Einstein proposed his general theory of relativity, it caused an uproar, though it was received with scepticism by his colleagues: his theoretical predictions had to wait for their vindication until the solar eclipse of 29 May 1919 provided the necessary proofs. Even so, his Nobel Prize for Physics (1921) was not awarded for his new interpretation of the universe.

And so, introducing his *Traitée de Métapsychique*,[30] Charles Richet could justly claim:

Everything of which we are ignorant appears improbable, but the improbabilities of today are the elementary truths of tomorrow. Among the discoveries which by reason of my advanced age I have seen developed under my own eyes, so to speak, I will take four which in 1875 would have seemed absurdly inadmissible:

1) The voice of an individual speaking in Paris can be heard in Rome (Telephone).
2) The germs of all diseases can be bottled and cultivated in a cupboard (Bacteriology).
3) The bones of a living person can be photographed (X-rays).
4) Five hundred guns can be taken through the air at the speed of 180 miles an hour (Aeroplanes).

Anyone who uttered such audacities in 1875 would have been thought a dangerous lunatic.

(1922, p. 9)

First hypotheses about the unconscious

Occultism and in particular Spiritualism became implicated in what was "a veritable land-rush to claim 'ownership' of the unconscious, alternately known . . . as the subconscious": the driving impulse behind turn-of-the-century spiritualism was for the most influential thinkers "the urge to extend the domain of science into an exciting unknown that transcended materialist boundaries," not a denial of rational truth but an "idealistic faith in the expansive, humanistic relevance of science."[31]

The Interpretation of Dreams by Sigmund Freud (1900), *L'Inconnu et les problèmes psychiques* by Camille Flammarion (1900), and *Des Indes à la Planète Mars* by Théodore Flournoy (1899) all appeared at the start of the century.

In *L'Inconnu* – subtitles: *Manifestations de mourants. Apparitions. Télépathie. Communications psychiques. Suggestion mentale. Vue à distance. Le monde des rêves. La divination de l'avenir* – Camille Flammarion (1842–1925), founder and first president of the Société Astronomique de France, and also fascinated by spiritualism, had collected over 1000 episodes of manifestation by dead or dying individuals, spiritualist experiments, phenomena of clairvoyance, premonition, and thought transmission. On the basis of these testimonies, which he had requested from readers of the *Annales des Sciences Psychiques*, the *Petit Marseillais* and the *Revue des Revues* (in the same way that he would collect astronomical observations or atmospheric records for weather forecasts), Flammarion asserted the incontestable reality of telepathy and of telepathic, prophetic, and divinatory dreams, and his conviction about the existence of psychic forces of an unknown nature. We should remember that he was also a passionate devotee of literature and, along with Jules Verne and H G Wells, invented a new genre of scientific novel under the name of "science fiction."

Des Indes à la Planète Mars – subtitle: *Étude sur un cas de somnambulisme avec glossolalie* – by Théodore Flournoy (1854–1920) was a model for all those who later went on to study unconscious processes through the use of mediums. As a professor of philosophy and psychology, Flournoy studied the medium Catherine-Elise Muller (Mlle Hélène Smith) for more than four years and tried to explain some of the mediumistic phenomena as subliminal or subconscious perception and cryptomnesia: according to his hypothesis, a medium's trances revived forgotten memories and primitive or infantile stages of psychological development. For all his assiduous and passionate participation in séances, he was ironic about belief in occultism, and as far as telepathy was concerned, he regarded it as a still unexplained physical phenomenon, a communication between living beings that had nothing to do with souls and the occult.

In examining the language of the medium, the creator of structural linguistics Ferdinand de Saussure, professor of Indo-European languages and Sanskrit, and the Indo-European comparative grammar specialist Victor Henry collaborated with Flournoy. Henry analysed the text of the imaginary language of Mlle Hélène Smith, highlighting how, in her strange and original way of speaking, she defensively deconstructed the language she knew: in inventing Sanskrit or Martian, she systematically abolished the consonant "f," initial of *français*, "symbolic representative" of that language, known and familiar, which she had to avoid above all.[32]

In his dealings with the young medium, Flournoy underestimated his own role as her mentor and as initiator of the theosophical mystique which emerged in the linguistic inventions of her apparent glossolalia, just as he failed to consider the inevitable erotic investment aroused in her immature personality. When the book was published, Muller felt so violated and betrayed that she broke with her psychologist permanently.[33]

In its day, *Des Indes à la Planète Mars* was as successful as *The Interpretation of Dreams*. Although it was the Freudian hypothesis about the unconscious which passed into history – especially the method of unveiling the unconscious that Freud had conceived and the discovery of transferential infantile sexuality – for some time, the academic from Geneva and his studies of mediumistic phenomena competed with his Viennese colleague for fame and followers.[34]

At the end of the century, with his colleagues and friends Myers and James, Flournoy tried to construct a theory of subliminal consciousness which might enable him to organise and understand the highly diverse conceptions formulated hitherto about the *seconde condition*, including the hypothesis about hypnoid states proposed by Breuer and Freud in the *Studies*.

However, the boundary between psychological science and occultism was very thin. The Society for Psychical Research, which had initially promoted valuable research on hypnosis and automatic writing, had an occultist tendency, and much of its exploration into telepathy concentrates on the possibility of communicating with the dead. Crookes had discovered spiritualism after his younger brother's death, under the illusion of entering into dialogue with him, and for a long time afterwards remained under the influence of his amorous relationship with the young medium Florence Cook.[35] As for Myers, despite his researches in the subliminal mind, he became a passionate student of the paranormal because of his faith in survival after death.[36] Axel Munthe, his physician, tells how Myers, taking this conviction to the limit, made a solemn pact with his friend, James, that whichever of them was to die first should send a message to the other as he passed over into the unknown. But in 1901, when Myers died at the Hotel Costanzi in

Rome, while communications from his departed spirit began to be received in different places all around the world, there was nothing to be recorded by the friend who was on the spot. Sitting "on a chair by the open door, his note-book on his knees, pen in hand, ready to take down the message with his usual methodical exactitude," William James was overcome with grief and yet unable to find relief in that notebook, which stayed open and blank on his knees.[37]

Better results were achieved by W T Stead, the celebrated pioneer of investigative journalism who died in the wreck of the *Titanic*: following the dozens of editions of his collection *After Death, or Letters from Julia* (1897), letters received from his dead friend by means of telepathic automatic writing, in 1909 he opened a bureau of communication with the deceased.[38]

It is important to stress that, in the popular language of the period, the attribute "psychical" referred to the soul as a spirit distinct from matter, and as such, it evoked a sphere of interest that had not yet been clearly differentiated from the occult. Thus, at the end of the nineteenth century, the new terms psychotherapy and psychoanalysis did not sound like medical terminology but aroused a certain suspicion since they declared a kinship with occultism and exorcism.

The very title, *The Interpretation of Dreams, Die Traumdeutung*, sounded in German indecently like *Sterndeutung* (astrology), recalling the interpretations of dreams performed by seers and the booklets on sale at religious festivals.[39] Freud was moving, provocatively, onto a fertile but equivocal frontier.

It is no coincidence, and no surprise, that Myers, at a meeting of the Society for Psychical Research, should be the first English writer to consider the works of Breuer and Freud: he mentioned the *Preliminary Communication* only three months after it was published and, in March 1897, presented a summary of the *Studies on Hysteria*. In 1911, Freud agreed to become a corresponding member of the Society, commenting to Jung that "the *list of members* is most impressive."[40] A few years later, he also became an honorary member of the American Society (1915). It was for the *Proceedings* of the Society for Psychical Research that he wrote a paper on the topic of the unconscious in psychoanalysis where he clarified the descriptive, dynamic, and systematic significance of the term which he feared was still too ambiguous and which he intended to differentiate from subliminal consciousness.[41]

He declined subsequent requests to collaborate. Freud never committed himself to so-called psychical research and he never sat around a table to take part in séances. For many years, he kept up his interest in the topic of thought transmission, the only sphere he considered worthy of interest: his scientific perspective, rooted in observations made in the dissecting room, left no room for illusions about death. In *The Psychopathology of Everyday Life*, he wrote:

> To my regret I must confess that I am one of those unworthy people in whose presence spirits suspend their activity and the supernatural vanishes away, so that I have never been in a position to experience anything myself which might arouse a belief in the miraculous.

> (1901a, p. 261)

and in *Delusions and Dreams in Jensen's Gradiva*:

> the belief in spirits and ghosts and the return of the dead, which finds so much support in the religions to which we have all been attached, at least in our childhood, is far from having disappeared among educated people, and that many who are sensible in other respects find it possible to combine spiritualism with reason. A man who has grown rational and sceptical, even, may be ashamed to discover how easily he may for a moment return to a belief in spirits under the combined impact of strong emotion and perplexity.
>
> (1907, p. 71)

Unlike Freud's, the scientific journeys of his two first pupils and friends, the Swiss Carl Jung and the Hungarian Sándor Ferenczi, did take them through the field of spiritualism, which, in different ways, influenced their professional vocations and their interest in the psyche.

And while, despite his childlike exuberance, Ferenczi was as averse as Freud to any occultist view, Jung showed himself from the start to be thoroughly "steeped" in spiritualism.[42]

Notes

1 Wiseman, 2011.
2 Kerr, 1993, p. 51.
3 Phillips, 2001, p. 18.
4 Natale, 2016.
5 Natale, 2012.
6 1957, p. 48.
7 Natale, 2016.
8 See Prochnik, 2006.
9 Kerr, 1993, p. 51.
10 Natale, 2016.
11 See Conan Doyle, 1926.
12 Luckhurst, 2002.
13 1886.
14 Galton to Myers, 11/02/1889, in Luckhurst, 2002.
15 1886, p. xiii. In those same years, the similar and non-occult term of "empathy" (*Einfühlung*) was introduced in Germany in the fields of the arts (Vischer, 1873) and aesthetics (Lipps, 1903).
16 Evrard et al., 2017, p. 13.
17 Rhine, 1949, p. 27.
18 His death also seems to have been extraordinary: he fell into *trance* due to an exhausting performance at the Lambs Club in New York, was pronounced dead and subjected to an autopsy (Wiseman, 2011).
19 Clare Cavanagh, Stanislaw Baranczak, translators: Houghton Mifflin Harcourt, 2015.
20 La Cecla, 2006, p. 7.
21 Sconce, 2000 and Natale, 2012.
22 Maddox, 2006, p. 57.
23 Luckhurst, 2002.

24 Twain, 1884, 1891, 1895.
25 Milutis, 2006.
26 Douglas, 2004 in Natale, 2012.
27 Natale, 2012.
28 Lomas, 1999.
29 Marconi, 12/12/1932, Istituto Centrale Beni Sonori ed Audiovisivi.
30 *Métapsychique* stood for *Okkultismus* (German) and *Psychical Research* (English).
31 Prochnik, 2006.
32 See Canestri, 1983.
33 See Kerr, 1993 and Giacomelli, 2008.
34 Later Henry Flournoy, son of Theodore, and Raymond de Saussure, son of the linguist Ferdinand, both analysed by Freud, formed the first psychoanalytic nucleus of French Switzerland.
35 Luckhurst, 2002.
36 Myers, 1903.
37 1929, p. 372.
38 In Varèze, 1948.
39 Ellenberger, 1970.
40 17/02/1911.
41 1912b, a paper that Freud wrote directly in English, in an attempt to explain this concept "to English readers with English words" (Steiner, 1991).
42 Jones, *III*, p. 411.

Chapter 10

Jung, spiritualism and countertransference

The world of the dead

Jung, *poltergeist* phenomena, and séances

Gustav Jung (1875–1861) was born in Kesswyl in Switzerland to a family with an illustrious ancestry. His father was a Protestant theologian and pastor, and so were two paternal uncles. The family legend told that his paternal grandfather, a physician and freemason, whose name Carl had been given, was an illegitimate son of Goethe. His mother's family, which included no fewer than six pastors, was deeply immersed in religion but also in spiritualism: his other grandfather kept a chair in his study reserved for the spirit of his dead first wife, Magdalene, with whom he had a private conversation once a week at a fixed time, to the great displeasure of his second wife, Augusta.[1]

Augusta had suffered a long episode of catalepsy in childhood while tending to her sick brother and was thought to be dead, after which she acquired the gift of second sight.[2] As a child, her daughter Emilie, mother of Carl, had to sit behind her father as he wrote his sermons in order to protect him from the influence of evil spirits.[3]

Carl, born after his mother had lost a first child, a boy who died within a few days,[4] developed a precocious and brilliant intelligence at the expense of the affective sphere, with a deep split in his personality: he identified its roots in his parents' separation when he was three years old. At the time, his mother, who had started to suffer from depression, was admitted for a lengthy period to the Friedmatt psychiatric hospital where her husband was chaplain, leaving him with the sense that there was an "innate unreliability" in woman. Carl was put into the care of an elderly aunt and her maid, a woman who felt "very strange and yet strangely familiar," as he states in his memoir, and subsequently "came to symbolise . . . the whole essence of womanhood."[5] Night terrors, bad dreams, panic attacks by day, and a series of accidents that he suffered as a child – crossing the bridge over the Rhine Falls to Neuhausen, he already had one leg under the railing and was about to slip through: the maid caught him just in time – made him later think that he must have been dealing with an unconscious impulse to suicide.

There had certainly been an early suffering of the Self, the first signs of which were insomnia and a widespread eczema, a suffering which Winnicott interprets

DOI: 10.4324/9781003246473-11

as psychotic illness, related to the parental separation and to maternal depression[6] linked to the death of the first child.[7]

In the splitting of his personality, Jung felt himself to be reflecting the dual nature of his mother, with whom he had a disturbing fusional relationship. He had learned to recognise in her the signs of a second, nocturnal personality manifesting itself, unexpected and frightening, one that turned her in on herself and, paradoxically, put her in direct communication with her son's thoughts: "She would then speak as if talking to herself, but what she said was aimed at me and usually struck to the core of my being, so that I was stunned into silence."[8]

As Jung recalled in his memoirs, his mother seemed "like one of those seers who is at the same time a strange animal, like a priestess in a bear's cave. Archaic and ruthless; ruthless as truth and nature."[9] His father had functioned as a substitute mother for a long time: the parents slept separately, and the boy slept in his father's bed. But while Carl could trust his father more than his mother, or women in general, in the end what he sensed in his father was "reliability and powerlessness."[10] Carl very soon lost his faith in his father's authority, his religious faith, and the Church, and, feeling that he was a heretic, he lived wholly absorbed in a philosophy of his own where the experience of God represented a personal certainty, "one of the most certain and immediate of experiences," rather than a credo.[11]

Growing up practically as an only child – his sister, born when he was nine, seemed never to have existed for him – Jung was a solitary, closed spirit. He had passed through the bewilderments of adolescence with the constant sensation of isolation from his peers, though he was fully aware of his active withdrawal from relationships because of his fear of alienating himself from those secret and incommunicable, almost autistic spaces that were indispensable to him.

He suffered nervous crises for many months and stayed away from school. He began fainting; he learned how to simulate the blackouts at will in difficult situations and as an escape from his relationship with his father. As if balancing out these features of withdrawal, he noticed that he possessed a *participation mystique*" with others, a kind of telepathy and precognition which made him perceive ideas and realities that he should not have known about.[12] "My entire life consisted in elaborating what had burst forth from the unconscious," he wrote in old age,[13] "and flooded me like an enigmatic stream and threatened to break me."

Enrolling to study medicine at Basel University, Jung had soon doubted the truths being offered by scientific materialism and, mistrusting all his teachers, began a passionate reading programme of his own in the field of spiritualism, certain that he would find the fundamental meanings of nature and the human spirit in those phenomena. He obviously also engaged in hypnosis, discovering himself to be a skilled hypnotist, but it was through occultism that he began to be interested in the psyche: wondering if dreams might have something to do with the spirits, he revived the magical thinking of his childhood and the experiences of fusion with his mother's second personality.

In his autobiography, he recalls how, a few days after deciding on a career in medicine, he had two strange accidents at home in his mother's presence; two episodes marked by deafening bangs like pistol shots, after which it was discovered first that the round top of the walnut dining table and then the broad blade of the bread knife in an old cupboard had split. These abnormal events impressed themselves strongly on his imagination, convincing him of the reality of occult phenomena.

Shortly afterwards he heard of some relations who had been engaged for some time in table-turning – as he relates in his memoirs – during which the 15-year-old Helena Preiswerk, his cousin on his mother's side, produced somnambulistic states and spiritualistic phenomena. Curious about this, he started to attend these sessions: this experience, which "wiped out all earlier philosophy," made it possible to "achieve a psychological point of view."[14] He decided to devote his thesis to the psychology and pathology of occult phenomena, convinced that this unexplored field could provide him with a rich experience to illuminate the psychology of the unconscious.

In his thesis Jung described the occurrence of somnambulistic states in "SW," initially spontaneous and then self-induced, the appearance of a peculiar talent in impersonating and imitating acquaintances living or deceased, the manifestation of the ability to move in a harmonious and graceful way, as well as to speak fluently in a more refined language like literary German. He noted the extreme variability of trance crises, from ecstasy and catalepsy to the almost theatrical dramatisation of visions. He also described séances and the use of automatic writing or the psychograph (an upturned glass on which two fingers of the right hand rest, which moves along small cards with letters, laid out around them). He reported the evocation of a "guiding spirit" who declaimed complex genealogical constructions for subsequent reincarnations and preached a mystical science based on a system of "energies." Jung diagnosed a *déséquilibre* or unstable character in the girl, with a hysterical imprint linked to semi-somnambulism, a specific symptom of disturbed attention responsible for the splitting of personality and the appearance of automatisms and visions. Autosuggestion, hysterical reverie, and pathological lying were considered by Jung secondary compromises of the naturalness of trance states, characterised by sleepwalking performances that no conscious staging would be able to achieve.[15]

An important part of the work was devoted to what he called "heightened unconscious performances"[16] and to the topic of thought transmission, which was exactly what had prompted his interest in occultism in the first place.[17] Jung revealed the medium's ability to "read the thoughts" of the *agent provocateur* by means of table movement through the heightened reproduction of spontaneous movements and intentional vibrations (on the table or the glass) not subjectively perceptible by the other participants in the séance but which the medium received automatically. He speculated that the young woman had "a receptivity of the unconscious far exceeding that of the conscious mind"[18] caused by a sensitivity of a hysterical type. On this subject, Jung showed he was aware of Freud's theories about the mechanism of hysterical identification, quoting them specifically, and held views close to Freud's hypotheses about its aetiology.[19] In his conclusions, he explained the case of the medium as being linked to a disturbance of puberty and the emergence of repressed sexuality in the second, split-off personality where a dream of satisfied sexual desire was expressed, differing from true dreaming in its continuity and persistence.

Kerr informs us of what Jung omits, both in the thesis and in his memoirs: that he himself had initiated those séances and used hypnosis to induce trance states in his cousin, with whom he was also in love. Kerr writes, "None of the participants' fathers were told about the sessions, which were held at nightfall in the Klein-Hüningen presbytery that served as the Jung home."[20]

The arrival at *Burghölzli*

After graduating, Jung left Basel for Zurich in December 1900 to work at the *Burghölzli*. At that time, the cantonal hospital run by Eugen Bleuler was an almost obligatory training centre for European psychiatrists who came to stay in its complex of buildings located like a fortress on the hills overlooking the eastern districts of the city.

Bleuler himself lived there with his wife and children alongside his colleagues' families, and the patients in a strange monastically austere community dedicated to opposing alcohol and criminality lived side by side with the absolute disinhibition of madness.

In this remarkable context, a "psychiatric monastery,"[21] the clinical culture of Zurich centred on mutual self-observation in groups and had begun to employ not only hypnosis, which had been practised there since the time of Forel, but also the word association test invented by Galton and Freud's new method of dream interpretation, sometimes in combination. Experimentation had both therapeutic and diagnostic purposes aimed at identifying "complexes" – a term coined by Bleuler to indicate a nucleus of fixed ideas in association – both in the mentally ill and the sane, in the staff of the clinic and their family members.

It was an "ardent and enthusiastic" circle, as A A Brill described it, recalling his first impressions of his stay as a student at the *Burghölzli*. "I am sure that no such group of psychiatric workers ever existed before or since then."[22]

The subject of Jung's graduate thesis and his interest in telepathy appealed to Eugen Bleuler. Big, "handsome, and brimming with enthusiasm, with a voice and laugh so loud they filled the room,"[23] Jung quickly stood out among his colleagues with his magnetic presence, keeping his students spellbound by his temperament and the wealth of his ideas, a real *spirit of fire*.[24] But he seemed more like a soldier than a man of science,[25] and in the hard penetrating gaze of the youthful photographs, there is a hint of arrogance, presumption, even sadism.[26] While exercising a strong attraction over men, he would later repel them with his "flashes of hostility" and "his coldness and lack of consideration."[27]

With the director's support, Jung spent some time in Paris, also at the Salpetrière, where Pierre Janet had succeeded Charcot and was conducting experiments in word association and automatic writing in his Psychological Laboratory. On his return, Jung set up house in an apartment at the *Burghölzli* with his rich young wife Emma, née Rauschenbach, who came from a family of industrialists in Schaffhausen, and in 1903, he set up a Laboratory for Experimental Psychopathology where he experimented with the Word Association Test.

Based on the reaction test devised by Galton and perfected by G Aschaffenburg in the Kraepelin school, the method used a chronometer regulated to a fifth of a second to quantify a subject's resistance in their separate reactions to a list of stimulus words. Jung's job was to test the theory of the weakening of associated links between ideas in *dementia praecox*, which was the reason Bleuler renamed it "schizophrenia" (from the Greek σχίζω – split, 1911). After publishing the results of his experiments and constructing the statistical basis for the theory of complexes and for the new diagnostic method, in 1905, Jung gained a professorship in psychiatry and became an *Oberarzt*, in effect the deputy director of the *Burghölzli*.

From 1904, Bleuler had begun an intense epistolary relationship with Sigmund Freud with the aim of learning the method of interpreting dreams from him directly in order to use it with patients and medical staff and on his own dreams.[28]

In 1906, Jung also began a correspondence with Freud, and in 1907, he made his first visit to Vienna: by this time he had two daughters, was beginning to be known for his work on the associative experiment, and had recently published *The Psychology of Dementia Praecox*. Aged 32, 20 years younger than Freud, but already qualified in German and academic psychiatry, Jung was in a position to project psychoanalysis beyond Vienna and its Jewish circle, and give it an international dimension. Zurich not only aided the dissemination of Freud's ideas in the world of scientific psychiatry but provided psychoanalysis with its official institutions. As early as 1908, with Bleuler's support, Jung set up the first psychoanalytic Congress in Salzburg and, under Bleuler's and Freud's direction, created the *Jahrbuch für Psychoanalytische und Psychopathologische Forschungen*; in 1910, he would be among the founders of the Internationale Psychoanalytische Vereinigung[29] and elected as its first president.

But, above all, pupils started coming from the *Burghölzli* to Vienna: many promising psychiatrists who would be numbered among the pioneers of psychoanalysis had clinical experience of the new therapeutic technique in the Swiss hospital and came into contact first with Jung before meeting Sigmund Freud in person. Through the *Burghölzli* passed Max Eitingon – who, as a student, was the first to travel to Vienna with a patient for a consultation, even before Jung – the Hungarian Sándor Ferenczi, the Englishman Ernest Jones, the German Karl Abraham, A A Brill (a New Yorker of Austrian origin), the Dutchman Van Emden, the Russian Nikolai Ossipov, and the Poles Herman Numberg and Eugenia Sokolnicka (one of the founders of the *Paris Psychoanalytical Society*, SPP). Obviously there were also many Swiss, such as Alfons Mäder, Emil Oberholzer, Franz Riklin, Jakob Honegger, and not least Ludwig Binswanger, whom Jung brought with him to Vienna in 1907.

As we will see, this group should naturally also include the Russian doctor Sabina Spielrein, who entered the *Burghölzli* as a patient in 1904, was treated by Jung, graduated in medicine, and joined the Vienna group in 1911, later becoming one of the first women to practise psychoanalysis in Russia. But we must not forget the doomed genius Otto Gross, the Austrian neurologist and philosopher who was Kraepelin's assistant in Munich and later became an enthusiast for psychoanalysis: hospitalised in Zurich for detoxification from cocaine and opium, he was briefly – before absconding – treated by Jung in the hope of beginning an analysis in Vienna and was diagnosed with *dementia praecox*. His notions about sexuality and the need to free it from social rules would have a certain influence on Jung, on the Sabina Spielrein affair, and perhaps on Jung's difficult relationship with Freud.

First visit to Vienna

The first meeting in Vienna between Jung and Freud in February 1907 constituted an event in the full sense of the term: the elation was still present in Jung's recollections at the age of over 80:

> We met at one o'clock in the afternoon and talked virtually without a pause for thirteen hours. Freud was the first man of real importance I had encountered in my experience up to that time, no one else could compare with him. There was nothing the least trivial in his attitude. I found him extremely intelligent, shrewd, and altogether remarkable. And yet my first impression of him remained somewhat tangled; I could not make him out.
>
> (1961, p. 149)

Admiration was mixed with a certain shyness because the Swiss was not able to allow himself too much intimacy in his relationships and, after initial enthusiasm, often withdrew from friendships.

Jung had arrived in Vienna with his wife, bringing Ludwig Binswanger (1881–1966), who had graduated with him, as a companion to provide moral support but also as an illustrious pupil, heir of a psychiatric dynasty, with whom to impress Freud: his grandfather had founded the Bellevue Sanatorium in Kreuzlingen, on Lake Constance, and his uncle was a professor of psychiatry who ran the clinic in Jena.[30]

"No great amount of imagination is needed to understand how joyfully and gratefully I answered 'yes'," Binswanger writes in his reminiscences of Freud, "when Jung one day surprised me by asking whether I would accompany him and his wife on their (first) visit to Freud in Vienna."[31] He called again the following week and from then on established a warm friendship with Freud and his family: despite following his own theoretical path – he would be the founder of *Daseinanalyse* – he joined the IPV in 1910 and remained a member throughout his career, maintaining an unchanged and affectionate relationship with the professor.[32]

By contrast, Jung's relationship with Freud was troubled from the start. With his intuition, Freud had immediately "read" Jung's unconscious. Binswanger tells us that

> the day after our arrival Freud questioned Jung and me about our dreams. I do not recall Jung's dream, but I do recall Freud's interpretation of it, namely, that Jung wished to dethrone him and take his place.
>
> (1957, p. 361)

As Binswanger recalls it, this was all part of the natural and friendly atmosphere that Freud could create with his aversion for formality and etiquette. Jung, however, was very ill at ease and felt dangerously fascinated. His dream – "I dreamt that I saw you walking beside me as a very, very frail old man"[33] – indicated the rapidity of his transferential cathexis but also its fleeting character: even before getting to know Freud, he has met him in a dream and is already preparing to lose him. The characteristics of this approach were connected to certain circumstances which motivated his visit – his responsibilities in the treatment of Sabina Spielrein – but also the reactivating of traumatic situations from childhood: six months later, with great difficulty, Jung confided in Freud that he had inspired a veneration in him like a "religious crush" accompanied by an "undeniable erotic undertone," that it was "disgusting and ridiculous," an "abominable" feeling rooted in his having as a boy suffered a sexual assault by a man he had previously admired:

> Even in Vienna the remarks of the ladies ("enfin seuls," etc.) sickened me, although the reason for it was not clear to me at the time. This feeling, which I still have not quite got rid of, hampers me considerably. Another manifestation of it is that I find psychological insight makes relations with colleagues who have a strong transference to me downright disgusting. *I therefore fear your confidence*. I also fear the same reaction from you when I speak of my intimate affairs.
>
> (Jung to Freud, 28/10/1907)

In the encounter with the other, Jung seemed afraid of losing his boundaries and, in attempting to restore them, he found himself prey to an erotic excitement dense with feelings of persecution. In order to defend himself, he tended to experience much of life in a dreamy state: as he put it himself, his relationship with external reality was secondary, frequently filtered through visions and "extremely vivid hypnagogic images."[34] That Jung had suffered traumas and abuse is quite evident from his autobiography, but he never again mentioned that specific episode of childhood molestation, neither in his memoirs nor in other contexts. We cannot rule out the possibility that he had made it up in order to rationalise the unease caused by the professor's direct impact on him, in a similar way to another waking fantasy going back to that first visit to Vienna: the occasion when Minna Bernays had taken him to one side and confessed her concern about her own relationship with Freud.

In 1953, Jung told Eissler that Minna made a giant transference onto Freud, of which he was not insensible. On being asked to clarify, he replied: "Oh, a liaison!? I don't know to what extent! But, my God, we know very well how it is, don't we!?"[35] Later, in 1957, he treated the liaison as certain in an interview with his friend Billinsky, but according to Jung's biographer Bair, this was an interpretation on the latter's part.[36] This revelation, made by Jung more than 50 years after the event, would fuel rumours of an affair between Freud and his sister-in-law, about which historians are doubtful. Freud's way of life, inspired by a work ethic and an almost ascetic sobriety, is hard to reconcile with this idea of sexual promiscuity. As we will see, it was in fact Jung – the unconscious does not lie – who would later organise his life into a stable bigamy.[37]

Jung was certainly curious about this foreign Jew, his world and his cohabitation with his wife and sister-in-law. During the visit he was also struck by the beauty of Sophie Freud, so much so that he later justified his erotic yielding to Sabina Spielrein – what he would call his "compulsive infatuation" for the beautiful Jewess – as a displacement of his first transference onto the professor and his daughter.[38]

In the relationship that many of Freud's followers had with him, sooner or later the fantasy emerged of sharing a woman with him, winning his daughter, or offering him their mistress. During that visit in 1907, Freud himself emphasised Sophie's resemblance to Jung's wife and interpreted the dream reported by Binswanger as an ambivalent desire to marry his daughter Mathilde (the eldest, less beautiful than Sophie). In 1908, when he first meets Ferenczi, Freud will fantasise about his marrying Mathilde, whereas in 1914, he will firmly oppose Jones's courtship of Anna.

What is obvious is that Jung left Vienna deeply disturbed by Freud's ideas about sexuality, seduced and intimidated by the "*numinosum*" which seemed to transform Freud's face:

> It was only the emotionality with which he spoke of it that revealed the deeper elements reverberating within him. Basically he wanted to teach – or so at least seemed to me – that, regarded from within, sexuality included spirituality and had an intrinsic meaning. But his concretistic terminology was too narrow to express this idea.
>
> (1961, p. 152)

Jung was already confused by the experience he was going through with Spielrein. He soon proposed the use of a terminology other than "sexual" for the libido, a less offensive and impudent, more innocuous one,[39] and, despite continuing to be a representative of psychoanalysis for several years, he was always nurturing his own heresy as he had done towards his father's religion. When he finally started openly keeping his distance, Freud's attempt to dissuade him would be absolutely counter-productive:

> I can still recall vividly how Freud said to me: "My dear Jung, promise me never to abandon the sexual theory. That is the most essential thing of all. You see, we must make a dogma of it, an unshakable bulwark." He said that to me with great emotion, in the tone of a father saying, "And promise me this one thing, my dear son: that you will go to church every Sunday." In some astonishment I asked him, "A bulwark against what?" To which he replied, "Against the black tide of mud," – and here he hesitated for a moment, then added – "of occultism."
>
> (1961, p. 150)

In reporting the episode, Jung would accuse the sexual theory itself of occultism, deploring the fact that, seeking the "*deus absconditus*" in the depths rather than on high, Freud was running away from the "mystical" part of himself and rejecting everything that philosophy, religion, and the nascent parapsychology could say about the soul. On sexuality, he added:

> I had grown up in the country, among peasants. . . . Incest and perversions were no remarkable novelties to me, and did not call for any special explanations. Along with criminality, they formed part of the black lees that spoiled the taste of life.
>
> (*ibid.,* p. 166)

For his own part, Freud was far from indifferent to the occultist promptings of his illustrious pupil from their first meeting, and, perhaps so as not to be left behind, he took the opportunity of a new edition of *The Psychopathology of Everyday Life* to insert into the last chapter, left somewhat suspended, a section on prophetic dreams to which he attached great importance. In the new section (1907), he declared that, far from rejecting phenomena such as "presentiments, prophetic dreams, telepathic experiences, manifestations of supernatural forces and the like," he was convinced that many of the available observations could be clarified on the basis of incipient knowledge of unconscious processes without leading to doubts about "the coherence of things in the world."[40]

Later he briefly referred to the episode of a "Premonitory Dream Fulfilled" observed in 1899 and gave a personal vignette of the time when he received the title of professor, explaining that strange but quite common coincidence in which we meet the person we are thinking about in an unconscious "rendezvous," as a product of dream-work in action.

Easter 1909: Jung's spiritual complex and Sabina

Jung's second visit to Vienna shortly before Easter 1909, again accompanied by his wife, is an excellent illustration of how the question of occultism immediately came into play. The troubling and tragicomic clash between the two great men happened as soon as Freud named Jung the heir to his legacy. In return, Jung asked Freud's opinion about precognition and parapsychology, subjects that he had promised himself he would bring up with Freud.

The professor's brusque reaction unleashed a "curious sensation" in the Swiss; thoughts like "flames suddenly flaring up,"[41] along with a series of *poltergeist* phenomena:

> It was as my diaphragm were made of iron and were becoming red-hot – a glowing vault. And at that moment there was such a loud report in the book-case, which stood right next to us, that we both started up in alarm, fearing the thing was going to topple over on us. I said to Freud. "There, that is an example of a so-called catalytic exteriorization phenomenon." "Oh come," he exclaimed. "That is sheer bosh." "It is not," I replied. "You are mistaken, Herr Professor. And to prove my point I now predict that in a moment there will be another such loud report!" Sure enough, no sooner had I said the words than the same detonation went off in the bookcase. To this day I do not know what gave me this certainty. But I knew beyond all doubt that the report would come again. Freud only stared aghast at me. I do not know what was in his mind, or what his look meant. In any case, this incident aroused his mistrust of me, and I had the feeling that I had done something against him.
>
> (1961, p. 155)

Those who had the opportunity to know Jung in person were unforgettably struck by the impact of the fascinating side of his personality and by his manner of speaking, which, unlike his writings, was a bubbling rush of ideas in free association: in his conversation, the most profound, the most subtle, and sometimes the most paradoxical views followed one another with an incomparable ease and rapidity. His fascination, often likened to that of a magician,[42] affected Freud too, leaving him troubled by Jung's speculations and by those twin reports like gunfire in his own house, in the second-floor library of his studio.

The impression only faded once the Swiss had left, leaving Freud irritated at having allowed himself to fall under the spell of such wizardry.

In the letter written straight after the visit, Jung boldly brought the subject up again, mentioning the treatment of a patient in whom some "first-rate spiritualistic phenomena" were emerging.[43] But he expressed a certain disappointment:

> When I left Vienna I was afflicted with some *sentiments d'incomplétude* because of the last evening I spent with you. It seemed to me that my spookery struck you as altogether too stupid and perhaps unpleasant because of the Fliess analogy. (Insanity!). . . . If there is a "psychoanalysis" there must also be a "psychosynthesis" which creates future events according to the same laws.
>
> (2/04/1909)

He ungraciously ended by declaring that the evening spent with Freud had "most happily" freed him from the oppressive sense of his authority. Freud replied promptly:

> It is strange that on the very same evening when I formally adopted you as eldest son and anointed you – *in partibus infidelium* – as my successor and crown prince, you should have divested me of my paternal dignity, which divesting seems to have given you as much pleasure as I, on the contrary, derived from the investiture of your person.

And returning to his paternal role:

> Now I am afraid of falling back into the father role with you if I tell you how I feel about the poltergeist business. . . . I decided to continue my observations after you left, and here are the results. In my first room there is constant creaking where the two heavy Egyptian steles rest on the oaken boards of the bookshelves. That is too easy to explain. In the second, where we heard it, there is seldom any creaking. At first I was inclined to accept this as proof, if the sound that was so frequent while you were here were not heard again after your departure – but since then I have heard it repeatedly, not, however, in connection with my thoughts and never when I am thinking about you or this particular problem of yours. (And not at the present moment, I add by way of a challenge.)
> (16/04/1909)

After making it clear that his willingness to believe had been caused by the magic of Jung's presence, in an attempt to diminish the blow inflicted on him, Freud allowed himself to confide some of his own superstitions:

> Some years ago I discovered within me the conviction that I would die between the ages of 61 and 62. . . . Then I went to Greece with my brother and it was really uncanny how often the number 61 or 60 in connection with 1 or 2 kept cropping up in all sorts of numbered objects, especially those connected with transportation. This I conscientiously noted. It depressed me, but I had hopes of breathing easy when we got to the hotel in Athens and were assigned rooms on the first floor. Here, I was sure, there could be no No. 61. I was right, but I was given 31 (which with fatalistic licence could be regarded as half of 61 or 62), and this younger, more agile number proved to be an even more persistent persecutor than the first. From the time of our trip home until very recently, 31, often with a 2 in its vicinity, clung to me faithfully.
> (*ibid.*)

Freud later presented his analysis of this conviction, which arose in 1899 following the coincidence of two evens, the publication of *The Interpretation of Dreams* and his being assigned the telephone number 14362.

The common element that had struck him was 43, his age at the time:

Thus it was plausible to suppose that the other figures signified the end of my life, hence 61 or 62. Suddenly method entered into my madness. The superstitious notion that I would die between the ages of 61 and 62 proves to coincide with the conviction that with *The Interpretation of Dreams* I had completed my life work, that there was nothing more for me to do and that I might just as well lie down and die. You will admit that after this substitution it no longer sounds so absurd. Moreover, the hidden influence of W Fliess was at work; the superstition erupted in the year of his attack on me.

You will see in this another confirmation of the specifically Jewish nature of my mysticism. Otherwise I incline to explain such obsessions as this with the number 61 by two factors, first heightened, unconsciously motivated attention of the sort that sees Helen in every woman, and second by the undeniable "compliance of chance," which plays the same part in the formation of delusions as somatic compliance in that of hysterical symptoms, and linguistic compliance in the generation of puns.

Consequently, I shall receive further news of your investigations of the spook complex with the interest one accords to a charming delusion in which one does not oneself participate.

(*ibid.*)

Jung took his time about replying: in a rather formal manner, he declared himself in agreement about not giving too much credit to impressions and, with regard to his own "spook complex," he gave an assurance of not having "gone over to any system yet." No comment on their relationship or on Freud's reflections about it.[44] Jung's feeling of liberation was highly premature: all the more so when, a few weeks later, he was compelled to ask Freud for help with Fräulein Spielrein, who had written to the professor asking for a meeting.

The dangerous fascination of the "beautiful Jewess"

> DON GIOVANNI: *The poor girl is mad, my friends,*
> *leave me alone with her,*
> *perhaps she will calm down.*
> *(Aside to Donna Elvira)*
> *Softly, softly! People are beginning*
> *to gather around us.*
> *Be a little more prudent,*
> *you will become an object of gossip.*
> (L Da Ponte, W A Mozart, *Don Giovanni, or The Rake Punished,* 1787)

In the autumn of 1977, the Jungian analyst Aldo Carotenuto, who had for some years been interested in a Russian woman doctor cited by Paul Roazen as a pupil of Jung and analyst of Jean Piaget,[45] was informed about the rediscovery of a package of documents in cellars of the *Palais Wilson* in Geneva, the former head-quarters of the Institute of Psychology where Spielrein had worked immediately after the war. The documents consisted of the letters which the young woman had exchanged with Jung and Freud, and her diary (1909–1912). The rediscovery enabled a reassessment of the personality and thought of this psychoanalyst – whom we would otherwise only know from a reference in *Beyond the Pleasure Principle* (1920b) – together with her role in Jung's early career and his friendship with Freud.[46] The clinical and relational tangle that was being brought to light could be compared to the previous one between Breuer and Freud: the drama of the confidences between colleagues was played out again, with a request for help and an about-face with regard to the sexual theory.

Sabina Spielrein was a Russian Jew who during adolescence had begun to suffer from a grave form of hysteria with schizoid features. She had been a precocious and sensitive child who, returning home after spending some of her childhood away from the family, had established a very turbulent relationship with her mother and even more so with her impulsive and violent father. At the age of 16, her disturbances had been aggravated by the death of a younger sister. In August 1904, she was taken by her parents to the *Burghölzli*, where she spent several months and was treated by Jung using a procedure based on psychoanalysis.

He was 29 at the time; she was 19 and was his first patient. He did not use the couch but made her sit on a chair while he sat on another chair behind her, perhaps as his mother had done to protect his grandfather from evil spirits.[47]

After a while, the analysis also took on the characteristics of teaching and sci-entific collaboration: in the meantime, Sabina had enrolled in the Faculty of Medi-cine and in 1905 helped her analyst to obtain his teaching post. In September of the same year, on the insistence of Sabina's mother, who wanted her to be treated by Freud, Jung wrote a letter of introduction which described her as a "highly intelligent and gifted person of greatest sensitivity" but lacking "any sense of appropriateness and external manners," which he "attributed to Russian peculiari-ties."[48] The letter remained among the documents in her clinical file because the young woman stayed in therapy with him until he gradually changed:

> Four and a half years ago Dr. Jung was my doctor, then he became my friend and finally my "poet," i.e., my beloved. Eventually he came to me and things went as they usually do with "poetry." He preached polygamy; his wife was supposed to have no objection, etc., etc.
> (Spielrein, letter to Freud never sent, *The Diary*, 11/06/1909)

When the Swiss decided to get in contact with Freud,[49] he was also motivated by the intensity of his passion for this patient and the difficulties involved in treat-ing her. The first subject addressed in his correspondence with the professor was indeed the sexual genesis of hysteria, on which he expressed his own reservations: not convinced of the new method's therapeutic potential, he was reluctant to use the term "transference," which Freud was suggesting, to distinguish what Jung was still calling "certain personal rapports" in the treatment.[50]

As early as the second letter, Jung was asking for an opinion on the case of Sabina.[51] From the depth of his own experience, Freud tried to encourage him:

> You are probably aware that our cures are brought about through the fixation of the libido prevailing in the unconscious (transference), and that this transference is most readily obtained in hysteria. Transference provides the impulse necessary for understanding and translating the language of the ucs.; where it is lacking, the patient does not make the effort or does not listen when we submit our translation to him. Essentially, one might say, the cure is effected by love.
>
> (6/12/1906)

Apart from this initial exchange, and a later hint in which Jung confided a female patient's erotic fantasies about him,[52] the case of the "young Russian" was not brought up again. She did, however, appear in Jung's writings, in *The Psychology of Dementia Praecox* (Jung, 1907) and *The Freudian Theory of Hysteria* (1908). Only in the spring of 1909, shortly after the highly charged Easter visit just described, did Jung complain to Freud of being persecuted by a complex, and he mentioned a scandal he had become involved in with a patient "solely because I denied myself the pleasure of giving her a child," he explained. (It should be noted that, after two girls, in November 1908, the Jungs had had a son born to them, Franz Carl.)

> I have always acted the gentleman towards her, but before the bar of my rather too sensitive conscience I nevertheless don't feel clean, and that is what hurts the most because my intentions were always honorable

concluded Jung in the letter of 7/03/1909.

Freud was ready to believe that his pupil was the subject of a slander and, in his reply – after announcing that he had accepted an invitation from Stanley Hall at Clark University in Worcester – and perhaps so as not to jeopardise the American project,[53] he generously offered Jung his solidarity: "such are the perils of our trade," he wrote, quoting *Faust*, "In league with the Devil and yet you fear fire?" and took the opportunity to be remembered affectionately to Jung's illustrious grandfather.[54]

The situation was a good deal more complex and was starting to threaten Jung's family and professional life.

That spring, warned by an anonymous letter – perhaps from Emma Jung herself – Sabina's mother had written to Jung from Rostov thanking him for his work with her daughter but also asking for an explanation of what had been going on. Instantly alarmed, Jung had rudely and scornfully replied that the absence of a fee did not allow a clear distinction between treatment and friendship, and announced, "My fee is 10 francs per consultation." He then harshly withdrew from the girl who, in her resentment, reacted aggressively and broke off the sessions.

Jung was now living in a state of fear that Sabina Spielrein would make his conduct publicly known: indeed, he was convinced that she had already started to spread rumours about him now that she was working as an assistant in the hospital where she had been treated and was preparing her thesis under Bleuler's supervision. It was an unbearably menacing situation. First of all, he tried to get his marriage back on track, with the complication that at this time he was analysing both his wife and his little daughter Agathli, born in 1904 and, let's note, like Spielrein, at the height of her oedipal cathexis and curiosity about birth. Jung wrote to Freud also about his daughter:

> My Agathli continues merrily with her discoveries. . . . The act of birth is now fully understood, as the little one announced in an amusing game. She stuck her doll between her legs under her skirt so that only the head was showing, and cried: "Look, a baby is coming!" Then, pulling it out slowly: "And now it's all out." Only the role of the father is still obscure and a subject for dreams. I'll tell you about it in Vienna.
>
> (7/03/1909)

And a few days later:

> My Agathli's achievements are guaranteed original: she has never heard of Little Hans. . . . A great outcry this morning: Mama must come – I want to go into your room – what is Papa doing? But Mama won't have her in the room. "Then you must give me some sweets." Later, when we had got up, Agathli hops in, jumps into my bed, lies flat on her stomach, flails and kicks out with her legs like a horse: "Is that what Papa does? That's what Papa does, isn't it?"
>
> (11/03/1909)

The fact that Jung had resigned from the *Burghölzli* in January in order to devote himself to his private practice and was preparing to move to the new house at Küsnacht on the eastern shore of the lake might have given ambiguous confirmation to the suspicions about his misdemeanours and fed the scandal which, given the Swiss people's rigorous morality, would certainly have arisen.[55]

And so, while Jung was bringing out the first issue of the *Jahrbuch*, in which he officially introduced psychoanalysis and its sexual theory to the scientific world and made himself its guarantor (Freud's new case study of "Little Hans" appeared in it), he was talking less and less to students and patients about the subject of sexuality and was already modifying Freudian theory in his own way. In order to avoid confronting his own difficulties and the ethical violations he had allowed himself, he was cutting his ties with both Zurich and Vienna.[56]

The "spiritualist" Easter visit to Freud, followed by a holiday with his wife in Italy – always a go-between bringing couples back together! – coincided with the leaving of the *Burghölzli*. The compelling interest in catalytic exteriorisation phenomena was not independent of the crisis in which Jung was finding himself: focusing his conflict with Freud on occultism was a distraction from the heated and thoroughly concrete subject of Sabina. The matter became impossible to defer in June when Freud received a letter from the Russian doctor after she had abruptly ended her relationship with Jung: Spielrein was looking for help and, wishing to protect both her former analyst and herself, asked the Master to act as guarantor and mediator, in the hope of gaining acknowledgement rather than justice.[57] This time Freud did not underestimate the situation: better informed about the young woman, but still attempting to recover his authority, he behaved very indulgently towards Jung. In his letter to the Swiss, he named the phenomenon of "countertransference" for the first time:

> Such experiences, though painful, are necessary and hard to avoid. Without them we cannot really know life and what we are dealing with. I myself have never been taken in quite so badly, but I have come very close to it a number of times and had *a narrow escape*. I believe that only grim necessities weighing on my work, and the fact that I was ten years older than yourself when I came to ΨA, have saved me from similar experiences. But no lasting harm is done. They help us to develop the thick skin we need and to dominate "countertransference," which is after all a permanent problem for us; they teach us to displace our own affects to best advantage. They are a "*blessing in disguise.*"
> (7/06/1909)

By using English words, Freud seemed to want to cool the intensity of the scandalous emotions in play. Offering his pupil a justification, and flattering him with masculine complicity, he added:

> The way these women manage to charm us with every conceivable psychic perfection until they have attained their purpose is one of nature's greatest spectacles. Once that has been done or the contrary has become a certainty, the constellation changes amazingly.

Absolved from responsibility, Jung was now ready to declare himself Freud's son and heir and, having also been invited to Worcester, prepared to join Freud on the journey to America. While they waited, after a second exchange of letters with Spielrein, Freud gave Jung further reassurances about the countertransference:

> Remember Lassalle's fine sentence about the chemist whose test tube had cracked: "With a slight frown over the resistance of matter, he gets on with his work." In view of the kind of matter we work with, it will never be possible to avoid little laboratory explosions. . . . In this way we learn what part of the danger lies in the matter and what part in our way of handling it.
> (18/06/1909)

Meanwhile, Sabina Spielrein had got back in touch with Jung: to obtain the apology due to her and to her mother, and to ascertain that he really had given the professor all the facts. The Russian doctor was trying to avoid an irreparable breach: "My dearest wish is that I may part from him in love," she wrote in a letter addressed to Freud but kept in her diary.[58] And her mother, coming to Zurich to clear things up, was very careful not to put either her daughter or the incautious analyst in a compromising situation with the institution and with Bleuler.

Only at this point did Jung decide to confess to Freud that he had lied in trying to pass off the patient's revelations as hysterical fantasies: "a piece of knavery," he now admitted, fearing that in revenge the woman was already exposing him to scandal. He concluded the letter by reiterating his friendly feelings towards Freud: "I am looking forward very much to America" (21/07/1909).

For his part, Freud had already offered Spielrein an apology for having misconstrued the subject of her proposed visit: "Please accept this expression of my entire sympathy for the dignified way in which you have resolved the conflict," he wrote to her on 24/06/1909.

So, in June 1909, the Spielrein affair seemed to have faded away, and the noisy Easter spirit appeared to have been a benign delusion of Jung's.[59] Freud thought he had reasserted his authority and weight of experience, but, exchanging ambition for devotion, had let himself be seduced by his pupil once again. Stimulated by Jung's problems, he was not only able to fine-tune the technical question of transference and countertransference but also to identify and reflect on a particular element of occultism, its "Uncanny" (*Unheimlich*) effect.

The two questions, incest and occultism, thus arose in an intimate association before jointly constructing the embryo of what will be recorded in the history of scientific thought under the very precise name of countertransference.[60]

By becoming lovers, Carl and Sabina had not only broken the ethical rules of treatment but believed they were "kindred spirits," convinced that they knew each other's unspoken thoughts. In violating the limits imposed on intense analyst-patient infatuation, the first boundary to crumble would be exactly that intrapsychic one between self and object, and indeed the experience of psychological fusion would have preceded physical involvement.[61]

Spielrein's analysis had involved Jung at a deep level, putting him back into contact with unresolved elements of his relationship with the feminine. Like his falling in love with his cousin during the period of the séances, the encounter with Sabina's transference had reactivated his disturbing relationship with his mother and, immersed in incestuous and traumatic fusion, he had let himself be led by countertransferential illusions in his swerve towards occultism. While Sabina Spielrein was writing in her diary, "I was able to read Dr Jung's thoughts both when he was nearby and *à distance*, and he could do the same with me," or "He said then that he loves me because of the remarkable parallelism in our thoughts; sometimes I can predict his thoughts to him,"[62] Jung was regressing to his previous mysticism – the prophetic quality of dreams and the reading of thoughts – and, by detaching himself from the reality of the body, he was allowing himself to deny sexuality and his own responsibilities.

The crisis with Spielrein, as well as the conflict with Freud, pushed Jung to the edges of human experience and plunged him "into the potentially ever-present human side-world of premonition and strange coincidence."[63]

When Freud and Jung left for the United States, the situation was still under control: their friendship seemed to have emerged from it unscathed, the contributions to the second volume of the *Jahrbuch* were being assembled (including the case study of the "Wolf Man" which Freud had presented at Salzburg), and Stanley Hall was awaiting his famous visitors in Massachusetts. Now Sándor Ferenczi entered the scene, accompanying Freud to America and in his new explorations of the occult.

Notes

1 Kress-Rosen, 1993.
2 Moreau, 1976.
3 Fodor, 1971; Jung, 1961.
4 Brome, 1978; Eissler, 1993.
5 Jung, 1961, p. 21.
6 1964.
7 Eissler, 1993. According to Miller (2009), Jung's mother had had three stillborn children before his birth and "had withdrawn into a world of ghosts and spirits" (p. 24).
8 1961, p. 69.
9 Ibid., p. 70.
10 Ibid., p. 21.
11 Ibid., p. 84.
12 Ibid., p. 70.
13 *The Red Book*, 1913–1930, (1957), epigraph.
14 Ibid., p. 135.
15 1902, p. 47.
16 Jung, 1902, p. 80.
17 Kerr, 1993.
18 Jung, 1902, p. 80.
19 Ibid.
20 1993, p. 50.
21 Ibid., p. 39.
22 In Ellenberger, 1970, p. 796.
23 Miller, 2009, p. 26.
24 Binswanger, 1957, p. 361 e 1956, p. 11.
25 M Freud, 1958a.
26 Which makes us understand the fact that "this man has dedicated the best years of his life to introspection by immersing himself totally, perhaps without return, in dreams, in ecstasy, desperately looking for an ever deriding and elusive god" (Lopez, 1979, p. 235).
27 Kerr, 1993, p. 53; Freud to Frenczi, 30/05/1912.
28 The friendship between the two men continued throughout their life, even after Bleuler's exit from the IPA (Falzeder, 2005).
29 IPV, now IPA, *International Psychoanalytical Association*.

30 Bertha Pappenheim was hospitalized in the Kreuzlingen Sanatorium. Nietzsche was treated in Jena in 1889–90.
31 1957, p. 1–2.
32 "All my scientific development, in positive as well as in negative respects, has been determined by my efforts to formulate the philosophical as well as the scientific bases for psychoanalysis" (Binswanger, 1957, p. 23).
33 Jung to Freud, 02/11/1907.
34 Jung, 1961, p. 256.
35 Eissler interview 29/8/1953, Borch-Jacobsen and Shamdasani, 2012, p. 356.
36 Roudinesco, 2014, p. 473.
37 In Jung's memoirs, the chapter relating to the episode was cut (Gabbard, 1995).
38 Jung to Freud, 4/06/1909.
39 Jung to Freud 31/03/1907.
40 1901a, p. 281.
41 1961, p. 152.
42 Shamdasani, 1998, p. 5.
43 According to Kerr, he was referring to Sabina Spielrein (1993).
44 12/05/1909.
45 Spielrein analysed Piaget for a few months in 1921 and supported him in presenting a paper at the IPA congress in Berlin in 1922 (Roazen, 1975).
46 Carotenuto, 1980.
47 Kerr, 1993.
48 25/11/1905 in Carotenuto, 1980, p. 101.
49 In the spring of 1906, he sent him *On Psychophysical Relations of the Associative Experiment* (Jung, 1907).
50 Jung to Freud, 5/10/1906.
51 "At the risk of boring you, I must abreact my most recent experience. I am currently treating an hysteric with your method. Difficult case, a 20-year-old Russian girl student, ill for 6 years" (23/10/1906).
52 6/07/1907.
53 Balsamo and Napolitano, 1998.
54 9/03/1909.
55 Appignanesi and Forrester, 1992; Kerr, 1993.
56 Kerr, 1993.
57 4/06/1909.
58 10/06/1909, in Carotenuto, 1980, p. 116.
59 Kerr, 1993.
60 Balsamo and Napolitano, 1998.
61 Gabbard, 1995.
62 Diary, 1909 and 9/11/1910, see Carotenuto, 1980, p. 109 and p. 33.
63 Kerr, 1993, p. 213.

Chapter 11

Ferenczi, the unclassifiable

The sultan and his "clairvoyant"

In the spring of 1933, Ferenczi's pupil, Clara Thompson,[1] whose analysis in Budapest was dramatically cut short by his death, recalled the "brilliant intuition which enabled him to guess at the patient's thoughts almost before the patient himself knew them":

> Possessed of a genuine sympathy for all human suffering, he approached each new case with an enthusiastic belief in his ability to help and in the worthwhileness of the patient. . . . he tried more and more to apply psycho-analysis to very difficult cases.
>
> (1964, p. 66)

In the same year, Sigmund Freud began his eulogy for his friend with this remarkable anecdote:

> We have learnt by experience that wishing costs little; so we generously present one another with the best and warmest of wishes. And of these the foremost is for a long life. A well-known Eastern tale reveals the double-sidedness of precisely this wish. The Sultan had his horoscope cast by two wise men. "Thy lot is happy, master!" said one of them. "It is written in the stars that thou shalt see all thy kinsmen die before thee." This prophet was executed. "Thy lot is happy!" said the other too, "for I read in the stars that thou shalt outlive all thy kinsmen." This one was richly rewarded. Both had given expression to the fulfilment of the same wish.
>
> (1933, p. 227)

In this way, the Master noted his bitter satisfaction at having outlived his pupil, but this time, the Sultan had lost the best of his soothsayers. Freud went on to recall how in Worcester, strolling together in front of the Clark University building before the conference, he invited Ferenczi to propose the subjects he should talk about, and half an hour later Freud would improvise on the

DOI: 10.4324/9781003246473-12

suggested topic. Their collaboration instantly became a form of shared thinking. He added:

> We spent together in Italy the summer holidays of several successive years, and a number of papers which appeared later under his or my name first took form there in our talks.
>
> (*ibid.*, p. 298)

Despite this, Ferenczi would be one of the few pupils who was able to create and maintain his own differentiated and original area of theorisation on the margins of Freudianism, proceeding sometimes in isolation and sometimes in collaboration with barely orthodox analysts (as Freud's former pupil Otto Rank became, deciding to found his own school in 1925, and as the "wild" analyst and healer Georg Groddeck was from the start). In particular, Sándor Ferenczi became a theorist of technique, unlike Sigmund Freud, who only published a few brief articles on the subject and never produced the long-awaited "handbook."

But was Ferenczi ready for anything in his analytic work?

He certainly exerted himself to the extreme, like nobody before him. When, late in life, he came to experiment with "mutual analysis," Anna Freud herself recognised a special clinical ability in him:[2] what would in other hands have degenerated into clumsy and manipulative improvisation took the form of disciplined and honest boldness in the Hungarian, venturing into territories where he was fully aware of the risks. Freud did not conceal his dissent:

> One problem alone absorbed his interest. The need to heal and to help had become imperious. Probably he had set himself aims that to-day are not to be reached with our therapeutic means.
>
> (1933, p. 299)

A restless and tormented spirit, "prescient innovator of all modern trends" in psychoanalysis,[3] Ferenczi was ahead of his time. After his death, a barrier of censorship was raised by Ernest Jones and, indirectly, by Freud himself: against his technique and his theory, but also against him as a person.

The controversy between Freud and Ferenczi represented a trauma for the psychoanalytic community but also a fulcrum for its development, a place for comparison and for defining its tradition, since it gave a voice to distinct and apparently opposed elements of the subject's psychic experience and its relation to the other.

Paul Roazen would write:

> Ferenczi's last period of work, such as his rich "Confusion of Tongues between Adults and the Child" paper, served to highlight an essential aspect of Freud's teachings: for there is an altogether surprising gulf between what adults see and experience as opposed to what children appear to feel. The

poet in Ferenczi enabled him to recapture for science something essential of the imaginative life of the child.

(2003, p. 82)

To forget Ferenczi is to wrong Freud, who repeatedly said that all analysts can call themselves Ferenczi's pupils:[4] he was in fact the teacher of many great psycho-analysts, from Anna Freud and Melanie Klein to Balint, Winnicott, and the British independents, as he was for a number of American analysts and even Bion. Some of them were his patients; others learned from him without realising it via that thought transmission that passes through the analytic generations.

Disappearing and reviving in a sort of "karstic propagation"[5] after crossing unsuspected subterranean strata, it was only in the fifties that Ferenczi's "theory of counter-transference," with its rich vein of innovation, flowed back into the mainstream and found its original place in dialogue with Freud's "theory of trans-ference."[6] Lacan, among the few who remembered the Hungarian, mostly to try to challenge the emerging current running through Ferenczi and Balint to English psychoanalysis, was able to pun on the French pronunciation of his name: *faire ainsi* (do thus) but also *faire un ksi* (cross him out).[7]

He has achieved the posthumous fame that is the destiny of the "unclassifiable ones" that is, "those whose work neither fits the existing order nor introduces a new genre that lends itself to future classification."[8]

A psychoanalyst "of a restless mind"

Sándor Ferenczi (1873–1933) was a typical member of the Budapest intelligentsia at the end of the nineteenth century. Like so many Jews who played key roles in European culture, his family came from Galicia. His father, Baruch Fraenkel, had migrated when very young to Hungary, where he became passionate about the independence movement and took part in the Hungarian uprising of 1848.

Fraenkel had been allowed to make his home in Miskolcz, a small town in the northern region of Tokai, where he had opened a bookshop with an attached print-ing press, publishing house, and concert agency, a small business which prospered for a long time and managed to survive both world wars under the management of the few of his heirs who had escaped the Holocaust.[9]

Miskolcz was where Sándor was born, the 8th of 12 children and his father's favourite, acquiring his interest in culture, study, and research but also his pro-gressive and independent outlook.

Clara Thompson also reports:

Of a restless mind, he sought always to satisfy his curiosity, a curiosity which did not rest with mastering such problems as how to make fireworks and how to discover the inner structure of the eye from reflections on his own eyeglasses, but sought always to master the secret of people – why they did what they did, what they really felt. He sought in his father's bookshop for the answer. No bookshelf was too high for him, and he often read while still sitting on the top of the ladder. So quiet and lost to his surroundings would he become that once the clerk actually started to carry the ladder away, not realizing that he was perched there.

(1964, p. 65)

Ferenczi's relationship with his father seemed to compensate him in part for not having felt loved by his mother, whom he experienced as distant, hard, and unjust. In his case, too, there had been another child who died young and perhaps a maternal depression.[10] Sándor had inevitably cultivated resentments, fantasies of retaliation, and fears of abandonment: at least until his father's death, which came when he was 15.

The large family had fostered Ferenczi's development of a sociable and friendly temperament, but, as a middle child, he had to struggle from the outset against what Freud called his "powerful brother complex."[11] He himself came to intuit how his insatiable curiosity about the world had its origin in his childhood desire for knowledge aroused by his mother's series of pregnancies. Ròza Eibenschütz, this remarkably energetic mother who was also active in the community – she was chair of the Jewish women of Miskolc[12] – had grown up in Austria and spoke fluent German as well as Hungarian.[13] She passed her bilingualism on to Sándor, along with her emotional ties to Vienna where, after finishing his secondary education, he went to live with an aunt in order to study medicine.

Sigmund Freud and his Hungarian pupil had not only the German language and their Jewish roots in common but also their families' migration from Eastern Europe, the experience of the Empire's multiethnicity, a secular spirit, a sense of humour, cultural interests, and a scientific education from the same teachers. Not least, they shared an interest in hypnosis which in Ferenczi's case manifested itself well before his clinical practice when, while still at his high school, he loved to experiment with his schoolfriends or the assistants in his father's bookshop.[14]

The two men's friendship shared a further link in the names of their respective brothers: *Zsigmond* Ferenczi (elder) and *Alexander* Freud (younger).

Despite these affinities, Ferenczi would not come into contact with psychoanalysis until relatively late. He ran into Freud's work almost symptomatically: while still a student in Vienna, he had picked up *On the Psychical Mechanism of Hysterical Phenomena* by Breuer and Freud (1893) and had instantly dismissed the Freudian hypotheses on sexual aetiology.[15]

In 1897, after graduation and two years of military service, Ferenczi returned to Hungary and made his home in Budapest.[16] Here the young physician – already a neurologist, psychiatrist, and expert in forensic medicine – became an assistant at the old *Ròkus* Hospital, the one where, in the mid-nineteenth century, Ignaz Philipp Semmelweis, the "saviour of *mothers*," had been the first to understand the value of asepsis in obstetrics. Confined to the poor people's ward, though it had been his wish to devote himself to the neuroses, Ferenczi had to deal with the most diverse pathologies and especially with venereal diseases, treating prostitutes and homosexuals, whose marginalisation he took to heart.

His curiosity led him into the field of occultism during almost the same period that Jung was experimenting with his cousin in Basel: spiritualism was a focal point of debate at the time, and from 1897 to 1899, Ferenczi had many opportunities to take part in séances organised in Buda at the house of his friend E Felletàr, a professor of pharmacology. This is the period of the anecdote about the medium – the professor's young niece – who was given a piece of paper on which Ferenczi had written, "What is the person I am thinking about doing at this moment?" and replied, "He has just got out of bed, he is asking for a glass of water, and he is falling back again . . . dead." Instantly remembering that he had an appointment with a patient, Ferenczi rushed to the house to find that the vision had proved correct.[17]

The strange episode aroused his curiosity but did not influence his critical judgement, and his first publication, devoted to spiritism (1899), testifies to his scientific interest in the unconscious.

Ferenczi and the hidden treasure of Spiritualism

Let's leave it to Ferenczi himself to explain how, in 1898, the idea of writing that first article came to him:

> As I had no chance to perform psychological experiments, I had begun to experiment on myself. I had tried to explore the so-called "occult" phenomena, amongst others. It happened once that, following a dinner with friends, I walked through the permanently locked gates of the "Little Ròkus" well after midnight. I went to the junior physicians' room and engaged in "automatic writing," frequently discussed by spiritualists of the time. Janet had already published interesting observations on this phenomenon. It occurred to me that it was already very late; I was tired and somewhat aroused emotionally. All these circumstances favoured the exploration. So I picked up a pencil and, holding it loosely in my hand, I pressed it against a piece of blank paper: I decided to let the pencil move "on its own," letting it write whatever it wished. Senseless scribbles came first, then letters, words (some of them strange to me), and finally whole sentences. Soon I was having a real dialogue with my pencil – I asked questions and got quite unexpected answers. Being young, first of all I demanded answers to big theoretical problems, but later more practical questions occurred to me. Finally the pencil suggested the following: "Write an article on spiritualism for Gyógyászat, the editor will be interested" . . . The next day I wrote my first medical paper entitled "Spiritism." I started off from the phenomenon of automatism which I had observed in myself. My conclusion (which has been generally accepted since then but now based on more sound evidence) was that so-called occult phenomena do not contain anything supernatural and should be viewed as manifestations of the functioning of the unconscious.
>
> (1917, p. 430)

The episode shows a Ferenczi who enjoyed playing at a sort of naïve self-analysis, which nevertheless brought about an important change in his life, since it put him in touch with the surgeon Miksa Schachter of the Hungarian Medical Society, editor of the journal *Gyógyászat* (Therapy).

A Jew with strong religious sentiments and moral principles to match, but liberal and generous, Schachter not only accepted Ferenczi's article but encouraged him to write again, on scientific matters such as public health. Schachter became a teacher and father to him, welcoming him into his family as Sigmund Freud would later do.

That first article, "Spiritism," highlights the freedom and courage of Ferenczi's mind: he expresses his astonishment at the conviction held by "educated and intelligent men who give no evidence of hysteria," such as Crookes, Lombroso, Du Prel, and other celebrated scientists, who had turned themselves into prophets of spiritualism, and exhorted the anti-spiritualists to study this social phenomenon in order to draw new psychological knowledge from it. Anticipating what would be the main theme of his psychoanalytic research, he intuited that in these experiences a mechanism of Ego fragmentation was present:

> It is quite possible that the greater part of spiritistic phenomena is based on the same *simple or multiple split in mental functioning*, one function alone being focused in the mirror of consciousness, while the rest is carried on automatically and unconsciously. This would explain how a medium automatically, unconsciously and without the intent to deceive may guide the coin over the alphabet to form intelligent words.
>
> (1899, p. 144, my italics)

It was the psychological viewpoint that aroused his curiosity; his ambition to find a scientific explanation for the reality and authenticity of telepathy would later have a significant influence on Freud. This is how Ferenczi concluded his article:

> the movement of spiritism may lead to similar findings as did alchemy. (Those who were intent upon making gold made very useful chemical discoveries.) Or the same thing will happen to the spiritists as to the heirs of a crafty peasant who were told, on their father's deathbed, that treasure was hidden on the land. As a result, the sons have turned the soil upside down, and reaped the reward of a plentiful harvest. The alchemic gold, the hidden treasure of the spiritist may bring us a rich harvest, beyond anticipation in a little cultivated field of science, the psychology of the human mind.
>
> (1899)

At the time, his interest in hypnosis, non-conscious processes, and "subjective" truths, as well as dreams, did not include any reference to psychoanalysis. On the contrary – as he would enjoy admitting[18] – having become a regular contributor to *Gyógyàszat*, he refused to review *The Interpretation of Dreams* because, having skimmed it, he hadn't thought it worth mentioning. Instead, he had reviewed the contemporary work by the Italian psychiatrist Sante De Sanctis, *Dreams* (published in German in 1901), a less revolutionary and scandalous text, and wrote a Hungarian translation of the final chapter "Miraculous Elements in Dreaming," devoted to an examination of prophetic or telepathic dreams.[19]

However much he was influenced by the puritanism of his mentor Schachter, who would not have appreciated such a bold treatment of sexuality, Ferenczi himself nursed a certain repugnance towards the subject and a deep prejudice against women.[20] This initial resistance was followed by an almost morbid insistence on the theme of sexuality in his subsequent correspondence with Freud and in his self-analysis, where old fears would emerge about masturbation and resorting to prostitutes.

Compared with Jung, who was physically imposing and good-looking, with a richly resonant voice and who, in his good moods, dazzled with vigour and energy, exercising a remarkable fascination over women,[21] Ferenczi, who was only a few years older than the Swiss – and 17 years younger than Freud – was neither so attractive nor as sunny. Anguished by nature, in his photos, the features of his slightly swollen face suggest an almost foetal softness and, together with his psychosomatic sensitivity, make him look like an aquatic creature forever in search of a liquid habitat he could at last dive into. Women rarely liked him, whereas he appeared instantly lively, brilliant, and appealing in male company: whether there were 2 people present or 20, Ferenczi was at the centre and the focus of attention with the impassioned way he spoke.[22]

As he neared the age of 30, he began a relationship with a married but separated woman eight years older: Gizella Palos, née Altschul, came from a family he had known since his childhood in Miskolcz and had also become a relative when his younger brother Lajos had married one of her daughters.

This first love of Ferenczi's answered his deep inner needs so fully that it remained the most important relationship of his life. Beset by doubts, dissatisfactions, resentments, and nostalgia, Gizella would always be at his side: for over 20 years as his lover, from 1919 as his wife. But Ferenczi, who would nurse a tortured regret at not having married a younger woman and having no children, had never been in any hurry to start a family and had lived a bachelor existence for years, visiting prostitutes, living in hotels, and spending a lot of time in cafés writing poetry and absorbing himself in the most varied subjects in the company of writers and artists, at least until psychoanalysis captured his interest.

The encounter with Freud: a postponed transferential appointment

It was through Jung that Ferenczi came into contact with Freud. It happened that his colleague Philippe Stein, whom he had met in 1905 at a congress of the anti-alcohol association, told him about the new method developed in Zurich for measuring mental functioning by means of word association: Ferenczi was immediately fired up by this and began experimenting with the chronometer test on the responses of whoever was at hand in the cafés of Budapest, whether they were writers, poets, or painters, waitresses, or the women who cleaned the toilets.[23] He went to Zurich with Stein to meet Jung and may have had some experience of analysis with him. Having now devoured all the available psychoanalytic literature, in February 1908, at the age of 35, bearing a letter of introduction from the Swiss and again accompanied by Stein, he made his first visit to Berggasse. Ferenczi was enraptured, almost thunderstruck by the new friendship, to which the Master responded with equal intensity and warmth. The Hungarian soon became an external correspondent of the Viennese Wednesday Society but more especially a permanent[24] and intimate confidant of Freud.

In the lectures he gave at the Royal Society of Medicine on his return to Budapest, Ferenczi addressed the mechanism of repression, trying to understand why he had initially felt such aversion for the theory of the sexual origin of the neuroses and had not given a thought to establishing whether it might contain some elements of truth.[25] Later, the analysis of a dream allowed him to connect the previous exaggerated strictness in morals and intolerance of obscene words to the humiliation of having been detected by his mother when he was six and a half years old, "writing down on a piece of paper a dictionary, so to speak, of all the obscene expressions he knew." The severe punishment that followed resulted in "a lack of interest in erotic matters for many years after."[26]

This start reflected his resistance to the impetuosity of his deepest emotions. It is likely that the first glance at *Studies on Hysteria* had instantly impregnated Ferenczi with Freudian theory and he had deferred the experience, not deleted it, so as to work through it inwardly with a steadily emerging rather than magical understanding.[27] The transferential appointment with Freud changed his whole life, not just his professional vocation: it was a falling in love, an unhappy passion in which he repeated his relationship with a father whose favourite he had been but who weakened him with his indulgence. What emerged, as Jones mercilessly described it, was his insatiable need to be loved.

Freud was won over by the Hungarian's vivacity and talent: there was none of the Wagnerian intensity of the encounter with Jung, but the offer of unconditional friendship, faithful intimacy, and scientific commitment had the instantaneous familiarity of the disciple meeting the teacher.[28] In April, at the first psychoanalytic congress in Salzburg (where he presented the paper *Psycho-Analysis and Education*), Ferenczi was already being invited by Freud to spend the summer holidays with his family in Berchtesgaden. Here he easily won the friendship of Freud's eldest son, Jean Martin, and prompted Freud to fantasise that he might marry Mathilde.

Ferenczi became the loyal companion of his scientific journeys and recreational travels, splendidly occupying the place left vacant by Freud's brother Alexander and sister-in-law Minna.

In the years before the war, the two friends always spent the late summer holidays together – in 1910 in Sicily, in 1911 at Collalbo, in 1912 in Rome, in 1913 at

San Martino di Castrozza – not to mention the Easter weekends or overnight stays that Ferenczi would make, whenever possible, in Vienna. To begin with, there were no women who might divide them.

In 1909, as they were getting ready to plan their first vacation together, having had to give up the idea of Sicily because of the earthquake that had devastated Messina in December 1908, they were thinking of falling back on Egypt when a renewed offer from Stanley Hall persuaded Freud to accept his invitation and give five lectures at Clark University in Worcester, in the United States. That Ferenczi should come too was a matter of course.

G Stanley Hall (1844–1924),[29] rector of the university set up in 1889 with funds from the industrialist J Clark, wanted to celebrate the 20th anniversary of its foundation with an international conference dedicated to progress in the social sciences. He succeeded in gaining Freud's participation by increasing the fee (3000 marks, the same sum that he had offered to Wilhelm Wundt from the beginning) but still more by postponing the date of the celebrations to late summer. Freud and Ferenczi then learned that Jung had also been contacted with a view to his holding three sessions on the experiments in association. The late request, some months after the second invitation to Freud,[30] did not stop the Swiss finding one of the last first-class cabins on the same ship as his colleagues.

The swift and splendid *George Washington*, belonging to *Norddeutsche Lloyd*, was due to leave Bremen for New York on 21 August 1909.[31] *Big George*, as the steamer was nicknamed, and the *ménage à trois* of dream analysis and mutual interpretation on the crossing – a marathon of free associations along the bridge[32] – unleashed Ferenczi's infantile complexes and was the start of the break-up for Jung. Freud seemed to foresee this on the day before their departure.

Notes

1 1964, p. 66. With K Horney, H S Sullivan, E Fromm, and F Fromm Reichmann, in New York, Clara Thompson was at the origin of modern intersubjective psychoanalysis.
2 To Andreas-Salomé, 22/11/1931.
3 Aron and Harris, 1993, p. 1.
4 1933.
5 Martin Cabré, 2005.
6 Heimann, 1950.
7 See Granoff, 2000.
8 Arendt, 1968, p. 155.
9 Carloni, 1989.
10 Martín Cabré, 1997.
11 Freud, 1923a, p. 268.
12 Carloni, 1989.
13 Haynal, 2002.
14 Carloni, 1989.
15 Ferenczi, 1908a.
16 Mèszàros, 1993, in Aron and Harris, 1993.
17 Fodor, 1979, p. 106.
18 Balint, 1968.

19 Gyimesi, 2018; Mèszàros, 2008. De Sanctis's text dealt the characteristics of dreams in animals, children, old people, adults, and various mental illnesses. Speaking of telepathy, he reported that he had even read about a case of telepathy in a dog (1899, p. 58).
20 Mèszàros, 1993.
21 Kerr, 1993 and M. Freud, 1958a.
22 Roazen, 1975.
23 Balint, 1968.
24 Sabourin, 1985.
25 1908a.
26 1911, p. 142.
27 Ibid.
28 Rodrigué, 1996, *I*, p. 512.
29 Student of W James, founder of the American J. of Psychology, first president of the American Psychological Association.
30 According to Roazen, there was another guest who refused (1975).
31 It seems that the recently inaugurated German steamship was given that name, dear to Americans, with the aim of making it easier for European emigrants to enter the United States during the inspection at Ellis Island. At the outbreak of the Great War, the *George Washington* was at the Hoboken terminal, on the Hudson, opposite Manhattan: when the United States entered the war, she was requisitioned for transporting troops and on 13/12/1918 she set sail for Europe with President Woodrow Wilson on board.
32 Rodrigué, 1996, *I*,

Chapter 12

A journey to America

Three men and an eventful, mutually analytic crossing: the outward journey...

On 20 August 1909, in Bremen, Freud and Jung met again for the first time since the Easter visit and the exchange of letters about Sabina Spielrein. The subjects of occultism, the sexual implications of treatment, and the conflict with authority formed part of their luggage for the journey.

Given their violent feelings of rivalry, sharpened by the invitation to America, a confrontation between the two men was inevitable.

Freud treated his pupils to lunch at the *Essighaus*, an old restaurant on the harbourside, and with Ferenczi's help persuaded Jung to break the *Burghölzli* rule and toast their departure. Immediately after this success – though Jung had not always been an abstainer: in fact, he was known to his school and drinking companions as "the Barrel"[1] – the professor had the first of two fainting fits that afflicted him in the presence of the Swiss.

> Jung mentioned, to our great satisfaction, that he has decided to give up his abstinence, and he begs us to encourage him. We toast with an excellent wine. Whether because I drank too fast or was disturbed by the sleepless night, it happened that when I was eating the salmon I broke into a bad sweat with a feeling of faintness. I had to forgo the other courses of the meal. I didn't trust myself to drink any more. Jung will now take care of the drinking for me. Of course, the whole attack was soon over.
>
> (Freud, travel diary, Bremen, 21/08/1909, early morning)[2]

As had happened to him in the past during a conflict with Fliess, the passing out may have been a consequence of the effort to repress the heated emotions stirred up by his pupil. It seems that Freud resented the insistence with which Jung, not content with taking him to see the perfectly preserved mummies in the so-called Bleikeller under Bremen Cathedral, then talked enthusiastically during lunch about the "peat-bog corpses," mummified bodies from the first centuries AD, rediscovered in the late nineteenth century under the peat in certain north

DOI: 10.4324/9781003246473-13

European marshlands. Freud perceived this as a direct attack on him. The situation was further complicated by the skeletons in Jung's cupboard, that disturbing nucleus of primitive, entirely unconscious instinctuality, which aroused the fear that the pupil not only wished to engage in conflict with his teacher but was ready to seduce him and then wipe him out, as he had done with Spielrein.[3]

For Freud, it was not just a reaction to Jung's violent parricide, which he did not fail to interpret: as he became the protagonist of a success story, the fantasy of being "wrecked by success" made Freud relive his own confrontation with his father. Only a few hours earlier, Freud had arranged 20,000 marks' worth of life insurance against the perils of the journey ("Ferenczi did the same for ten thousand"),[4] but there was no insurance that could protect him from the re-emergence of oedipal conflict. America was the confirmation of that longed-for victory which was still able to take him by surprise.

After the first fainting fit, the rest of the journey was thrilling. The first sight of the *George Washington* had been breathtaking: once on board, almost drunk and disoriented by that immense, elegant steamer, Freud found himself triumphantly welcomed into his cabin by a bunch of orchids, a gift from his patient Elfriede Hirschfeld,[5] which adorned his desk for the whole crossing. A moment later, he found his cabin steward intently reading *The Psychopathology*, a coincidence worthy of the "Lubitsch touch." This was what it meant to be famous![6]

The days at sea flew by, despite the clocks having to be put back by three quarters of an hour every morning: while the other passengers devoted themselves to every possible entertainment, moving between the saloons in the cafeteria and the deckchairs on the bridge, the three colleagues played their own enjoyable and highly distinctive game of mutual analysis.

As they entered the grand harbour at New York, it was with a foretaste of success that Freud turned contentedly to Jung and observed that the Americans had no idea what was going to be revealed to them. The Swiss was quick to note, with some scorn, the intensity of Freud's ambitious desires, and when Freud replied frankly that he was the humblest of men, the least ambitious in the world, his pupil increased the dose by denouncing this as a further presumption.[7]

Once they had disembarked, after the relative quiet of the sea voyage, the short stay in New York turned out to be hectic. Guided by Brill, the travellers packed everything into a few days: from the Lower East Side, to Chinatown, the night excursion to Coney Island, the tour in Central Park, the visit to the Metropolitan Museum and Columbia University. They did not miss a late evening concert of Apache music and shopping at Tiffany's. Joined by Jones, they all went together to the cinema for the first time.

It all took a toll on their bodies and especially their stomachs. During a two-way walk along Riverside Drive, with the view of the Palisades in front of him, Freud again showed Jung his fragility, with an embarrassing inability to hold his urine.[8] Jung had to help him, but took the opportunity to reiterate the interpretation of his regressive exhibitionism.

As he remembered years later, Jung was disappointed and resentful because during the crossing and in the full immersion *à trois* in the analysis of dreams, slips, and free associations, Freud had reserved himself the privilege of not revealing all his own thoughts about a dream, with the justification that he did not wish to lose his authority. This lack of reciprocity and the fact that Freud was placing "personal authority above truth"[9] would constitute a first pretext for the break.

It seems that Freud's dream was about his wife and sister-in-law, and it can be imagined that, when asked for details, he might have had reservations about replying: he would have found it difficult to let himself go by confiding in Jung, who had behaved so improperly in his own married life.[10] However, on the subject of the castration complex, Freud was ready to let himself be interpreted and calmly took the Swiss bombardments on the chin. More excited than he wanted to admit, the words with which he recalled that experience reveal a sense of alienation, an inability to believe his own eyes, similar to what he had felt in 1904 at the Acropolis:

> At that time I was only fifty-three. I felt young and healthy, and my short visit to the new world encouraged my self-respect in every way. In Europe I felt as though I were despised; but over there I found myself received by the foremost men as an equal. As I stepped on to the platform at Worcester to deliver my *Five Lectures on Psycho-Analysis* it seemed like the realization of some incredible day-dream: psycho-analysis was no longer a product of delusion, it had become a valuable part of reality.
>
> (1925, p. 52)

It is no wonder that, in confirmation of Jung's interpretation about ambition and his remarkable incontinence, during the ceremony awarding him a degree *honoris causa*, Freud found it unusually difficult to hold back his tears, as was noted by Ferenczi.[11]

It has to be said that the welcome to New York was a bit disappointing at first: "Interviews with the reporters gave little trouble, and the only account in next morning's paper baldly announced the arrival of a certain 'Professor Freund [sic] of Vienna'."[12] There weren't even the big requests for consultations that he had hoped for. Still, the Clark University conferences were making some noise, and the newspapers in Worcester and Boston gave prominence to Freudian theory, and with fewer criticisms and distortions than he had expected.

Despite the exceptional nature of the occasion, Freud had followed his usual habit at the university and left home without preparing a text. Instead, after discussions with Ferenczi, he gave his conferences with the freshness of improvisation, in his clear and engrossing style. He gave his lectures in Clark's art library, a simple medium-sized room lit by a large skylight and packed with people, speaking in German because he did not feel he had sufficiently mastered English, but in such a quiet voice that his eminent listeners, whose attention he had seized straightaway, had to rearrange their chairs into a semicircle in order to follow him more easily.[13]

At that time, German was the international language of science: William James knew it well, as did Stanley Hall and J Jackson Putnam, who, like many other American physicians, had completed their studies in Europe. Freud began by illustrating Breuer's cathartic method, his own theory of hysteria and repression, the technique of free association and the interpretation of dreams, slips, and mistakes and concluded with the hottest topic of all: infantile sexuality and transference. It seems that what most struck the audience was not his attack on conventional sexual morality but the revolutionary way he talked about sexual behaviour, in a simple and precise manner that no one had ever used before in medicine.[14]

Jung's conferences dealt with the association experiment, the theory of complexes, and also the infantile psyche. In his exposition, which he later published under the title *Psychic Conflicts in a Child* (1910), he presented some observations about the analysis of a four-year-old girl, his own daughter Agathe, introducing the concept of introversion and underlining the adaptive utility of poetic fantasy in adolescence and childhood. Oddly, the pseudonyms he chose for her and her little sister were those of Freud's daughters Anna and Sophie.

The *Boston Evening Transcript* published a surprisingly reliable summary of the lectures and even a credible account of an interview with Freud, held by Stanley Hall at his home in Worcester.[15] It should be noted that the arrival of psychoanalysis in America had coincided with the renewed popularity of suggestive psychotherapies: in search of an answer that was not the diagnosis of hereditary degeneration offered by official psychiatry, many neurotics turned to Christian Science and to forms of healing obtained by prayer.

The interview began by addressing the Emmanuel Movement, a current local phenomenon led by the psychologist rector of the Episcopal church of Boston's Back Bay and the Rev Samuel McComb.

Freud was at pains to stress that his discipline had not connection whatever with any form of religion and, on this occasion – unlike what he would assert in the twenties about "lay" analysis (Freud, 1926) – defended the importance of a medical training for psychotherapists. He commented:

> I can easily understand that this combination of church and psychotherapy appeals to the public, for the public has always had a certain weakness for everything that savors of mysteries and the mysterious, and these it probably suspects behind psychotherapy, which in reality has nothing, absolutely nothing, mysterious about it.
>
> (Albrecht, 1909, p. 335)

Even so, historians report that the Worcester congress also included unofficial updates about occultism and "psychical research." America was also in the grip of a feverish mania for spiritualism:

> Across the nation, in the homes of the poor and disenfranchised and of the highly educated, prosperous elite alike, séances were being staged with a zeal suggesting that if the participants didn't act now, the dead might forever hold their peace.
>
> (Prochnik, 2006, p. 141)

Freud had the opportunity to discuss it with William James, a then-prominent figure of American psychology, who had been one of the presidents of the Society for Psychical Research. Although very ill – he had an attack of angina during the walk together[16] – James had come on purpose to meet Freud and bring him at the request of Stanley Hall a copy of his opus on Mrs Piper, the most influential medium of the age, at least in New England, with the results of his experiments on extrasensory perception.[17]

James had been interested in Freud's work since 1894[18] and had introduced his friend Putnam to psychoanalysis, bringing Putnam with him from Boston to welcome the illustrious guest. Unlike James who, despite his scientific loyalty to the facts, believed in telepathic, trance-based communication from the dead, Putnam was not so interested in his studies of the medium. He held out against the seduction of the supernatural even when contemporaries for whom he had the greatest respect embraced it, just as he held out against the new forms of Christian faith-based healing,[19] but he was inextricably entangled in the national conversation about psychic experience himself and eager to discuss it with Freud and Jung. His personal interest in mediums could be traced back to an ancestor involved in the Salem witch hunt.[20]

At a meeting organised at Hall's house, James exhibited his experiences of Piper's mediumship. Even her detractors agreed that "Piper was not a fraud in the usual sense in which the term applied to mediums":[21] she did not stage shows and did not make substantial profits ($20 per session for a limited audience). From a modest family, Leonora E Piper was 50 years old in 1909. There seems to have been a traumatic childhood experience at the origin of her medium ability, a sleigh accident that had caused her recurrent gynaecological problems.

Over the years, Piper's sessions had taken on a stable form that James carefully described.[22] The woman sat in an armchair and chatted with her sitters. In front of her stood a table on which lay three pillows. Conversation was "superficial," trying to empty her mind: today we could compare it to free associations. Gradually she appeared more sleepy, the physiological parameters were all slowed down, and slowly she went into a trance and lay down. Her hand took up a pencil and began to write at the prompting of her control, or she spoke in the voice of the dead visitor.

The whole thing would last between an hour and a half and two hours. While inside the trance and "possessed by the dead," the medium would be engaged in a form of conversation with her sitters, who would ask her questions or stimulate her with words that were very reminiscent of the test of verbal associations and that helped them in trying to interpret her communications.

The exit from the trance was even slower and Piper's face went through various contortions similar to those Charcot studied in hysterics, expressing pain and repulsion, sometimes griding her teeth or weeping.

In 1910, Stanley Hall's pupil, Amy Tanner, referring specifically to Freudian theories, in her *Studies on Spiritism*, came to conclusions completely opposite to those of James, linking Piper's mediumship with traumatic events of her child-hood that had affected her health since adolescence.

Tanner was of the opinion that Piper suffered from "dissociation," a divided self triggered by traumatic injury in puberty and subsequent shocks to her sexual physiology. Despite various gynaecological pathologies, the woman had married and had children: according to Tanner, the dissociation at the base of mediumistic abilities was clearly connected to her state of health and improved with the meno-pause to the detriment of the sessions.[23]

James's departure did not end Freud's exposure to the American fever for spiri-tualism. On Freud's last night at Clark, Stanley Hall arranged another exchange on the occult. This time, the colleagues had the opportunity to examine a medium directly. The one presented by Hall had already been studied by James, but he hoped that the session would give him the chance to observe how Freud used psy-choanalytic theory to dismantle the psychic statements of the woman. In reality, it was a rather different case from Piper, in which the complex of the young woman was so immediately transparent that Freud and Jung were in agreement and did not take long to extract a confession from her that her mediumship was a fraud designed to attract the attention of a young man.[24]

After completing the scientific tasks in Worcester, the American adventure ended with a visit to Niagara Falls and, after a bumpy ride on a rickety horse-drawn carriage, a stay at Putnam Camp in the wilderness of the Adirondacks, at the invitation of Putnam himself. This cultural colony in the isolated High Peaks had been founded in the mid-1870s by Bowditch, Putnam, and James, representa-tives of the so-called Boston School, a tightly knit group of psychologists, phi-losophers, neurologists, and psychiatrists who ranked among the most erudite and progressive thinkers of their day. It was intended to host distinguished artists, scholars, and literati for a return to contact with wild nature as a response to the stresses of a rapidly industrialising society.[25]

The stay at Putnam camp offered Freud the opportunity to try out the true American wilderness: for the first time, his ability to walk and climb mountains was put to the test by the climate and the harshness of the High Peaks.

William James had not received a completely positive impression of Freud, though he had certainly appreciated Jung, and a few days after the Worcester meeting he wrote to his friend Flournoy:

> I hope that Freud and his pupils will push their ideas to their utmost limits, so that we may learn what they are. They can't fail to throw light on human nature, but I confess that he made on me personally the impression of a man obsessed with fixed ideas. I can make nothing in my own case with his dream theories, and obviously "symbolism" is a most dangerous method.
>
> (28/09/1909)[26]

Freud, by contrast, continued to be favourably impressed by his meeting with James, and it was through his influence that Freud agreed to become a corresponding member of the Society for Psychic Research in 1911 (Jung had been a member since 1908).

Ernest Jones quickly made his puzzlement known and put Freud on guard:

> in spite of the good names in it, the society is not of good repute in scientific circles. You will remember that they did some valuable work in the eighties on hypnotism, automatic writing etc., but for the past 15 years they have confined their attention to "spook-hunting," mediumism, and telepathy, the chief aim being to communicate with departed souls. . . . It does not seem that your researches lend much support to spiritism, in spite of William James' ardent hope.
>
> (17/03/1911)

In conclusion, Jones cynically added, since William James had died in the interim: "Poor James. One hasn't even the consolation of thinking that he knows better by now."[27]

... and back again

At ten o'clock on Tuesday 21 September 1909, the *Kaiser Wilhelm der Grosse* of the Norddeutscher Lloyd line weighed anchor to bring the three psychoanalysts back to Europe. As Jung wrote to his wife, on that foggy morning, the ship left behind the towering whitish and reddish heaven-storming towers of New York City on the left and the smoking chimneys and Hoboken docks on the right, and to the rhythmic and piercing sound of the foghorns, the city's profile was soon lost to view; not long afterwards the big swells of the ocean began.[28]

The return voyage was more arduous than the voyage out, and the ship, a smaller one than the *George Washington*, and not nicknamed *Rolling Billy* for nothing, had to face a day-long storm.

> Yesterday there was a storm that lasted all day until nearly midnight. . . . The objects in my cabin had all come to life: the sofa cushion crawled about on the floor in the semidarkness; a recumbent shoe sat up, looked around in astonishment, and then shuffled quietly off under the sofa, a standing shoe turned wearily on its side and followed its mate. Now the scene changed. I realized that the shoes had gone under the sofa to fetch my bag and my briefcase. The whole company paraded over to join the big trunk under the bed. One sleeve of my shirt on the sofa waved longingly after them, and from inside the chests and drawers came rumbles and rattles. Suddenly there was a terrible crash under my floor, a rattling, clattering, and tinkling. One of the kitchens is underneath me. There, at one blow, five hundred plates had been awakened from their deathlike torpor and with a single bold leap had put a sudden end to their dreary existence as slaves. In all cabins round about,

unspeakable groans betrayed the secrets of the menu. I sleep like a top, and this morning the wind is beginning to blow from another side.

(Jung to Emma, Kaiser Wilhelm der Grosse, 25/09/09, 1961, p. 370)

There were internal uproars, too: the extreme daily exertions in analysis had been no joke and had changed the mood.

Jung was sleeping badly, and he resumed his open hostility towards the professor: he had already cast a shadow over the sexual theory in his second American lecture. Having barely tolerated Freud's refusal to reveal all the ideas associated with his dream on the outward crossing, Jung decided he too would ignore the rule and upon returning introduced a deliberate lie into the associations to a dream of his own.

This was a long and complicated dream, rather gothic, as Jung's dreams often were, more like fantastical daydreams:

I was in a house I did not know, which had two stories. It was "my house." I found myself in the upper story, where there was a kind of salon furnished with fine old pieces in rococo style. On the walls hung a number of precious old paintings. I wondered that this should be my house, and thought, "Not bad." But then it occurred to me that I did not know what the lower floor looked like. Descending the stairs, I reached the ground. There everything was much older, and I realize that this part of the house must date from about the fifteenth or sixteenth century. The furnishings were medieval, the floor were of red brick. Everywhere it was rather dark. I went from one room to another, thinking "Now I really must explore the whole house." I came upon a heavy door, and opened it. Beyond it, I discovered a stone stairway that led down into the cellar. Descending again, I found myself in a beautifully vaulted room which looked exceedingly ancient. Examining the walls I discovered layers of brick among the ordinary stone blocks, and chips of brick in the mortar. As soon as I saw this I knew that the walls dated from Roman times. My interest by now was intense. I looked more closely at the floor. It was of stone slabs, and in one of these I discovered a ring. When I pulled it, the stone slab lifted, and again I saw a stairway of narrow stone steps leading down into the depths. These too, I descended, and entered a law cave cut into the rock. Thick dust on the floor, and in the dust were scattered bones and broken pottery, like remains of a primitive culture. I discovered two human skulls, obviously very old and half disintegrated. Then I awoke.

(1961, pp. 158–159)

The interpretation of this dream was to become fundamental for the development of Jung's thought, on a par with the dream of Irma's injection for Freud. In the analytic work done during the crossing, the "two skulls" aroused the professor's interest. He kept returning to them and urged the Swiss to unearth a *desire* in relation to them. As fate would have it, in this case, too, the question revolved around corpses in a cellar: in order to avoid a repeated and predictable interpretation of parricidal and fratricidal fantasy (and to provide material that would keep Freud busy without disturbing him, as in Bremen), Jung concealed his real thoughts and lied that he associated the two skulls with "my wife and my sister-in-law." In his memoirs he explains the lie thus:

> After all I had to name someone whose death was worth the wishing! I was newly married at the time and knew perfectly well that there was nothing within myself which pointed to such wishes.
>
> (*ibid.*, pp. 159–160)

Sometimes lies aren't false, and Jung was so unaware of his hostile feelings towards his wife and his lover that he had wiped out the memory of the affair with Sabina Spielrein, which he had still not worked through: Eissler interprets the two skulls as an expression of anger and vindictive desire towards Emma and Sabina, and the association to the "remains of pre-historical creatures" as a way of evading guilt through a shift from the personal biography to the ancestors (1993).

The "lie" enabled Jung to avoid an outright clash (he convinced himself that Freud had not understood anything of his dreams) but nevertheless brought to light some truths about himself and perhaps also a provocative allusion to the never completely analysed dream that Freud had had on the outward journey (and which did indeed involve a wife and sister-in-law).

Winnicott claims that when Jung decided to lie to Freud, withdrawing from an analysis that would probably *not* have cured him (and "would have involved aspects of psycho-analytic theory that are only now, half a century later, beginning to emerge as a development of psycho-analytic metapsychology"), he did something profoundly sane and maturational. By being able to take responsibility for a lie, he was becoming whole – "a unit with a capacity to hide secrets instead of a split personality with no place for hiding anything" – and taking a first step towards the construction of that more highly evolved psychic mechanism, repression, which he did not yet have available to him.[29] Today we would say that Jung could not yet afford a "repressed unconscious," nor a stable boundary within which to define himself: for this reason, in the encounter with the other, he was exposed to unrepresentable and potentially traumatic sensations that each time reopened a split.[30]

While Jung, *Young* in name and in fact, was not ready to have a teacher, Freud had nevertheless done him a service without realising it. But that was not all. In the mimetic mirroring between the two men, whose personalities were like the obverse and reverse of a coin, Winnicott suggests that "Freud's famous fainting turns" in Bremen, provoked by Jung, could tell "the same story the other way round": Freud's regression from his predominant and rigid defensive mode of repression to splitting, with a blackout comparable to Jung's fainting in the presence of his father.[31]

Let it be clear that Winnicott does not intend to propose diagnoses but to reflect on the relationship that each of the two men was able to establish with the more "crazy" and "true" part of his own self and on their mutual incompatibilities. He warns:

> Freud's flight to sanity could be something we psycho-analysts are trying to recover from, just as Jungians are trying to recover from Jung's "divided self", and from the way he himself dealt with it.
>
> (1964, p. 450)

Jung would have to reflect on this important dream in solitude, starting out on an inner journey that would keep him withdrawn into himself for a long time while continuing to show apparent friendship and respect in his correspondence with the professor. The difficulty of this experience of self-cure, which especially occupied him after the official break with Freud, appears evident when we read the part of his memoirs relating to his childhood, the only part written in his own hand: it is Winnicott again who interpreted it as the autobiographical description of childhood schizophrenia and at the same time of the heroic effort he made thereafter to heal himself.

In *Memories, Dreams, Reflections* (1961), Aniela Jaffé presents in the form of an autobiographical narration the material collected in the interviews with Jung, who was then over 80: the memoirs contain very few references to external events or to private life, since Jung was interested in reporting only his internal events. "My life is a story of the self-realization of the unconscious," he stated in the Prologue. Though very reluctant to write about himself, Jung unexpectedly became passionate about the work of reconstructing his early childhood, to the point of directly engaging in fixing his own memories as a child, almost with a physical need to write. Winnicott considers this part of the text the final fruit of that incessant work of integration towards the acquisition of a whole Self, which had accompanied Jung since the split of his basic personality dating back to those years (1964).

A highly dramatic childhood memory which Jung locates at the age of four and calls "extraordinarily interesting" may help us to grasp the form of his early suffering, which Freud wrongly tended to interpret as neurotic and conflictual.

In his associative reconstruction, Jung cites the episode amid his evocation of the glowing, snow-covered mountains near Zürich, the Uetliberg, an unattainable land of dreams, and the incomparable splendour of the glistening water of Lake Constance (symbol of the split-off and idealised aspects of the first experience with his mother) and the memory of when, restive, feverish, unable to sleep, his father carried him in his arms, singing old songs (when a general eczema had appeared, as a psychosomatic defence against his mother's deep depression and the threat of disintegration suffered by his Ego). And this is the scene:

> strangers, bustle, excitement. The maid comes running and exclaims, "The fishermen have found a corpse – came down the Falls – they want to put it in

the washhouse!" My father says, "Yes, Yes." I want to see the dead body at once. My mother holds me back and sternly forbids me to go into the garden. When all the men had left, I quickly stole into the garden to the washhouse. But the door was locked. I went around the house; at the back there was an open drain running down the slope, and I saw blood and water trickling out. I found this extraordinarily interesting.

(1961, p. 7)

If we interpret it in the same way as a dream, the memory suggests that the child-ish curiosity about the drowned man shut in the washhouse arose from the search for something of himself that remained dead, unborn, left in the maternal womb by his father. The identification with his mother required this traumatic reality to stay split, unseen, and denied, not thought about, but little Carl had succeeded in circumventing the prohibition and in confronting the emblem of a violently impressive primal scene which, in spite of being concealed and locked away, was continuing to arouse curiosity and excitement. The "corpse in the washhouse" would appear to constitute the first form of that unconscious fantasy about his own missing birth (the mythical waters of childbirth bring a drowned man, not a newborn baby in a basket) and about the maternal depression which had so troubled his childhood, a fantasy that would re-emerge in the analytic relationship with Freud: both at the start of the voyage in the mummies of the "lead cellar" of Bremen Cathedral and the peat-bog corpses, and during the crossing in the dream-representation of the two skulls that Jung decided to lie about.

Freud's role in the failed analysis of his pupil and in Jung's withdrawal should not be overlooked.[32] Jung might have recognised his authority, and accepted his interpretations, if Freud had grasped his fragility and had been firm about his behaviour with Spielrein from the beginning, presenting him with a limit. Jung had at first feared him as an inquisitor and then regarded him as an accomplice, as had happened with his father. Once again he felt unprotected against an incestu-ous and fusional maternal image. Freud had not been helpful to him because he too had corpses of childhood mourning to deal with, the death of Julius, and these were the basis of his fainting.

It will be vitally important for Jung to construct a psychological theory capable of helping him contain his excitatory and terrifying fantasy. In order to do this, he will have to detach the psyche from the body and from sexuality – the very reali-ties to which Freud, by contrast, needed to anchor it – and to locate its roots in antiquity in a spiritual and impersonal "collective unconscious," which he consid-ered more profound than the Freudian one. Inhabited by the spirits of the ances-tors, the Jungian unconscious will be built as a metaphorical house of the dead, with many floors, to be explored by searching for archetypes – collective primi-tive and potential symbols – and for remote states of consciousness.

In order to understand the incompatibility between Freud and Jung, it can be suggestive to compare the two paradigmatic forms – one more conflictual, the other more associated with mourning – assumed by the infantile sexual fantasy and curiosity about the primary scene, at the centre of their personal psychic development and the construction of their theories: in the first, the childhood memory of crying desperately in front of the empty cellar with the subsequent finding of the mother, and the Irma dream; in the second, the memory of the corpse in the washhouse and the dream of the house with the two skulls in the cellar.

From the *Kaiser Wilhelm der Grosse*, aware of the inner journey he was gearing up for, Jung wrote pensively to his wife:

> The beauty and grandeur of the sea consists in our being forced down into the fruitful bottomlands of our own psyches, where we confront and re-create ourselves in the animation of the "mournful wasteland of the sea." Now we are still worn out from the "torment of these last days." We brood over the past few months, and the unconscious has a lot of work to do, putting in order all the things America has churned up within us.
>
> (22/09/1909)

***

Sándor Ferenczi had experienced a different but equally tempestuous crossing. The unexpected addition of Jung to the voyage had forced him to share Freud's attention: as a spectator of the American events, staying in the background but available to support Freud's weakness, an old feeling of jealousy and rancour had inevitably started to come over him, kept in check by his apparent docility in accepting second place and welcoming his sibling-rival.

On his return, he would state that:

> America is like a dream. On the whole, it went as I expected – I had much more reward and satisfaction from the company on the voyage than from what I engaged in over there. But I am extraordinarily glad to have made the trip.
>
> (Ferenczi to Freud, 5/09/1909)

Freud noted that, on leaving New York, the Hungarian had forgotten to return the key to his room in the Hotel Manhattan, taking it with him as a souvenir. He interpreted the slip as Ferenczi's having wanted to keep the *(Frauen)zimmer* [prostitute] secret so that no one else could enter it/her until he returned. Freud's pun captured Ferenczi's desire for exclusivity and the infantile attack on the mother who betrays.[33] What is certain is that, as a first outcome of the journey, Ferenczi decided not to keep his lover under wraps any longer and to introduce her to Freud. Back in early summer, shortly before their departure, he had dared to hint to Freud that he was on holiday "in pleasant company,"[34] and during the crossing he had frequently spoken about his girlfriend, using the tragic and romantic pseudonym Frau Isolde, with which she appeared in one of his dreams. There had been some negative comments and some interpretations of the difference in their ages. In spite of this, during the winter, Ferenczi resolved to introduce her to Freud and to submit to his judgement.

The development we are about to witness will see Ferenczi facing the situation with Frau G and his plan to marry her, on the one hand, and the new passionate search for the "occult" and thought transmission, which he will share with Freud, on the other. As soon as they had set foot in Europe, the two men would in fact start to pay visits together to other fascinating and disturbing *Frauen*: clairvoyants.

***

And finally, what do we say about Freud and the souvenirs he brought back from America along with the Chinese jade cup purchased at Tiffany's and the small brass porcupine that was given to him, whose quills he used to hold letters waiting to be answered? On his return to Europe, though seemingly reinvigorated by the company of his young pupils (for a while, he shaved off his beard), he said he had been sorely, viscerally tried by the trip: for many months he complained of digestive problems – his "American colitis" – and even claimed that the voyage had caused a deterioration in his handwriting.[35]

"America is a mistake; a gigantic mistake, it is true, but none the less a mistake" he used to say.[36]

For him, too, the trip had entailed an inner turmoil which Jung had tried to interpret and Ferenczi had confined himself to observing in silence. The Hungarian pupil would later think back to the problems manifested by the Master (the fainting fit in Bremen, the incontinence on Riverside Drive, and the ease with which he could feel moved) judging them to be hysterical symptoms arising from the conflict with his chosen heir. In the *Clinical Diary*, he acutely noted a bit of analysis that Freud had shared with him:

> dying as soon as the son takes his place, and regression to childhood, childish embarrassment, when he represses his American vanity. (Possibly his contempt for Americans is a reaction to this weakness, which he could not hide from us and himself. "How could I take so much pleasure in the honours the Americans have bestowed on me, when I feel such contempt for Americans?") Not unimportant is the emotion that impressed even me, a reverent spectator, as somewhat ridiculous, when almost with tears in his eyes he thanked the president of the university for the honorary doctorate.
>
> (1932, 4/08/1932)

Of course, Freud's intolerance for America became proverbial; it included the language, since, despite his knowledge of English, their version of it had been incomprehensible to him. The revival of this intolerance by his sister Anna's yearly visit from New York leads some people to think that it also had to do with jealousy and envy of his brother-in-law Eli, who had made himself rich. Before deciding to practise in the Vienna General Hospital, Sigmund had considered the possibility that he and Martha might emigrate with them.[37]

"America is only good for making money," he later claimed ungratefully and, like many intellectuals of the time, prided himself on despising every novelty from the New World: whatever was "*Echt Amerikanisch*" was frivolous and superficial.

However, something else needs to be said.

Once within sight of New York harbour and the famous Statue of Liberty, Freud immediately said, "They don't know we're bringing them the plague!" (we owe this legend to Jacques Lacan);[38] on his return to Europe he himself seemed to have been infected by one of the American mistakes and was bringing home a new interest in occultism.

"If Freud thought he could escape the occult by playing meteorologist to the erotic tempest in America he was mistaken."[39] After meeting James, he was fascinated by research into mediums and not in the least inclined to leave the field clear for James or Jung but ready to make his own sorties even into this new territory on which he certainly had something to say. And he would have Sándor Ferenczi as a companion.

The parallels between psychic sessions and psychoanalytic sessions could not have escaped Freud in descriptions of Piper's mediumship. The effort to empty the mind before receiving the dead was not so different from the surrender of ego-based control over consciousness that the analyst undertook. The content of the control conversations during a trance seemed to derive from the same depths of the underworld that Freud invoked as a source in the epigraph of *The Interpretation of Dreams*.[40]

The similarities were many and quite obvious.[41] Prochnik thinks that

> The phenomenal popularity of spiritism in turn-of-the-century United States helped prepare Americans for the dynamics of the psychoanalytic relationship, foreshadowing the special kind of listening and talking involved in Freud's practice.

But he also states:

> It's unlikely, however, that Freud would have felt confident enough at this point about the stature of psychoanalysis to suggest that the similarities between the two activities proved that psychoanalytic matters were at the core of the sitter's trance. It's more plausible that he would have felt afraid that the parallels would cast aspersions on his own methodology.
>
> (2006, p. 155)

Arriving in Bremen harbour at midday on 29 September 1909, after six weeks of living together and mutual analysis, the three travellers were not the same. The group broke up, but something "new" had been set off between them, something to do with "thought transference." Not only had they started to share an interest in "telepathy," but the analytic dialogue had opened up communication from unconscious to unconscious.[42]

The Swiss went straight back to Zurich, alone.

Freud immediately missed him and began unconsciously looking in the crowd for his light-coloured hat with the black band: "The day after we separated an incredible number of people looked amazingly like you," he wrote.[43] As if to console himself, he prolonged the journey for a few days with Ferenczi: they stopped in Hamburg and in Berlin, where they began their scientific collaboration into the occult by going to consult Frau Seidler, a clairvoyant who claimed she could *read* the contents of a sealed envelope when blindfolded. The two men agreed with Ferenczi's hypothesis that the woman was able to read the thoughts of the person consulting her and that, on this occasion, she had read the Hungarian's thoughts about the recent journey.

Notes

1 Oeri, 1977.
2 In Freud E., 1961.
3 Carotenuto, 1980.
4 Jones, *II*, p. 61.
5 Freud's historical "grand-patient" (Falzeder, 1994b).
6 Jones, *II*.
7 Clark, 1980.
8 Ferenczi, 04/08/1932 p. 184.
9 Kerr, 1993, p. 266.
10 Carotenuto, 1980.
11 1932.
12 Jones, *II*, p. 62.
13 Albrecht, 1909.
14 Conci, 2000.
15 On 11/09/1909 by Albrecht (1909). Kerr (1993) considers it a fake.
16 Freud, 1925a.
17 Clark, 1980; Prochnik, 2006.
18 Hale, 1971.
19 Prochnik, 2006.
20 Silver, 2006. Putnam, the first American professor of neurology and representative of New England's philosophical idealism, would become one of the leading exponents of psychoanalysis in America. Backed by Jones, he would be among the founders (1911) and presidents of the American Psychoanalytic Association.
21 Prochnik, 2006, p. 153.
22 Ibid.
23 Ibid.
24 Kerr, 1993, pp. 241–242, Prochnik, 2006,
25 Prochnik, 2006.
26 In Kerr, 1993, p. 245.
27 At the same time, while in Canada, perhaps in reaction to the feeling of exclusion, Jones began to write, in German, his essay on the nightmare and medieval superstitions (1912), quoted by Freud in the "Forsyth case."
28 22/09/1909, in Jung, 1961, p. 369.
29 1964, pp. 451–452.

30 In this regard, see also Saban's different interpretation, 2016.
31 Ibid., pp. 450 and 452.
32 Carotenuto, 1980.
33 *Zimmer* (room) and *Frauenzimmer* (prostitute), Freud to Jung 4/10/1909.
34 22/07/190.
35 Jones, *II.*
36 Ibid., p. 67.
37 Sigmund to Martha 15/04/1884, in Schwartz (1999).
38 Lacan reported that he had heard it from Jung (Rodrigué, 1996).
39 Prochnik, 2006, p. 141.
40 Ibid.
41 See Tanner, 1910.
42 Rodrigué, 1996, *I*, p. 514.
43 4/10/1909.

Chapter 13

The Danaan gift

The clairvoyant who reads Ferenczi's mind

After Freud had left, Ferenczi stayed on in Berlin for further meetings with the clairvoyant and managed to involve his elder brother Zsigmond who lived there, persuading him to question her some more on his behalf. As soon as he got onto the train back to Budapest, he set to work writing up his further observations for sending to Freud, who had in the meantime asked for information about Frau Seidler from the publisher Hugo Heller (a member of the Wednesday club and the only person who was brought into the secret: his popular salon was filled with rumours about this clairvoyant).[1]

The Hungarian's notes seemed to be a continuation of the American analytic work into which the woman had remarkably inserted herself. Questioned by Ferenczi about a letter from Freud, she guessed not only the recent voyage but the pupil's unconscious thoughts about his teacher: his unbounded admiration and also the lack of a complete rapport in the past few weeks caused by his jealousy of Jung. Ferenczi declared himself convinced that the clairvoyance worked by "psychic induction" and did not rule out the possibility of "a kind of ecstatic hyperaesthesia for minimal expressive movements," for those thoughts that we inevitably reveal in our words, movements, and so on. While reassuring Freud that he had no intention of becoming a believer in occultism, he was wondering about the reality of clairvoyance.[2]

By return of post, without picking up on his pupil's declarations of reverence and jealousy, Freud agreed that in some mysterious manner the woman was able to "read" thoughts with her eyes, "just as you read this letter," he wrote. And on divinatory faculties, he added, "You really don't need to regret that you didn't ask about the future. That always forms itself anew, even the dear Lord doesn't know it in advance."[3] After several days' reflection, he returned to the topic, saying he was convinced that the process of reproducing thoughts was based on very elementary psychic work. Ruling out the hypothesis of an exceptionally sensitive capacity for mimicry in the clairvoyant, he instead maintained that she might be "quite an imbecilic, even inactive person," and for this very reason capable of simply reflecting back what would otherwise be "suppressed through her own intellectual activity."

DOI: 10.4324/9781003246473-14

I subscribe to your interpretation that she guesses the thoughts, perhaps the ucs. thoughts of the experimental subject – with the corresponding misunderstandings and convergence of a kind of distortion in the transition from one psyche into another. So she seems to have interpreted the images of ships and traveling symbolically in connection with a death, as perhaps she might have done correctly in other cases, because she could not have known that she had a real traveller to America in front of her.

(Freud to Ferenczi, 11/10/1909)

In order to achieve a correct assessment of Frau Seidler's responses, he suggested that Ferenczi take into account the inevitable distortions of transmission linked both to the Pythia's misreadings and the resistances of the client, in this case Ferenczi.[4]

It was clear to Freud that they were not venturing into the field of occultism: the year before, at meetings of the Wednesday club, he had discussed some examples of telepathic phenomena which he had attributed to natural causes.[5] His presupposition was that such transmission of thoughts was not "a Ψ phenomenon" but a "a purely somatic one" and "a novelty of the first rank." Urging the most absolute silence – "We want to initiate Jung at a later date" – he concluded his dense letter with an ambiguous expression of gratitude to Ferenczi:

I am almost afraid that you have begun to recognize something big here, but we will encounter the greatest difficulties in exploiting it.

(11/10/1909)

Their plans for study concentrated on the unintelligible ways in which the conscious and unconscious thoughts of the clairvoyant's client, the spoken and/or written words, might awaken visual impressions in her psyche. Ferenczi set to work with a battery of tests to probe the woman's capacity for picking up stimuli from the other person's brain and instructed his brother in Berlin to present her with a series of letters – unsigned, written in German and Hungarian, and also a sealed envelope, the contents of which would be unknown even to his brother – including two missives addressed to Frau G: one sent by a former lover,[6] the other being one of his own. He seemed to want reassurance from the clairvoyant about his unconscious feelings for Gizella Pàlos and how his brother Zsigmond, and also Sigmund Freud, would accept his relationship with her.[7]

The research into thought transmission was beginning to weave itself into his transferential investment in the Master and in his lover.

The matter was complicated by the fact that, after returning to Budapest, Ferenczi continued to analyse himself and also undertook to treat Gizella, rather as Jung was doing with his wife and daughter.

His gushing enthusiasm had no limit: exhausted by his hunger for authenticity and mutual revelation, by the cruelty which, using this alibi, he allowed himself to inflict on the woman, and worn out too by the experiments in telepathy that he was starting to do with her, the Hungarian soon confessed:

> As far as my feelings toward Frau Isolde are concerned, I must say that the confession that I made to her, the superiority with which, after some reluctance, she correctly grasped the situation, and the truth which is possible between us makes it seem perhaps less possible for me to tie myself to another woman *in the long run,* even though I admitted to her and to myself having sexual desires toward other women and even reproached her for her age.
>
> (26/10/1909)

Ferenczi was venting onto Gizella his disappointment at Freud's first negative reaction to the difference in their ages, but immediately defended her:

> Evidently I have too much in her: lover, friend, mother, and, in scientific matters, a pupil, i.e., the child – in addition, an extremely intelligent, enthusiastic pupil, who completely grasps the extent of the new knowledge.
>
> (*ibid.*)

The invariable recourse for a man who is a convinced proponent of women's mental inferiority and suffers from a small-penis complex.[8]

In the two men's copious correspondence, the topic of telepathy – and those mysterious women who see and speak like ancient and powerful matriarchs – now formed a constant accompaniment: there was almost always a reference or hint about this closely guarded secret at the end of their letters or in a PS, but for the moment Jung was left out.

"In collaboration with Ferenczi I am working on a project that you will hear about when it begins to take shape," Freud wrote to the Swiss.[9] In the same letter, he anticipated an intuition he had about the mystery of Leonardo da Vinci's genius, and his inversion, which he intended to develop by applying psychoanalysis to his biography, a piece of research that had begun to engage him intensely in the attempt to elaborate his own bond with his pupils and delayed the writing-up of the American conferences. In this regard he added a note to *Three essays on the Theory of Sexuality*:

> the future inverts, in the earliest years of their childhood, pass through a phase of very intense but shortlived fixation to a woman (usually their mother), and that, after leaving this behind, they identify themselves with a woman and take *themselves* as their sexual object. That is to say, they proceed from a narcissistic basis, and look for a young man who resembles themselves and whom *they* may love as their mother loved *them*.
>
> (1905b, n. deleted 1910, p. 144)

Meanwhile Jung was being cagey and, briskly acknowledging his debt ("The analysis on the voyage home has done me a lot of good"),[10] said he was busy researching the traces of the nuclear complex of neuroses in mythology. In reality, he was finding it hard to conform to Freudian terminology, casting doubt on the concept of libido, and he intended to understand individual fantasy on the basis of myth.[11] This was behind the silence in his letters on the complexity of analysing his wife and the weight of her jealousy:

> Analysis of one's spouse is one of the more difficult things unless mutual freedom is assured. The prerequisite for a good marriage, it seems to me, is the license to be unfaithful.
>
> (30/01/1910)

And he freely announced that, "by design and after mature reflection," his wife was pregnant again.

Ferenczi sent Freud the documentation from his second series of experiments on Frau Seidler. The data supported the hypothesis of "reading": what's more, in her reading, the clairvoyant mixed the thoughts of other people with her own. "I find *the most striking* thing," wrote Ferenczi, "to be the candid confession of not knowing in the case of the converted letter which said nothing and which my brother had never read."[12] While he was still busily instructing Zsigmond on how to administer further tests (coded letters, which would not reveal their contents even on a normal reading) and even protesting when his brother did not make his own resistances available for analysis, the tireless Hungarian was contacting another clairvoyant, Frau Jelinek, this time in Budapest.

Child of a poor family and entirely uneducated, she had become a medium at the instigation of a hypnotist. "If you want to stay healthy, you have to tell the fortunes of people who come to you and let yourself be put to sleep by your husband for this purpose. If you don't do this, you will die" – this was the injunction imposed on the unfortunate woman by the doctor who had in some way "cured" the symptoms of somnambulism and hysteria which had appeared shortly after her marriage.[13]

After a face-to-face conversation with Frau Jelinek, this time without sealed envelopes to mediate the experience, Ferenczi was left highly impressed.

He had asked the medium four questions (about one of his sisters, his work, his relationship with Jung, and his relationship with Freud) in which he revealed his need for a guide, and the woman had answered in a way that stirred up strong emotions in him, feelings to which he gave himself without restraint, justifying it as the "success of the experiment." His last question to the medium – "What can you tell me about my Viennese friend?" – had received the curious response:

> You should remain true to him. A blockage [Sperrung] (sic!) has, to be sure, occurred, owing to the intervention of a third party, but you should not stop

sending him the letters, reports. Not only is he useful to you, but also you to him; for that reason never let go of him. Your trust will be very much strengthened until you find recognition.

"The strange thing is what she says about your relationship to me," commented Ferenczi to Freud, and later:

> Frau Seidler already told me (word for word): "You draw much from his (?), but he, too, could draw much from you."
>   Frau Jelinek says: "Not only is he useful to you, but also you to him."
>   Ideas like that are – consciously – quite foreign to me. It is possible that I can have them unconsciously. Consciously I very often have ideas of small-ness; in the ucs. they can correspond to a colossal desire for greatness and megalomania.
>
> (20/11/1909)

The two clairvoyants' "predictions" make it possible to recognise what was going on inside Ferenczi, in relation to Freud and Gizella: his measuring himself against his parents' idealised greatness, his always feeling inadequate, and the pressing need for reciprocity. In his self-analytic working-through, Ferenczi was becoming aware of the problem of the small child's confrontation not only with the paternal penis but also the frightening, great maternal vagina.

Freud readily highlighted the novelty of his pupil's suggestion but only to set it aside. In his thinking, the figure of the father had to remain fundamental: he considered the comparison with the mother as a secondary element. "I only believed in the comparison with the father's penis" which has filled the vagina, he replied.[14] Freud never became able to talk and theorise directly about the mother and the maternal vagina.

Winter came and, when Freud arrived in Budapest for a consultation, Ferenczi plucked up the courage to introduce him to his lover. This time he obtained Freud's blessing: the professor thought highly of "Frau G," as he called her in their correspondence, and he withdrew the concern about her age that he had expressed on the ship. Both of them wrote about her to Jung, Freud first:

> I no longer have to feel sorry for him. She is splendid, a woman who has only recently stepped down from the summit of feminine beauty, clear intelligence and the most appealing warmth.
>
> (2/12/1909)

Ferenczi straight afterwards:

> I wrote Jung a long letter in which I confessed candidly about my "brother complexes" and explained that guerrilla war cannot be the tactic of choice in psychoanalysis; someone must lead, and this one person, besides you, is by

nature Jung himself. I also told him (probably in order not to garner any more sympathy) that you have altered your view about the "dead-end" that I have gotten into.

(7/12/1909)

The patient who reads Ferenczi's mind

As the transference evolved, Ferenczi began to tell Freud about an urgent and barely manageable demand for exclusivity and, from a misunderstood feeling of sincerity ("if one *could tell everyone the truth*"),[15] he felt bound to confess his own acute jealousies in detail: of Jones, Abraham, Brill, the Master's guests in Vienna, and even the sacrosanct Saturday evening tarot games. Naturally, Jung remained the principal object of his always-tormented "brother complex" in those self-analytic "diary pages," which, like an adolescent, he included in his letters to Freud alongside the enclosures about telepathy.[16]

As he became aware of Ferenczi's demand for unlimited friendship, Freud no longer felt like committing himself to a passionate relationship like the one with Fliess, nor does he seem to have been keen on getting involved in his pupil's analysis. He paid great attention to Jung, having identified him as his heir, while taking the Hungarian for granted and expecting absolute devotion from him. And indeed, the loyal and eager Ferenczi procured ancient statuettes for Freud to add to his collection, showed himself to be flattered by the fact that Freud valued some of his ideas and asked Ferenczi if he could use them without citing him, and was always available to spend a weekend in Vienna.

After the initial enthusiasm, the correspondence contains hints about the research into thought transmission and about shared reading or plans for exploration, but it reveals a less engaged Freud, "still not one hundred percent" in his health, who leaves Ferenczi to play around with telepathy and is taken up with more serious matters.[17] He has to work on his Leonardo essay and the American conferences but is also defining his formulation of the Oedipus complex as a nuclear complex of the neuroses, a distinctive theoretical and clinical concept of psychoanalysis. The use of the term "complex" borrowed from Jung (Freud would subsequently prefer the term "conflict," so as to emphasise the dynamic elements in play) clearly signals the process of competition and mutual assimilation between teacher and pupil in this period.[18]

In February 1910, having consulted famous psychiatrists in vain and stayed at various European clinics, a young Russian, the 23-year-old Sergei Konstantinovich Pankeyev, turns up in Berggasse [the case known as *The Wolf Man* (1914c)], giving Freud the clinical opportunity to reiterate to Jung the importance of the infantile factor in the initial stage of forming the adult neurosis. Comparing notes with the Swiss is fertile and prompts Freud to try out new themes. After bringing Freud's attention to Jensen's novella *Gradiva: A Pompeiian Fantasy*, Jung now shows him the book of memoirs by Daniel Paul Schreber, president of the Dresden Court of Appeal, an autobiographical account of a paranoid breakdown, and seems to be inviting Freud to compete with him on the topic of *dementia praecox* and *paranoia*.

We should note that, alongside the internal conflicts, there were the pressing concerns of external politics since German psychiatry, represented in Munich by Kraepelin, had launched an attack on psychoanalysis and on the Zurich Clinic. A prominent group of neurologists decided to boycott the departments where the Freudian method was being applied ("the clinicians' conspiracy"),[19] and very soon, a pupil of Kraepelin, Alois Alzheimer, whose judgement was among the most authoritative, used his new journal to make a systematic refutation of all published psychoanalytic work, declaring it unscientific.[20] In order to ensure the survival of his discipline, Freud had initially thought of joining the International Order for Ethics and Culture founded by Forel, into which many currents of psychotherapy were flowing, but he set the plan aside because it didn't guarantee that psychoanalysis would be able to preserve the distinctiveness of its identity and its theoretical roots.

It was becoming a matter of urgency to build an autonomous organisation, fully identified as "Freudian," which would regulate the admission of new members and have its own publishing house and an official journal for disseminating, updating, and circulating its ideas. The Hungarian's talent was put at the disposal of Freud's programme: Ferenczi did not hesitate to act as his "paladin and secret Grand Vizier."[21] And so in March 1910, at the second Psychoanalytic Congress held in Nuremberg, while Freud spoke about the future of psychoanalysis, addressing the problem of the countertransference for the first time and the self-analytic preparation of the physician (1910b), Ferenczi immediately followed this by looking to the past, reprising the history of the movement with the aim of moving the proposal to set up an International Psychoanalytical Association,[22] nominating Jung as president, and identifying Zurich as its base. An irony of fate, given his subsequent reputation as a heretic, that the founding of the psychoanalytic institution should be acknowledged by Freud as primarily Ferenczi's achievement.[23]

The Hungarian's ambivalence about the nomination of the crown prince was immediately apparent: his first move, which was to nominate Jung as president for life as well as giving him exclusive power to choose whether any publication could call itself *psychoanalytic*, provoked such strong objections in the Vienna group that it was immediately rejected.

In order to quell the ill-feeling, Freud promised Adler the presidency of the Vienna section and appointed him, together with Stekel, to edit a new journal, the *Zentralblatt für Psychoanalyse*, the mouthpiece of the local society. There were no rewards for Ferenczi, and indeed, after the Congress, Freud was no longer inclined to make the effort of travelling to Berlin to visit the clairvoyant.

The affective cost of Ferenczi's gesture of support for his rival was not slow to make itself felt, all the more so because Jung, seemingly nonchalant about these accolades, had accepted an invitation at short notice to hold a consultation in the United States and barely returned in time for Nuremberg. Alone and ignored, prey to "brother-envy"[24] but, above all, in the grip of his unconscious identification with Freud,[25] Ferenczi was unable to find solace in the small group of pupils that he was building up in Budapest or in his relationship with Gizella, for whom he was still both a secret lover and her analyst. His reluctance to make their tie official was compounded by the difficulties of divorce in Hungary and by the fact that Gizella would never leave Geza Pàlos until their daughters were both married. And while Magda, the younger one, was happily married to Layos Ferenczi, the elder, Elma, at 23, was still living with her mother and, apparently unaware of the relationship with Sándor, was beginning to cause concern because of her own romantic sufferings.

Ferenczi was now feeling the burden of difficult, infrequent, clandestine amorous encounters and having to deal with an increasingly insatiable need for support from Freud, which he presumed was heightened by the experience of professional analytic practice.

As he struggled to identify and master the countertransference, which he had called the "*energy expended against themselves*"[26] by psychoanalysts, torn between burdening Freud with his self-analysis or sparing him, he allowed a melancholy tone of resignation to appear in their correspondence in alternation with accounts of progress in his work on interpreting his own father complex, his "unjustified infantile desire to be the first and only one with the father."[27] In the background of these letters, infused with a tormented filial transference, Halley's comet was coming ever nearer and bringing its evil omens. The Earth would pass through its fearful tail of gas in May, and the statements of astronomers like Camille Flammarion, inflated by the newspapers, had excited the most impressionable minds. Ferenczi did not share their fears about the imminent end of the world,[28] but he wasn't sleeping well and was filled with fantasies about the longed-for late summer holiday, a kind of honeymoon promised by Freud, which would take them to Sicily, a destination they had had to abandon the previous year.

While they waited, he was beginning to notice that "telepathic" phenomena could emerge spontaneously from the analytic relationship: "things that correspond to one another that came to my mind and that of a patient," he wrote to Freud.[29] Those remarkable experiences were added to the "*Seidler dossier.*"

Just before their departure, Ferenczi enclosed the detailed account of his observations about thought transmission in analysis, drafted with a view to publication as a scientific article.

Entitled *A Few Recent Observations (On the Theme of Thought Transference)*, the manuscript – which travelled from Budapest to Vienna and back again and was never published – was not in fact "carefully preserved" but records a degree of distress: part of the letter is missing because half of the last sheet is torn off.[30]

This is how Ferenczi begins:

> It has already frequently occurred to me that a patient, a homosexual with strong resistances and equally strong ucs. transference to me, sometimes includes things in his associations that occupy me with particular intensity. But I remained for a long time with this uncertain impression (which I also,

incidentally, once communicated to Prof. Freud). Lately (in the face of resistance and transference which was becoming more and more intense) these correspondences were accumulating, so I decided to pursue the matter and note down precisely the individual occurrences.

(17/08/1910)

What we find there are six observations of thought-induction (five with a homosexual patient and one with a patient whose diagnosis is not indicated) and for each instance the patient's associations and the analyst's converging ideational contents.

A fantasy of the patient as soon as he lay down on the couch provokes concern in the analyst about having earlier used the couch for a sexual encounter (July 25); the image of a kind of mannequin, which appeared in the patient's associations, corresponded to the impressions left in the analyst the previous evening by reading about psychosis (July 26); the patient's fantasy about the name of a city, and a woman, a wife in velvet pants, had reminded Ferenczi of a scene from the same morning, in which riding with the wife of a colleague, he finally noticed her pregnancy (August 3); and so on.

In all these experiences, it is the patient who acts as the "thought reader" and Ferenczi notes that "the patient's 'thought-reading' associations always come at the beginning of the analysis."

In the conclusion which, as already mentioned, is incomplete, Ferenczi states:

> These things have no influence at all on the progress of the analysis. The patient immediately associates to them things that concern him. If something in this transference is true, then the transferred words have no different effect from that of Jung's stimulus words.
>
> Everyone "introjects" them according to his own needs.
>
> My homosexual is a first-rate masochist (his cruelty, naturally, is lodged behind it). This masochism (possibly) enables him to apperceive impulses toward which others are unreceptive. I project the stimulus words ucs. he *introjects* them. [I act like a man, he like a woman; he is, of course, a homosexual.]
>
> (*ibid.*)

The fifth example is worth noting because it allows us to catch, in *statu nascendi*, the preconscious form being taken by the content of Ferenczi's unconscious complex: in this indirect way, reflected in a patient's fantasies, we see the first appearance of the Hungarian analyst's erotic transference to Gizella's daughter, Elma Pàlos. We read:

> Patient sees (in fantasy) a girl lying on the dissecting table [as if in a caricature of "Le Rire," where a naked woman who is about to be operated on is surrounded by doctors]. A doctor holds her from below, binds her with a

cambric bandage as wide as a hand; another doctor operates on her. "My Rosmarin, my Georgine" [Hungarian song]. She wriggles her legs (feet?). Her upper body is separated from her lower body. As if screwed in.

Self-analysis [I won't repeat the patient's analysis]:

On the same day I was assisting my friend Schächter with a surgical intervention on Frau G.'s elder daughter. The poor girl became infected about three months ago during the extraction of a wisdom tooth, and the pus had to be surgically emptied four to five times. She causes us much worry. Cosmetically not affected for the time being, but the danger is always there. During the last-mentioned operation she was sitting on a chair; Dr. Schächter in front of her; her legs between those of Dr. Schächter. I was standing behind her and was holding her head. After the surgery I made her a gigantic bandage; the cambric bandage was very wide (wider than the breadth of a hand), they were laughing about my excess of zeal. During the surgery she was wriggling her legs.

(*ibid.*)

The patient's fantasy mirrors what Ferenczi is trying to address with the help of his teacher, Schächter-Freud.

Like an oracle, it seems to show him that he is seeking a place in the scene between the couple, his maternal tendency, and, above all, his emerging desire for Elma which will be so decisive for his future. Ferenczi seems unaware of it and does not elaborate on it in his interpretations. The patient's fantasy and the scene experienced by the analyst strikingly reprise the motif of the dream of Irma's injection and the episode of Eckstein's operation (including the detail of the long strip of gauze). Freud and Ferenczi really seem to be profoundly entwining their thoughts and destinies, even though Ferenczi was apparently the more engaged of the two.

In making the final arrangements for the holiday, Freud comments on Ferenczi's observations, claiming that they finally shatter all doubts about the existence of thought-transference.

However, he recommended preserving "the secret long enough in the maternal womb" since it was difficult news to receive, given the fear that it was likely to arouse.[31] Ferenczi replied:

I am a little frightened by the fact that you view my observations on thought transference as proven. . . . I know that it is a Danaan gift when one gets such ideas, that is to say, has such experiences.

(23/08/1910)

The Palermo incident, or the interpretation of paranoia

The two men left at the end of August with a very brief stay in Paris. They lunched at the *Café de Paris*, and in a few hours, Freud tried to show Ferenczi, who was in the city for the first time, everything that was precious to him there: the Paris of the Salpêtrière and Charcot, the Louvre with its archaeological museum and its Italian art department. Fascinated by the mysterious character of Leonardo da Vinci and his genius, in the essay he had just published Freud had attributed Leonardo's thirst for knowledge to his exceptional capacity for repressing and sublimating his primitive instinctuality (1910d). And perhaps identifying with Leonardo, denying that he put his homosexual tendencies into practice, and portraying him as surrounded by young pupils but sworn to abstinence, Freud set off on his vacation with Ferenczi.

And now the long-awaited Italian voyage: Rome, Naples, and finally Palermo.

The inconveniences, but also the intoxication and novelty of the journey; the heat and wild enchantments of Sicily; the splendour of the colours, scents, and sights; the little aggravations of living together; and perhaps the comparison with the previous, exciting American expedition and the weight of Jung's absence . . . all of this led the two friends inexorably towards the collision that had been brewing for some time.

The conflict became evident in Palermo on the evening when, staying in the old palazzo of Weinen's Hotel de France in Piazza Marina, the two colleagues settled down to work together on the book suggested by Jung, *Memoirs of My Nervous Illness* by Daniel Paul Schreber, president of the Dresden Appeal Court (1903), a text which resonated powerfully with the intensity of paternal and filial passions and could certainly have emphasised the affects in play between the two men.[32]

We know that the president's crisis began in half-sleep, when he had the thought about how "beautiful" it could be to be a woman who submits to copulation.[33] Ferenczi wanted to be loved and cared for with the despotic neediness of a small child towards his mother but also with the expectations of a woman from her lover. Freud had already interpreted these demands as linked to the early loss of Ferenczi's father and attributed his urgent "need to help" patients to the same cause.[34] Ambivalent about the intimacy being demanded, he was bored by the excessive dependency displayed by his friend, by what he regarded with irritation as his feminine passivity,[35] and also began to show a certain withdrawal of affective engagement from the clinical work that he had been so passionate about in the past.

We do not really know what happened in Palermo. We know the version that Ferenczi gave to Georg Groddeck in 1921 when, after his mother's death, he put himself in Groddeck's care:

> For a very, very long time now, I have taken pleasure in maintaining an aloof reticence, and I like to cover up my feelings, often even with close friends. . . .
>
> On several occasions I let myself be analysed by him [Freud] (once for three weeks, another for 4–5 weeks, and for years we've travelled together every summer.) But I never felt free to open myself totally to him. He had too much of this "prudish respect" he was too important for me, too much of a father. The result: in Palermo, where he wanted to do this famous work on paranoia (Schreber) with me, I had a sudden attack of rebellion. I jumped to

my feet, on the first evening of work, because he wanted me to take down his dictation. I told him that to have me simply take down his thoughts was not writing a paper together. "So this is what you are like?" he said, taken aback. "You obviously want to do the whole thing yourself." That said, he now spent every evening working on his own.

(Ferenczi to Groddeck, 24/12/1921)

Let's suppose that on this occasion, Freud had intended to repeat the Worcester experience and prepare his text together with this friend. In Sicily, Ferenczi rebelled. In the days that followed, they journeyed laboriously through Agrigento and Syracuse, tormented by the sultry, paralysing arrival of the sirocco and under the threat of a suspected cholera outbreak in Palermo.

They only mentioned it on their return and by letter:

> The beautiful days that I spent in your company, the ideas that were stimulated in me at the time will, I hope, have a favorable influence on my activity. . . . Still, I am sorry that you had in me a travel companion who is still so much in need of education. You will probably think that I have begun to subject the events of our living together, the manner in which I reacted to them, to extensive self-criticism. . . . I hope, in spite of this, . . . you can forgive the unavoidable "chimpanzee," who so often thwarted good intentions.

(Ferenczi to Freud, 28/09/1910)

> You will believe me when I say that I think back about your company on the trip only with warm and pleasant feelings, although I often felt sorry for you because of your disappointment, and I would like to have had you different in some respects. . . . you certainly expected to wallow in constant intellectual stimulation, . . . I would have wished for you to tear yourself away from the infantile role and take your place next to me as a companion with equal rights

(Freud to Ferenczi, 2/10/1910)

> I usually tend more toward modesty and self-deprecation – at least I always see my actual smallness. – But don't forget that for years I have been occupied with nothing but the products of your intellect, and I have also always felt the man behind every sentence of your works and made him my confidant. . . . So I am and have been much, much more intimately acquainted and conversant with you than you could have imagined. . . . During the trip I played the ridiculous and certainly very repugnant role of one who is misunderstood (like the Sicilian asses, perhaps) – and I was waiting for your accommodation in order to be able to tell you all this. . . . If you had scolded me thoroughly instead of being eloquently silent!

(Ferenczi to Freud, 3/10/1910)

Why didn't I scold you and in so doing open the way to an understanding? Quite right, it was a weakness on my part; I am also not that ψα superman whom we have constructed, and I also haven't overcome the countertransference. I couldn't do it, just as I can't do it with my three sons, because I like them and I feel sorry for them in the process.

(Freud to Ferenczi, 6/10/1910)

The "Palermo incident" certainly derived from personal affective impulses but also from various complexes, later reflected in theory. We can historically locate here the beginning of the fundamental difference which will come to distinguish the entire development of Ferenczi's thought and practice from Freud's, a difference of perspective which Ferenczi was very eager to maintain in his constant effort to balance and complete the master's theorising.

Freud (1911a) used the *Memoirs* to write an extraordinary interpretation of President's Schreber's paranoia which – after *Dora*, *Little Hans*, *The Rat Man*, and *The Wolf Man* – completed the great series of case studies: it was the first attempt to explain the psychoses using the analytic method, in an entirely revolutionary manner compared to the psychiatry of the time and to Bleuler (1911) himself.

The paranoid and hallucinatory delusion which developed in Paul Schreber consisted of the conviction that his doctor, God/father, had to emasculate him in order to turn him into a woman, sexually abuse him, and leave him for dead: the patient called this conspiracy "soul-murder."

Freud interpreted it as Schreber's psychotic way of re-investing affectively in external reality from which he had earlier withdrawn his libido in order to preserve his body: the delusional symptom was an attempt at healing which protected him from fragmentation and schizophrenic lack of contact, and – coming from outside and being transformed into its opposite – restored that homosexual desire for his father which the patient had tried to eliminate from inside himself. The triad of narcissism, unconscious homosexuality, and paranoia became distinctive of the primitive mechanisms of psychosis. In this hypothesis, Freud was also expressing his deep fear of anything to do with passivity, helplessness, and impotence in relation to the father but still more to the mother.

While still in Palermo, Ferenczi had set himself the problem of understanding the family context in which Schreber's illness originated, the possible elements in his environment and upbringing that could have provoked or contributed to the detachment from external reality and the narcissistic, self-preserving withdrawal: Freud would quickly confirm that the "bellowing" which the patient felt compelled to emit – and which he attributed to the "lower God, Ariman" – strikingly mimicked the behaviour of his father:

What would you think if old Dr. Schreber had worked "miracles" as a physician? But was otherwise a tyrant at home who "shouted" at his son and understood him as little as the "lower God" understood our paranoiac?

(6/10/1910)[36]

In his case study, however, Freud did not concern himself with Schreber's father nor even, as Gay (1988) and Quinodoz (2004) observe, with his mother. Though he referred to childhood experiences in trying to understand the serious symptoms of paranoia, he did not return to the abusive elements he had taken into consideration at the time of the *Studies*. The concrete relationship with the first caregiving environment now stood in the background, and Freudian investigation was fully focused on the subject's internal world and infantile drives. At the same time, in a note to *Formulations on the Two Principles of Mental Functioning*,[37] Freud explained how, in his hypotheses about individual psychic organisation, maternal caregiving and the presence of the parents during growth were premises that should be taken for granted. The presence of the other, and of his/her unconscious processes, were of secondary importance. Subsequently, Freud would nevertheless emphasise the distinctive interpretative sensitivity and the overvaluing of manifestations of the other's unconscious in certain patients afflicted by jealousy, paranoia, and unconscious homosexuality.

In 1958, H Searles endorsed this sensitivity and revealed the characteristic permeability of people with a serious mental illness to the other's unconscious processes, feeling their influence in their delirium without understanding its origin. It was only in the early 1970s, on the wave of Sixty-Eight, that without knowing it, some authors recovered parts of Ferenczi's theories about the part played by family members in the aetiology of pathology. In *Soul Murder: Persecution in the Family*, Morton Schatzman (1973) connected President Schreber's paranoia to the educational methods employed by his father – a German orthopaedic specialist and educational theorist who suffered from obsessional ideas and homicidal impulses – and theorised the parents' responsibility in limiting their children's freedom by means of harsh discipline and violent moral and physical restrictions (the father's writings were to enjoy great success in Nazi Germany).[38] Schatzman went on to wonder if the *fantasies* of seduction brought by patients, which Freud had initially considered *recollections*, were in fact memories of the "sexual fantasies of parents" about them, expressed in turns of phrase (of the kind used by Dr Schreber), glances, facial expressions, and so on.[39] From Masson (1984) onwards, many others have questioned the possibility of an analyst's neutrality in the presence of traumatic events going back to the preverbal age or abusive parents, casting doubt on Freud's abandonment of the seduction theory and resurrecting Ferenczi's censored theory.

In the writing of the *Memoirs*, Freud had detected the part of Schreber which had succeeded in surviving the trauma and differentiating an Ego with a delusion. However, unborn parts of the Self had remained in the "bellowing," along with a still-undifferentiated reality of contagion and permeability: that affective *phonē* not articulated in language (an unheard cry that becomes a bellow, a matrix-area of the psychic boundary on which the subject and his speech are constructed) would be central to Ferenczi's thought. He would develop it fully in his last years when, returning to the concrete meaning of the trauma and abuse described by Freud in the *Studies*, he would insist on the importance of the primary affective reality of communication between parents and children.[40]

Freud and Ferenczi could not write together in Palermo and indeed never signed anything jointly. But from then on, they were unconsciously compelled to write together as "doubles":[41] an unconscious thought would develop contrapuntally in the theoretical and clinical paths that the two men followed, and in their intimate dialogue, it would take on a form and perspective that were only apparently opposed.

Once their correspondence was resumed after the return from Italy, Ferenczi sometimes surrendered to the fear of having disappointed Freud and compromised their friendship forever and at other times let resentment take hold of him, hungry with a crude, misunderstood sincerity which, in a free association of ideas, he wished could be reciprocated.

There were some attempts at explanation which only Freud considered exhaustive, repeatedly warning his pupil not to keep raising the matter and to drop his recriminations.

Just as he continued to be blind to his pupil's fragility, on the same occasion, he confided in the Hungarian that he had no deep inclination to help people because, he wrote, "I did not lose anyone whom I loved in my early years" (sic!). However, the associative thread of his thoughts immediately caused his fears of a predestined death to re-emerge, along with the old superstition connected to the break with Fliess, since he added, "Let us nevertheless firmly establish that I myself already decided quite a long time ago not to die until 1916 or 17. Of course, I don't exactly insist upon it."[42]

It would only be later, when he had to confront the crisis of the rivalry with Jung, after the second fainting fit in Munich in 1913, that Freud would reveal the trauma connected to the death of his younger brother Julius. For the moment, it was the episode with Fliess which represented that infantile traumatic reality in the present: "Not only have you noticed that I no *longer* have any need for that full opening of my personality, but you have also understood it and correctly returned to its traumatic cause," he confided to Ferenczi. Not appealing this time to the physical ills of aging or to the colitis that always afflicted him on journeys, Freud tried to justify to the Hungarian, and to himself, the emotional distance he had shown in Sicily. Like a lover who rejects a suitor because he's already bound to another,[43] he added:

> This need has been extinguished in me since Fliess's case, with the overcoming of which you just saw me occupied. A piece of homosexual investment has been withdrawn and utilized for the enlargement of my own ego. I have succeeded where the paranoiac fails.

And he goes on:

> As far as the unpleasantness that you caused me is concerned – including a certain passive resistance – that will go the way of memories of travel in general; small disturbances vanish through a process of self-purification, and

what is beautiful is left over for intellectual use. It was plain to see but also easily recognizable as infantile that you presumed great secrets in me. . . . Just as I shared with you *all* the scientific matters, I also concealed from you very little of a personal nature. . . . My dreams at the time were, as I indicated to you, entirely concerned with the Fliess matter, with which, owing to the nature of the thing, it was difficult to get you to sympathize.

(6/10/1910)

In response to Ferenczi writing, "Not everything that is infantile should be abhorred; for example, the child's urge for truth, which is only dammed up by false educational influences,"[44] Freud persisted in his withdrawal, trying to find a safe distance: "It occurs to me that a paralyzing influence emanated from you to the extent that you were always prepared to admire me."[45]

For two weeks, there were fewer letters between Vienna and Budapest and none of that intimacy which had previously united the two colleagues.

On 27 October, Freud hinted that he was bringing the Schreber case to completion and updated his friend on a new paper he had read to the Viennese group, about whom he complained: "It's getting more and more difficult to get along with these people. A mixture of shy admiration and stupid contradiction, *quand même,* not in the same persons." Ferenczi took this criticism to be directed at him and by return of post exclaimed, "Why didn't you mention the 'shy admiration and mute contradiction' in Italy? Everything could have turned out differently." In the postscript, he added a note about his researches into the connecting bridge between certain neuroses and thought transference: a clear sign of truce.[46]

The psychic work of the clairvoyant: two unfulfilled prophecies

At this point, it was Freud who returned to the theme of thought transmission, and with all his old bravura. In November, he sent Ferenczi a gift in the form of "a piece of news" for the dossier, granting Ferenczi the paternal rights to the subject.

This was an indirect observation of a clairvoyant who indisputably possessed the ability to read unconscious desire and to predict the past.

So listen. In Munich there is a court astrologer, a woman, who foretells the future with the aid of astronomical tables. One of my patients (Dr. Weil [Dr. phil.]),[47] a serious person not to be suspected of lying) had his future foretold by her. He gave her the birthday of his brother-in-law (his only sister's husband), and she thereupon produced not a bad description of him! What struck him most was the fact that she prophesied that the brother-in-law would undergo a poisoning from oysters or crayfish in July which would give his intestines trouble for quite a long time. The prophecy was made in January; it *didn't* come true in July, to be sure, but the commendable thing about it remains that the brother-in-law actually had experienced a poisoning from a lobster in July of the previous year. She thus simply prophesied the past again!!! My objection that this is the worst disgrace imaginable for a prophet made no impression on the storyteller, evidently because he so very much liked the prospect.

Another session in April of this year. He presented her with his own birthday and learned that there would be a death in this person's family in October; at the very latest, it would drag out until mid-November. It made a particular impression on him that she said *the same thing* upon hearing of his sister's birthday, even though she didn't know that it was his sister. He consciously related it to his father or uncle. Of course, again, only his brother-in-law could have been meant. Now, the timing fits the following context. A short time earlier he had set a date for his treatment with me for which *exactly the same thing* was arranged. He then came on October 10. Then he made the prophecy come true in this guise on November 15 by confessing to me that he was expecting his brother-in-law's death!!!

(15/11/1910)

Freud used this gift to win back his friend and encourage him to develop a parallel space for thinking and writing. Later, he exhorted him to hurry up and finish his essay on the role of homosexuality in paranoia[48] so that it could feature in the same volume of the *Jahrbuch* as his own work on Schreber, which was nearly ready.

Nevertheless, he had been struck by the observation of the prophecy.

Ferenczi responded to the offered rapprochement: he once again brought Freud in on the observations of thought-induction in analysis that he was continuing to collect,[49] asked him for the address of the astrologer in Munich with the idea of going to consult her, and announced his own "interesting news": "Imagine, *I am a great soothsayer, that is to say, a reader of thoughts*! I am reading my patients' thoughts (in my free associations)." Jokingly, he concludes, "When I come to Vienna, I will introduce myself as *court astrologer of the psychoanalysts*."[50]

Ferenczi's calling himself a "court astrologer," sometimes taken out of context as a symptom of madness, is a joke and also a provocation: we don't know how far it refers to the one true "astrologer," Jung, the favourite now absorbed in casting horoscopes for reading characters and destinies, and how much to his own condition as a soothsayer who must only give his sultan pleasant predictions (and really has no permission to follow different lines of thought).

However, the most important news passes unnoticed and is contained in a postscript where Ferenczi hints at an imminent trip to Vienna about "the matter of Frau G.'s daughter":[51] the storm is on its way.

Freud's reply is slow to come: Ferenczi has no idea of the emotional atmosphere he is stirring up in his correspondent. Freud, already thinking back to his friendship with Fliess because of the work on President Schreber's psychosis, cannot have forgotten his old enthusiasm for the cosmological fantasies of that "honorary astrologer'" who had later accused him of being a "reader of thoughts."[52]

Ferenczi, the medium, continues his hypotheses and sends him further observations on unconscious communication in analysis:

> if one is somehow upset and not quite with it (in regard to the experiment) – then one may be a good subject for emanation, but one is not suitable for perceiving. A calm mood is more favorable. The most sensitive are *those in love* (with the aid of their masochistic components). It is not beyond the realm of possibility that this "difference in voltage" and the thought transference conditioned by it also plays a certain role in ordinary life by influencing associations, and through them also mood, action, etc. Greater significance would accrue to it only in pathological cases. The perceiving person reacts to the transference with his own ucs. complexes, but he chooses among these only those closest to the ucs. complexes of the one posing the task.
>
> (2/12/1910, my italics)

When it finally comes, Freud's reaction throws water on the fire: he regrets having encouraged Ferenczi and asks "that you continue to research in secrecy for two full years and don't come out until 1913."[53] The Hungarian reassures him and presses on, satisfied that he can do without professional mediums:

> If the experiment succeeds, the percipient must be in a mood that is as calm and cheerful as possible, but still not too lively. If he is excited, then he is not suited for projecting the psychic rays. Impatience disturbs. The game reminds one of children's games of hiding; while associating, the questioner monitors what is good or inappropriate in what comes to his mind and seems constantly to give direction to the receiver. One sometimes sees outright how the receiver begins very distantly with his own complexes and is gradually led to the proximity of what is supposed to be guessed. A mute "warm" or "cold" is constantly being called out to him. Wonderful compromises come about between one's own complexes and the preconscious ideas of the questioner.
>
> (19/12/1910)

What are we to think of Freud? It is not surprising that, taking the opportunity of a trip to Munich to meet Jung, he planned his own visit to the "court astrologer," the famous clairvoyant whose Bavarian princes consulted her before any undertaking: a memory lapse made him forget Frau Arold's address and, seemingly disoriented, he asked Jung for help, breaking his silence about thought transmission. On 29 December, he updated Ferenczi in detail:

> Jung was again magnificent and did me a lot of good. I poured out my heart about many things, about the Adlerian movement, my own difficulties, and finally about my distress about thought transference. You should know that I had intended to write to the court astrologer, and had forgotten to take down her address, and in Munich I remembered only the street. I gave in

to such signs of inner weakness and initiated Jung into the matter, told him about your findings, my confirmation through that prophecy, and my proposal of a latency period until 1913. He laughed and admitted that he had been convinced for a long time and had himself initiated very substantive experiments, praised my caution, and declared himself ready for an agreement with you, if it should come to that. . . . I am glad he has such broad shoulders. I found this burden almost too heavy for me. Now, don't be jealous, but take Jung into your calculations. I am convinced more than ever that he is the man of the future. His own work has gone deep into mythology, which he wants to open with the key of libido theory. As gratifying as all that was, I still asked him to return to the neuroses at the proper time. That is the motherland where we first have to secure our mastery against everything and everyone.

The question of occultism worried Freud but drew him irresistibly. Only a few days later, forgetting his good intentions, he ran into the same error that he had warned his pupils against and ventured once again into the marginal area of the occult: and so he began the year by making Ferenczi the gift of a second unfulfilled prophecy:

Woman,[54] 37 years old, fallen ill with an obsessional neurosis since being informed by her husband that the cause of her childlessness lay in his azoospermia. Beginning with symptoms of anxiety in her 27th year. A year later (28th yr.) it was prophesied by a soothsayer in Paris from the lines in her hand that she would go through great struggles and have two children by her 32nd year. She consoled herself with that for a time; so now the prophecy is five years late.

Analysis. The struggles are clear from the situation. She had always wanted children; she, herself, was the eldest of five siblings. She vacillated at that time about whether she shouldn't leave her husband. Still, it is strange that the soothsayer told her this and revealed her wish to have a child directly, without questioning her further. But where do the time limit and the number two for the children come in?

My question: How old was your mother when you were born?

Answer: She was 30 when she was married. (Immediate correction:) she was thirty when I came into the world.

Question: How large is the age interval between you and your nearest sister?

Answer: One and a half years.

I: So, your mother didn't have any children at the age of 28. You consoled yourself by thinking, "I will be like *my* mother and already have two children at the age of 32."

Your struggles don't come into consideration as far as your mother is concerned?

She: No.

I: What does this fantasy further presume? That you separate from your husband or that he dies, so that you still have time, despite the year of mourning, not to stay behind your mother.

She: I always have the great fear that something will happen to him. When he wanted to leave yesterday evening, I had difficulties talking him into taking this morning's train. What if an accident happened with the train that I recommended!

(3/01/1911)

The observation is clear and precise: in this case, too, the interpretation of the occult phenomenon was indirect and, in the analytic situation, it found the frame through which it could best be appreciated: Freud never fully abandoned that secure base that he had constructed as his *Mutterland*. He confirmed the hypothesis that clairvoyants do not know the future but "predict the past" in the manner of dreams. Both prophecies underlined the fact that the fortune-teller knew how to give a voice to the enquirer's unconscious desire and – a surprising fact – could also insert precise details of his past: in the first case, the actual crayfish-poisoning survived by the patient's brother-in-law and rival the July before the consultation; in the second, the fact that the patient's *mother*, who had not given birth before the age of 30, had two children by the age of 32. In this second case, the foreseen facts were true for the patient's *mother*, in the sense that the prophecy would really have been the case if the clairvoyant had offered it to the mother, not the daughter, at the age of 27.

"Unfulfilled" prophecies were analytically true, therefore, since they captured the heart of the enquirer's unconscious structuring of time and gave them back a construction of their desire on the basis of incestuous desire in the first case, combined with a murderous wish and, in the second case, identification with the mother.

Freud's precious gift for Ferenczi's collection (he immodestly called it "perhaps the nicest piece that you have to date, as far as I am familiar with the material") was perhaps a way of seducing him and freeing himself from the subject.

But the coincidence of a second prophecy so soon after the first is remarkable: though it seems contemporary with this exchange of sentences as part of a lively and intimate, dialogue in a close, lively dialogue, it is not clear if it was a current or earlier observation by Freud. In any case, it reveals how taken he was with the topic, so entwined with his mother's expectations and his superstitions about death.

We know he continued to collect observations about the work of fortune-tellers and reappropriated these two prophecies for this purpose, since they appear in the first part of *Lecture XXX*, but over New Year 1911, he seemed quite keen to rid himself of the subject. The gift may have been intended to placate the Hungarian, who would have had good reason to resent the revealing of their secret to Jung, and enabled him to get closer affectively and even to extricate himself from the Palermo episode and light-heartedly to sound out his friend's state of mind:

> I will also ask you why your last letters, insofar as they were not about factual matters, revealed an elegiac character. I am awaiting the connection to our trip, but I will not let that stand.
>
> (10/01/1911)

Calmer and better understood in his role of sage counsellor, Ferenczi could share Freud's puzzlement about the airy mystico-theological speculations Freud had found Jung engaged in: "Your instruction that he should return to 'mother earth' will be very useful to him."

But at the end of the letter, he allowed himself, with the due detachment, to strike a note of jealousy: "I am sorry that we didn't strike the same tone in Italy that refreshed and gratified you in Munich" (3/01/1911).

Notes

1 Moreau, 1976.
2 5/10/1909.
3 6/10/1909.
4 11/10/1909.
5 4/03/1908, in Nunberg and Federn (1962).
6 The painter Victor Schramm, whose name had been the object of his forgetfulness, analysed aboard the *George Washington*.
7 14/10/1909.
8 And probably also premature ejaculation (Ferenczi, 1908b; Borgogno, 2010).
9 17/10/1909.
10 14/10/1909.
11 30/01/1910.
12 8/11/1909.
13 20/11/1909.
14 21/11/1909.
15 Ferenczi to Freud, 5/02/1910.
16 7/12/1909.
17 12/12/1909.
18 Kerr, 1993, p. 276.
19 Ferenczi to Freud, 22/03/1910. See also Falzeder and Burnham, 2007.
20 Freud to Ferenczi, 5/06/1910.
21 Freud to Ferenczi, 13/12/1929.
22 Ferenczi, 1911b.
23 Freud, 1923a, 1933.
24 Ferenczi to Freud, 5/04/1910.
25 Balsamo and Napolitano, 1998.
26 Ferenczi to Freud, 17/04/1910.
27 Ferenczi to Freud 5–17–27/04/1910.
28 Ferenczi to Freud, 18/05/1910.
29 27/07/1910.
30 17/08/1910 p. 215, footnote 11.

31 20/08/1910.
32 See Vigneri, 2001.
33 Schreber, 1903.
34 10/01/1910.
35 Freud to Jung, 24/09/1910.
36 As soon as Schreber succumbed to "thinking nothing" [*Nichtsdenken*], the "bellowing miracle" [*Brüllwunder*] appeared, "when my muscles serving the process of respiration are set in motion by the lower God (Ariman) in such a way that I am forced to emit bellowing noises"; Schreber, *Memoirs,* (Brabant et al. note 10, Freud to Ferenczi 12/10/1910).
37 1911b. The first title was *"Two Principles of Mental Functioning and Education."*
38 Two other sons suffered from mental illness, and the eldest son committed suicide.
39 1973, p. 117.
40 See Pierri, 2015.
41 Botella and Botella, 2001.
42 10/01/1910.
43 Brabant, 2003.
44 12/10/1910.
45 17/10/1910.
46 29/10/1910.
47 Willy Haas of Nuremberg, in analysis with Freud from September 1908 to July 1909, and from October to December 1910: philosopher of a phenomenological orientation, professor of political science, and orientalist. For a time he belonged to the IPV and became a founding member of the Munich psychoanalytic society (May, 2019). May corrects an error in the transcription of Freud's letter (Ibid., p. 50, n.).
48 Ferenczi, 1911c.
49 From sessions with a homosexual patient and an impotent patient and from the analysis of Gisela.
50 22/11/1910.
51 16/11/1910.
52 Freud to Fliess 9/10/1896 and 8/05/1901.
53 3/12/1910.
54 Elfriede Hirschfeld, who had begun the analysis with Freud in October 1908 and continued it, with various interruptions, for many years. He shared it with several colleagues: she had been treated by Janet; had a consultation with Bleuler; and was at times in treatment with Jung, Binswanger, and Pfister. The case figures in many of Freud's works and letters under various different pseudonyms (Falzeder, 1994; Skues, 2019).

Chapter 14

An epistolary novel

Ferenczi and incestuous countertransferential storms: from mother to daughter

In that same January of 1911, Fräulein Elma Pàlos set off a second incident involving Freud and Ferenczi.

As arranged, she arrived in Vienna accompanied by Gizella for a consultation with a surgeon about the scar left by that fateful abscess on her tooth. Ferenczi joined them over the weekend, officially as a relative and friend of the family standing in for the absent father. In fact, his marriage prospects were at stake, since Frau G, preoccupied by Elma's romantic difficulties, was not thinking of divorce: they were awaiting the professor's opinion about the girl.

Freud talked to Ferenczi about the young woman in a detached manner, proposing the hypothesis of a *dementia praecox*, which was the last thing that any of them had considered. Back in Budapest, surprised by the diagnosis and by the intensity of his own unhappiness about it, the Hungarian became depressed and tried to address the question from a professional viewpoint: accepting Freud's judgement, he assessed his disappointment as deriving from an "all too forgiving understanding," an inclination towards this "being induced by the patients," which he connected to unrecognised countertransferential impulses.[1]

In reply, Freud played down the situation: "The diagnosis says nothing about its practical significance. So let's hope for the best," but he ended his letter by compounding the problem. While praising the mother – "very nice; her conversation is particularly charming" – he did not refrain from making negative comments about the daughter:

> made of coarser material, participated little, and for the most part had a blank expression on her face. Otherwise, of course, there was not the slightest abnormality noticeable in her. The scar is really inconspicuous and gives good opportunities for her undeniable vanity.
>
> (8/02/1911)

Agreeing with the interpretation of Ferenczi's reaction ("it really branches out from the transference"), he overlooks the considerable unhappiness he had inflicted.

DOI: 10.4324/9781003246473-15

We can't help thinking that Freud had instantly and preconsciously identified the girl as a threat, his real rival in his pupil's devoted relationship with him, and had short-circuited an impulse of violent and unconsciously cruel jealousy, repressing Ferenczi's incestuous drive. In this connection, Balsamo and Napolitano speak of an incorrect diagnosis expressed in a Jungian nosography, one that would have signalled Freud's preconscious awareness that he was facing a repetition in Budapest of the furore between Jung and Sabina (1998).

Wounds from the past were re-opening (it had been another operation and another scar, Emma Eckstein's, that had compromised Freud's friendship with Fliess), but perhaps other conflicts were also fully present: those in his relationship with Anna, his 16-year-old daughter who was becoming a woman. Once again, unwittingly, Ferenczi presents Freud with something traumatic from which he withdraws.

It is also possible that Freud's unconvincing intervention had brought events to a head: his scorn showed all too clearly how much erotic potency the girl had. Certainly the impression it gave Ferenczi reveals how intimately involved he was and how seduced by Gizella's daughter, who, as she grew up, was clearly starting to develop an oedipal transference onto the young man who visited her mother, a fertile terrain in the mother-daughter conflict. Elma may have known nothing about the concrete reality of the relationship between the two, but she was certainly aware of her mother's feelings for him.

In the period following this, Freud's letters show him to be "completely capable of accomplishment" and "quite well, physically," almost hypomanic at his victory over Adler and Stekel – the "Max and Moritz" of the situation – who had at last resigned from the executive of the Vienna Society. He confined himself to a single hint on 12 March: "What is Frau G.'s daughter doing?"

Ferenczi was writing less and reflecting on the pupils' – and his own – transference onto the Master. Compared to Adler and Stekel, he declared himself to be "quite a simple "case" and free of that "neurotic character trait" which had made itself so obvious the previous summer in Sicily. But a few days later, at the end of the letter which he had hesitated to finish, he admitted to new, quite different personal problems: an erotic restlessness which he eased with fantasies about young women and failed attempts at unfaithfulness to Gizella, complaining that he felt "unshakably fixated" on her: "The equinoctial storms have fanned the flames of unsublimated erotism in me in the form of very youthful impetuosity."[2]

Distracted by his battles – regaining the editorship of the *Zentralblatt*, Bleuler's renewed coldness, the scandals being stirred up by Jones in Toronto, and, not least, disputes about where to hold the Congress which eventually took place in Weimar in September – Freud chose not to offer an opinion. After the hiatus, noting that Ferenczi was writing infrequently to him – "as if we had nothing more to say to each other" – he decided to return to the subject of the erotic restlessness to which he still owed a response and suggested a holiday in the South Tyrol so that they could discuss it in person.[3]

The Easter vacation, two days spent together in Bolzano, was only an apparent reconciliation, since Ferenczi did not have the courage to confide his "fantasies about marrying Elma."[4] When the correspondence resumed, the Hungarian hid behind his progress on thought-induction: some further experiences with Gizella, a gypsy woman's prophecy, and, among other things, a bizarre episode, almost a joke: his having guessed the name of a man encountered by chance on a bus, "Herr Kohn."[5] Freud was unstinting in his praise – "Your experience with Herr Kohn is also singularly beautiful" – as long as his pupil did not return to the subject of the "equinoctial storms."

> But are you certain that it isn't a cryptomnesia? Didn't you perhaps recognize the man from family resemblance? Is he supposed to carry around the visual picture of his own name with him while he rides the tram? Do you admit that this success also can't be explained by means of thought transference?
>
> (11/05/1911)

He was not by any means joking when he ended the letter with "Regards to you, uncanny one.": he was impressed by Ferenczi who, unlike Jung, could not be accused of mysticism. And indeed, building on his experiments in thought-induction, the Hungarian was probing into the psychoanalytic concepts of projection and introjection and looking for a connection between the constant arousal of the complexes and "the tendency to projection and *emanation*" of the complexes in paranoia, while the hysteric tends toward "introjection and complex *resonance*:"

> Still, I consider it possible that certain strongly feeling-toned ideational complexes are in a *constant* state of arousal, among them the determinations of one's own ego come into primary consideration. You know, of course, how one is inclined to hear one's own name being called out of all kinds of noises. [The joke in which someone calls out "Kohn" and punches an innocent Kohn who happens to stick his head out of a train window is also familiar to you.].
>
> (13/05/1911)

Ferenczi seemed to enjoy playing the magician but was still working on the terrain of analytic research. Freud found himself having to call on him for help in restraining Jung from his sorties into occultism, as well as the distortions that he was imposing on psychoanalysis (he had written the first part of *Transformations and Symbols of the Libido*).[6] "You should at least proceed in harmony with each other; these are dangerous expeditions, and I can't go along there."[7] But the Hungarian too was alarmed by the mystic turn being taken by the "secret arts" of the Swiss and suggested a meeting in order to find a common project. From now on, though, Jung was pursuing a solitary course.[8]

In the meantime, Adler also resigned from the International Association, and Stekel was preparing to do likewise. Faithful to the cause and to Freud, Ferenczi kept his feet on the ground scientifically, but in his personal life, he was vacillating. At the approach of summer, when he was already planning the late August holiday, the Hungarian, having merely hinted at a certain indecision in Fräulein Elma towards her two suitors and not raising the subject again with Freud, dropped his bomb and ended his letter of 14/07/1911 by saying,

> Frau G. thanks you for the greeting and returns it cordially. Just think, I decided to take her daughter (Elma) into psychoanalytic treatment; the situation, you see, was becoming unbearable. For the moment, the thing is working, and the effect is favorable.

Freud was concerned, and not fully in agreement:

> I wish you much practical success in the new enterprise with Fräulein Elma, but, of course, I fear that it will go well up to a certain point and then not at all.
> (20/07/1911)

Ferenczi retorted by playing down the concern:

> Elma's treatment is going along normally for the time being. In the meantime I will be able to report to you orally on her case.
> (24/07/1911)

How much, and in what terms, the two were able to talk about Elma's analysis during the two-week holiday in Collalbo and the days in Weimar, we do not know. Ferenczi's letters took up his usual topics: he added to his accounts of occultism, planned the visit to two mediums who worked "miracles" in a small north Hungarian town, suggested that Freud join the Society for Psychical Research. . . .[9] Only on 18 October came this hesitant mention of Elma:

> The analysis of Frau G.'s daughter (Elma) was already making very nice progress when one of the youths in whom she was (neurotically) interested (actually the only one who was worth anything) shot himself on her account a week ago. It is very questionable how the matter will go now.

Despite Ferenczi's denial, the image of the lover killing himself *"on her account"* seems to be a warning of an unhappy destiny awaiting the young woman. And in fact it was this tragic event which precipitated the situation. Only a month later, Ferenczi confessed to feeling erotically detached from the mother and to have fantasised for some time about a betrothal to the daughter.[10] He gave Freud to understood that he had these emotions under control and that he correctly connected them to the stage of the emancipatory journey he was undertaking, impelled by the young woman's transference. Focusing on his own difficulties with autonomy, he interpreted his erotic anxiety as a defence against acknowledging his profound tie of dependency on Gizella and on Freud himself.

In his reply to this declaration, Freud, rightly alarmed, called him "Dear Son" for the first time, said how sorry he was for Ferenczi's alternating sufferings of submission and revolt, invited him to be patient, and urged him to make himself as autonomous affectively as he was showing himself to be from a scientific viewpoint (and Freud also appealed here to the research on thought transference).[11]

The letter did not produce the desired effect.

What is still missing is the fatherly blessing. Fatefulness and Oedipal coincidences

After a couple of weeks, Ferenczi proclaimed that Elma had won his heart and that he intended also to make some decisions in relation to Frau G, who was fully abreast of the matter:

> Elma became especially dangerous to me at the moment when – after that young man's suicide – she badly needed someone to support her and to help her in her need. I did that only too well, even though I held my tenderness in check with difficulty for the moment. But the path was cleared and now, to all appearances, she has won my heart.
>
> (3/12/1911)

Freud asked him to break off the treatment without delay and make his way immediately to Vienna:

> Dear friend, First break off treatment, come to Vienna for a few days (Wednesday evening through Sunday (holiday in between) would be a good arrangement) don't decide anything yet, and give my regards many times to Frau G.
> (Freud to Ferenczi, 5/12/1911)

As in other cases, there are no writings to assist our narrative, and we can only intuit the content of the meeting. Ferenczi emerged from it feeling vaguely supported in whatever decision he might make but privately convinced of Freud's radical opposition.

Frau G received a formidable letter from Vienna, sincere and at the same time merciless, containing Freud's interpretation of the situation, the inevitability of the union between the two young people, his doubts about the couple's future development, and also his solidarity with what the woman must have been suffering:

> When, years ago, I first learned of the relationship that he had lodged himself in, I made a face and made it very clear to him that I wished something else for him. When I then became acquainted with you, I quickly learned

> to esteem you and was able to concede to him that, in comparison to other husbands and lovers, he possessed incomparably more than what he had renounced.
>
> (17/12/1911)

This was followed by a comparative silence: Freud waited for news without interfering, and Ferenczi sent the occasional rare medical bulletin about the states of agitation and unhappiness that he and Frau G were both suffering. On 30 December 1911, late at night, the Hungarian wrote Freud a letter wishing him a happy New Year and announcing his engagement and imminent marriage to Elma, not only with her mother's consent, but in fact under pressure from her: "What is still missing is the fatherly blessing." Only a few hours later came the counter-order: faced with some timid objections from the father, who had reminded Elma about her doubts and second thoughts on the threshold of a previous plan to marry, she had shown herself to be less certain and eager to marry Sándor.

On 1 January 1912, Ferenczi could only admit that "the issue here should be one not of marriage but of the treatment of an illness." With the agreement of Elma and her parents, he effectively demanded that Freud take her into analysis.

> My dear friend, how bitterly I feel being perhaps more perceptive and freer of illusion than others, and having to be right.

This is how Freud began the laborious letter in which he made clear all his doubts about the success of such a treatment. And he concluded,

> It pains me that I can't be with you now. I was depressed the whole time and anesthetized myself with writing – writing – writing.
>
> (2/01/1912)

As he witnessed the unfolding in Ferenczi of that "natural fate" that is the story of Oedipus, the inexorable development which psychoanalysis had accelerated, Freud was taken by surprise, even paralysed. He behaved more indulgently towards the Hungarian than to Jung and was unable to take responsibility for interpreting the incestuous and castrating impossibility of the trap Ferenczi had built for himself. Perhaps considering that his pupil would never be able to work through his old conflict of dependency on his mother and the vengeful urge of his unconscious homosexuality that "imperiously demands a child,"[12] Freud hoped at least to resolve Elma's neurotic transferential infatuation, doubting that she was capable of making Ferenczi happy.

Freud was worried more about his friend than about the young woman. He identified the main difficulty of the situation in the fact that Elma did not seem in a position to recognise, nor would probably ever be able to tolerate, the still-concealed fact that Ferenczi had "been her mother's lover in the fullest sense of the word."[13] In his opinion, this reality could only be overcome by someone with "a high degree of mental freedom" (sic) – Gizella, for example – but not by Elma who might go on to carry out "a revenge for her father."[14]

Agreeing to take Elma into treatment was the compromise that enabled Freud to take at least partial responsibility for his pupil's situation: the analysis was intended from the start to be brief, from the New Year to Easter, and would be concluded despite making substantial progress and the fact that the patient expressed the wish to continue.

During the course of the treatment, conducted in collaboration with Ferenczi – and without the girl's knowledge – the only legitimate diagnosis that Freud decided to make known to his friend was one of neurotic "infantilism":[15] but it was clear to Ferenczi that Freud had a low opinion of Elma and did not endorse his plan to marry her. As at their first meeting, so during the treatment: Freud never showed any positive attitude towards the young woman, going so far as to say that his feeling "utterly indifferent" was a favourable element in the success of the analysis.[16] Besides his jealousy of Sándor, he may have been paralysed in the face of the incestuousness and destructiveness of that inextricable entanglement, the confusion of boundaries, of couples, and of functions which, as we can recognise, characterise the traumatic nuclei of psychic life.[17]

In a state of torment, apprehension, melancholy, and remorse – internal battles in which he indulged without restraint – Ferenczi asked Freud to sort out the situation on his behalf and managed to restore the bond between them. He was hoping for the impossible, that Elma would resolve her paternal transference onto him and turn into the wife he hoped for and with whom he might start a family, while perversely maintaining his relationship to her mother whom he fatally wounded by wanting her back as a lover, only rarely imagining an encounter with an improbable other woman whose role would always be that of a replacement. The point would be reached where all he could do was spitefully and vindictively betray Gizella with prostitutes and even with her younger sister, Sarolta,[18] before seeking consolation once again in her maternal arms, realising that he couldn't manage without her.

After Elma's analysis in Vienna was over – and in Freud's opinion, its conclusion corresponded, as he had intended, with the definitive ending of the engagement – and after the two colleagues had spent an Easter holiday on the island of Rab, Ferenczi, going back on his previous resolution, decided to reassume the role of analyst to the young woman until the summer. In order to show that he had restored order to the situation and had it under control, he subjected himself and Elma to a complete ban on personal relations during the treatment, while secretly deluding himself that he would succeed where, in his opinion, Freud had failed, and that he would transform her into the fiancée of his desires. In letter after letter, Ferenczi told Freud that he had almost completely, definitely given Elma up and showed renewed enthusiasm for scientific projects. After reading a booklet about the "clever" horses of Elberfeld, he said he was ready to publish the paper on his experiments in thought transmission, and this time he received eager encouragement from Freud. But, all of a sudden, he abandoned these good intentions and, announcing that he had gone against his true feelings, declared that he was still expecting to marry Elma and imagining a happy future for himself, living with both his women.

The future turned out to be less happy, even though he did live with both women. At Christmas 1921, now married to Gizella, Ferenczi asked Groddeck for help, still describing his situation in the same terms:

> Your letter spurred me on to greater efforts; it helped me remove my mask in front of my wife, too – albeit partially. I spoke to her again about my sexual frustration, about my suppressed love for her daughter (who should have been my wife; indeed who in effect was my bride until a somewhat disparaging remark of Freud's prompted me to fight this love tooth and nail – literally to push the girl away from me). Oddly enough, with us these confessions usually end with me drawing closer to her again – overwhelmed by her goodness and forgiving nature.
>
> (25/12/1921)

Gizella, a lively, energetic, and cultured woman with many interests, who had managed to free herself from a marriage of convenience, witnessed this torment indulgently, took part in it compliantly, and indeed suffered it without taking sides. Perhaps the loss of her mother at the age of six[19] had left her affectively inhibited towards her difficult first child, who closely resembled her physically and, unlike her younger daughter, had developed a secretive, sad, and malicious nature, being spoiled by her father as his favourite. Seized with a sense of guilt, Gizella showed that she wanted to give way to her daughter, masochistically and cruelly wishing the engaged couple every happiness. On rare occasions, she made clear her intention to withdraw for good: in these situations Ferenczi plunged into real despair and revealed his inability to abandon the mother for the daughter, even on the sexual level, given that he was maintaining a virtually Platonic relationship with Elma.

A silent figure in the background, Elma's father, Geza Pàlos, whom she described as a "weak and passive man,"[20] after being left by his wife, denied her a legal separation without having the strength to react against her relationship with Ferenczi. Already turned in on himself, his premature deafness had accentuated his withdrawal into solitude. When consulted about the engagement, he cunningly used his timid objections to regain exclusive power over his daughter. Though opposed to psychoanalysis, he paid for the treatment with Freud and continued to remain in the shadows, only reappearing tragically on the scene nearly ten years later in the spring of 1919 by dying on the very day of Sándor and Gizella's wedding: "the connection is certainly striking," wrote Ferenczi to Freud.[21] The news of his death reached the bride and groom as they were on their way to the town hall. Like Groddeck, Ferenczi regarded it as no more than a tragic coincidence, linked in any case to the unhappiness of the separation and divorce: he doubted that Geza knew about the date of the wedding.[22] The widow of Lajos Levy, Ferenczi's friend and physician, would later tell Roazen that the coincidence, almost a sign of fate's hostility to the marriage, would in fact have been a suicide passed off as a heart attack with the aim of protecting Elma.[23]

Elma Pàlos, fragment of the analysis of a seduction

The suffering of young Elma Pàlos continued to be misunderstood and ignored at least until 1922, when Groddeck also took her into treatment and, for a while, tried to put an end to her cohabitation with her mother and Ferenczi.

In old age, more than half a century later, Elma tried to collect her memories in a letter to Michael Balint – Ferenczi's pupil and heir – during the long consultations that were necessary in order to bring about the publication of the compromising correspondence between Freud and Ferenczi which concerned her. She appeared overwhelmed by the love that Sándor had quickly manifested during the sessions, caught up almost against her will in the planned marriage that they had immediately confided to Gizella with no concern for her feelings, and abandoned to her fate by her father, a fragile and unhappy man in the grip of a sense of guilt:

> in spirit I was immature, self-conscious and desirous of love. . . . I was a young girl with a fiery spirit. . . . I was an evil seducer, I was only thinking about myself and did not care about my victims. But perhaps I was not evil at all, only the slave of nature!
>
> (7/05/1966)[24]

A victim, necessarily vital and seductive, Elma was paralysed by the clumsy, abusive, and mystifying actions of the adults who surrounded her.

Traumatised first of all by Gizella's difficult mourning for her own mother and by her parents' break-up, she was subsequently rendered incapable of fully working through her internal oedipal conflict and remained trapped in the couple. Her mother, whom she had never really possessed and from whom she could not separate herself, had put her continually into her own lover's arms, just as in the past Gizella had entrusted her to the "maternal" indulgence of her husband. Sándor had known her since she was a child; he had seen her grow up and, with a weakness like her father's, showed himself to be one of those conquests who are easy to ensnare.

In fact, Elma was seduced by Ferenczi and by her mother: as in many of these cases, the abuse was the least of the evils created by the pair, and the father's seduction was an extreme attempt to save his daughter's psychic life from her mother's coldness. We know that, in her tenacious desire for survival, Elma was even able to derive something good, or at least precariously defensive, from the analytic treatment with Freud, a further colonisation to which she submitted with docility and an understandable mistrust: but she never had a satisfactory married life or any children. She told Balint that she had soon realised she did not love Ferenczi very much and never really responded to his repeated attempts at re-establishing their initial relationship of affection and complicity.

She convinced Balint that there had never been a sexual relationship between them, and in this version of events, reconstructed at the age of 85, she added that the only man she could have loved had been her husband, but he had turned out to be a man too lost in his own dreams.

Elma met him the year after the events we are discussing at the international congress of the movement for female suffrage, held in Budapest in the summer of 1913, where he was working as a guide and interpreter since he spoke four languages.[25]

The journalist and art critic John N Laurvik, a handsome, fascinating American of Norwegian origin, probably made her hope that a man ready to take an unprejudiced view of women really could exist. They became engaged after a few days and he left for the United States with the aim of returning for the wedding a year later. The American fiancé turned out to be a dreamer, full of plans but unstable, unreliable, and sometimes violent: at one point he even got mixed up in the sale of forged paintings. Elma very soon began to think of leaving him. Back in Budapest for good in 1924, she never actually divorced Laurvik, but he kept promising to change, and she repeatedly went back to him.

These are the characters, as things stand in 1912: the various dice have been cast and the situation will not change substantially, not even after Elma's American marriage, Ferenczi's analysis with Freud, and his marriage to Gizella. Ferenczi continued to nurture troubled dreams of marriage and children with Elma, dreams full of regret and nostalgia, and to consider Freud responsible for the breaking-off of his engagement. Gizella clearly still thought that the right solution for Sándor was marriage to Elma: she urged him to make a decision when the engagement to Laurvik was in doubt,[26] and in 1916, she refused his first proposal, still hoping to ensure that her daughter would have a family.[27]

Even their closest friends were left in the dark about the hidden anxiety and pain underlying the apparently serene relationship of Sándor and Gizella and the deep roots of Elma's disturbances. It was only in 1927 that one of Ferenczi's intimate friends, Frédéric Kovàcs, while taking a cure at Baden Baden, learned about these secret destinies from Groddeck, who was by then very worried about the young woman's health. Kovàcs immediately confided in his wife:

At first Ferenczi fell in love with Gizella, then he got engaged to Elma; but the engagement was broken, and she then got married in America, and only later Ferenczi married Gizella. How about? And again, currently Gizella's greatest desire, and her project, is that Sándor divorces and marries Elma – she would not give up for another one, only for Elma – while she would be content to play the role of mother . . . is for this reason that Elma went from Paris to Berlin, that is, from Groddeck – who was there for a series of lectures – because she was very sick. . . . So it was this that had such evident effects on this poor, miserable, wonderful creature.

(8/01/1927)[28]

As late as 1927, Gizella was saying she was ready to divorce Sándor so that her daughter could marry him.

While declaring himself neutral, or strategically letting it be believed that he took the opposite view, Freud always defended the pairing of Sándor and Gizella; in 1917, like a father, he asked for her hand on his friend's behalf, and in 1919, he received the news of the marriage with satisfaction as the outcome of the analysis. In 1930, he was still able to defend Elma's disastrous American marriage, contentedly letting Ferenczi know that he had found her "beautiful and blossoming," "reconciled with her husband."[29]

The whole scene will remain fixed in its unhappy precariousness.

Lou Andreas Salomé was clear-sighted, as always, writing to Anna Freud about Sándor and Gizella – "the Ferenczis" – she argued that it was better to call them in the plural, both of them forming a single human being, which they had already been 14 years earlier, when Sándor introduced Gizella to her in Budapest (19/11/1926).

In this kind of blurring and mimetic trap, it is difficult, from the internal viewpoint of the various subjects and patients involved, to distinguish the victim from the seducer, and it is equally problematic to identify possible avenues for growth. It is certain that Elma helped Ferenczi to differentiate himself at least a little from Freud and Gizella, so as to achieve a marriage which occasionally caused him pain. Elma, an ordinary and inexpressive personality according to Freud, takes on the central role as one of the most important patients of his career. As Paul Roazen first wrote, her story is one of the most touching among the dramatic lives that became involved in the origins of psychoanalysis, and it only became known with the publication of Freud and Ferenczi's correspondence, which was postponed until 1998 for this very reason. Jones, who had had access to the letters, while missing no opportunity in his work to censure Ferenczi and denigrate his occultism and presumed incurable madness, maintained an absolute silence about them.

The open wound in Ferenczi's heart, a source of creativity

We may wonder how far the experience of analysing Elma Pàlos, like the encounter with Sabina Spielrein and the later treatment of Ernest Jones's partner, Loe Kann, had an impact on Freud's writings about technique in that period; how much it enabled him to highlight the problems associated with the countertransference but also to make them figure intimately, despite his protestations to the contrary, as inextricable and disastrous contingencies, consolidating in him that rigid position of defensive blockage which, after the first passionate period of work with hysterical patients and the first theoretical achievements, some readers recognise as early as the clinical fragment, *Dora*. The young Ida Bauer's problems emerged from a comparable incestuous situation with elements of seduction and abuse by her parents and parental representatives. That treatment had given Freud a way to recognise *a posteriori* the importance of the patient's negative transference and, at the same time, to acknowledge his own difficulty in identifying and interpreting it before Ida acted it out by prematurely ending the treatment.

Likewise, Freud's behaviour towards Elma seems from the start to indicate something very personal, countertransferential, and hard to manage.

While deceiving himself into the belief that he was not interfering in his friend's choices, he found himself inextricably involved and in his own way complicit in this insurmountable impasse affecting Elma's psyche and her whole existence: the opening up and immediate paralysis of that oedipal emancipation which Ferenczi, unlike Jung, would be strong enough to resolve without splitting. Despite this, and thanks to his subsequent analysis with Freud, Ferenczi kept himself fundamentally together, sane though certainly not "normal," capable of containing rather than expelling this vigorous and vital nucleus of pain inside him: an open wound in his heart, source of his creativity and his inexhaustible scientific and clinical ability, which began in these very years to manifest itself as an original and authentic source of research.

In a letter to Freud in summer 1912, Ferenczi was already aware of how his cruelty and harshness to Frau G might have derived from the infantile desire for revenge against his mother. It is no coincidence that he would be the first to highlight the mechanism of identification with the aggressor that Anna Freud would later describe as being one of the Ego's defences.[30]

Though with great ambivalence, in late 1912, after Elma's analysis had ended, Ferenczi began to consider having a real analysis, on the couch, with Freud; this took place in 1914 at the same time as Elma's wedding.[31] And so, in the attempt to escape and expel this pain, Ferenczi was able to blame Freud not only for having discouraged him from marrying Elma, obstructing his family romance and stopping him from having children, but also for not having done a good job of analysing him: all of this would bind him more indissolubly to Freud than anyone but Freud's daughter Anna.

Ferenczi's psychoanalytic research tells us about the place of the Mothers, the *Maelström*, to which he really descended and from which he was the first to engage in close dialogue on the subject with Freud.

While Freud explored the tragedy and conflict of the Oedipus complex and constructed the "theory of transference," Ferenczi resumed his research with a new initiative, venturing into the "ordeal" from which Breuer had retreated and which Freud had been able to transform into knowledge and theorisation: standing firm at the edge of the abyss, on the brink of dissociation, he worked on trauma and the pre-oedipal relationship, restoring to psychoanalytic understanding the part of reality that had been left behind and building his own "theory of countertransference" as a metapsychology of the analyst's unconscious processes.[32]

Beyond Good and Evil, with a Nietzschean creation of his own destiny, he would open the way to the understanding of the events preceding the oedipal tragedy, preverbal experience, and the primally maternal and female, the area he had tried to explore first through hypnosis and then with clairvoyants and mediums: the magical area where the infant experiences that "thought-reading" which precedes and supports the acquisition of language.

This would be Ferenczi's "great discovery": the deep significance of prophecies and thought transmission which Freud had foretold.

Sabourin writes:

> This is what the witnesses were thinking: those who knew Ferenczi describe him as a fragment of life in the pure state, accepting neither limits nor restrictions, heading in all directions at once . . . interested in everything with the same intensity, ready for all experiences. Others add that this vital excess had the flavour of desperation and death. This certainly fits well with Ferenczi's reputation for inveterate optimism; indeed, what is more desperate than the absolute refusal to abandon hope?
>
> (1985, p. 77)

Notes

1 7/02/1911.
2 17/03/1911.
3 2/04/1911.
4 14/11/1911.
5 3/05/1911.
6 *Wandlungen und Symbole der Libido* (1912), the original title of *Symbols of Transformation* (1967).
7 11/05/1911.
8 19/06/1911.
9 3/08/1911 and 11/10/191.
10 14/11/1911.
11 17/11/1911.
12 Freud to Gizella, 17/12/1911.
13 Ibid.
14 Freud to Ferenczi, 13/01/1912.
15 1/02/1912.
16 Freud to Ferenczi, 13/03/1912: "utterly" is in English in the original.
17 In his first diagnosis, Freud had grasped Elma's fragility, to which he had unconsciously reacted with a rejection, a countertransference prejudice rather than indifference.
18 Ferenczi to Freud, 18/11/1916.
19 Berman, 2004.
20 Ibid.
21 18/03/1919.
22 "the connection is certainly striking" (Ferenczi to Freud, 18/03/1919).
23 Roazen, 1998.
24 From Haynal's translation, in Berman, 2004.
25 Berman, 2004.
26 Ferenczi to Freud, 7/07/1913.
27 Ferenczi to Freud, 17/10/1916.
28 Frédéric Kovàcs to Vilma, in Fortune 2002, p. 116.
29 21/01/1930.
30 1936.
31 See Pierri, 2022.
32 Heimann, 1950.

Chapter 15

The Saturday goy
Getting to know Dr Jones

The *Welsh liar*

Up to now, Dr Ernest Jones has stayed in the background of the story: he has been quoted repeatedly as a source of testimony and historical reconstructions, but let's remember that he was one of the story's protagonists. This Welsh doctor became for Freud what Thomas Huxley had been for Charles Darwin: his *bulldog* – as he liked to call himself – and also his *alter Ego*.[1] For more than 20 years he was president of the IPA,[2] which he had helped to create, and also of the British Psychoanalytical Society for an equally long time; in January 1920, he had founded the *International Journal of Psychoanalysis* and the International Psychoanalytical Press. He would later promote the English translation of Freud's *opera omnia* in the *Standard Edition*.

In all respects a foreigner, Ernest Jones was the first of Freud's pupils not to be a native German speaker, and after Jung's defection, he was the only gentile in the circle of intimates. Often suspected of antisemitism in the small Viennese circle, even though his conduct made it seem more as if he had been *assimilated* into the Germanic and Judaic world, he undoubtedly functioned as the movement's *Saturday goy*, the man who undertook "unclean" actions: in his case, all the work of propaganda.

After he had written an unimaginably vast series of excellent, largely popularising psychoanalytic publications, at the request of Anna Freud and with the support of the International Psychoanalytical Association, in the early 1950s, he became Sigmund Freud's official biographer.

It must be admitted that, like all witnesses and biographers, he was pretty unreliable: suffice it to say that he earned the nickname of "the Welsh liar," in imitation of the more famous Lloyd George, well before taking on the task of historically reconstructing Freud's work, which brought him global recognition. It is to him that we owe the repression of the person and theories of Ferenczi, whom we could justly call Freud's *alter Id*.

Paul Roazen claims always to have admired Jones for the skilfulness of his historical inventions. On the basis that talking about him requires us "to activate all the resources of Christian charity," he presents Jones as a "master of manipulation

DOI: 10.4324/9781003246473-16

behind the scenes and of bureaucratic Machiavellianism."[3] Despite this, and bearing in mind the partisan nature of any interpretative narration, we cannot help liking him as a person and feeling grateful for his enormous and detailed work as an archivist and for the political tenacity he always showed in the movement's development and the promotion of psychoanalysis. In this context, Roazen recalls how in 1938, making all possible efforts to rescue Freud and the countless Jewish analysts in Eastern Europe from Nazism, Jones still found time to promote psychoanalysis even under the Third Reich after the purging of the Berlin Society, supporting the establishment of the notorious Göring Institute under the control of the Reichsmarshall's cousin.[4]

Ernest Jones's *Sigmund Freud: Life and Work* (1953–1955–1957) is a unifying origin myth addressed primarily to the diaspora of analysts who left central Europe after the war and were scattered across the United Kingdom and the Americas: his aim was to construct a historically correct Freud in terms of a psychoanalysis that was positivist, pragmatic, and regulated.[5] Since his narrative fictions were guided by noble and always emotional motives and by his obvious personal involvement in the events as they unfolded, we intend to trust him and his subjectivity. In telling Freud's life story, Jones was writing part of his own, and not only that: his is the testimony of a psychoanalyst, a pupil, the first to lie down on the couch as part of his training (note: it was Ferenczi's couch), to immerse himself deep in the Saga, in the genealogical tree of the tradition he was founding.

It is a rigorous testimony because, in writing his own legend, he knew it was analysts that he had to answer to first of all.

We will keep his version of events in mind because, while handing down institutional, formative, and official truths of psychoanalysis, it lets us intuit the censored and repressed ones between the lines. And we recommend the reading of his monumental work as still the fullest introduction to psychoanalysis and the understanding of its theories for anyone deciding to engage with this discipline, especially because it lends itself to stimulating new interpretative elaborations and enables us to *dream*, through its affective chiaroscuro of what is highlighted and emphasised and what is minimised, distorted, or mystified. As in the case of repression, this censorship preserves what it hides: "Secrets do not evaporate in their well-guarded hiding places, and the traces leading to them stand out as fresh as when they were made."[6]

The same oneiric density is not apparent in *Freud: A Life for Our Time* by Peter Gay (1988), Sterling Professor of History at Yale University, an extraordinary text but no less partial for that: even professional historians have personal and subjective motivations.

Peter Gay had more than one reason to identify himself with "his own" Sigmund Freud, a subject who occupied much of his historical research, as was the case with Roazen. Gay published a remarkable series of articles and books prior to the best-seller of 1988. Born in Berlin in 1923, Peter Joachim Fröhlich escaped Nazi persecution with his family in the spring of 1939 by taking the last steamer that would have recognised entry certificates to Cuba. Only by chance did he avoid boarding the fateful *MS St. Louis* which, after crossing the Atlantic back and forth, would bring its more than 900 Jewish passengers back to Europe. From Cuba, Gay fled to the United States where in 1946, at the age of 23, he changed his name (*Fröhlich* in German: cheerful, similar to Freud's *Freude*: joy) to Gay.

We can see how the emotionally determined censorship imposed by the psychoanalyst Jones is more significant than the omissions made by the historian Gay, who scandalously deletes Wilhelm Reich and has no room for Lacan,[7] to say nothing of thought transmission, a subject very dear to Freud and of such great interest as to make it impossible for a history of psychoanalysis to ignore it. Whereas Jones, despite his personal aversion to occultism, dutifully devotes a whole chapter to it, Gay despatches it in a few strokes, taking it for granted that telepathic phenomena and Freud's experiences of them have no substance, a thesis which the Welshman had dedicated so many pages and so much effort to maintaining.

"Jones testifies that Freud enjoyed telling stories of strange coincidences and mysterious voices," says Gay laconically, "and magical thinking had something of a hold on him, though it was never secure." And in commenting on Freud's first publication on the topic, *Dreams and Telepathy* (1925), he even remarks, "One wonders why Freud published the paper at all."[8]

Nevertheless, Ernest Jones could not share Freud's and Ferenczi's passion for thought transmission, which, besides making him feel excluded and jealous, caused him actual distress.

Unlike Jung and Ferenczi, our Jones was neither a visionary nor a dreamer. His childhood dreams were nightmares, and it is no coincidence that in 1910 he began to study mediaeval superstitions: with the aim of working through his childhood terrors of devils, vampires, werewolves, witches, and ghosts, he fittingly connected them to his oedipal images of his parents.

One of his first important psychoanalytic works, the monograph on the nightmare (1912), was written when he was living in Toronto, the time when Jung was immersing himself in astrology, and Freud and Ferenczi were deep in their investigations of clairvoyants and prophecies, as if Jones were trying to catch up with them and retaliate.

He had no illusions about this: in the grip of another kind of "occultism," he was necessarily "the man of true lies and true myths," a liar who firmly believed in his own lies and eventually managed to make some of them true, says Rodrigué (1996). Like Houdini, as he was creating his fictions, he maintained his solitary crusade against the occultism he attributed to Freud and by which he felt almost persecuted.

Difficult beginnings

Ernest Jones (1879–1958), born in the parish of Llwchwr in Wales, an eldest child and only son with three sisters, came from a modest social background: his father was a technician in the coalmines. His mother had had to wean him prematurely, and he consequently described himself as "a puny and ailing infant, with pronounced rickets and a not very happy disposition."[9] Dark-haired, small, and stocky as a terrier, he always had delicate health and a pallid complexion, and before the age of 30, he developed a rheumatoid arthritis which flared up repeatedly throughout his life.

From his autobiography, we learn about his night terrors but also about certain phobias, a fear of heights, for example (with the fear of falling or being pushed down), perhaps in reaction to his father's ambitious expectations and the intense conflict between the two of them. Born on New Year's Day, the celebration of his birthday had always been overshadowed and – so he tells us – he never liked the names he had been given, Alfred Ernest, after Queen Victoria's second son.

He tried to live as if he were his father's younger brother, never tolerating his authoritarian behaviour or accepting the idea of following in his footsteps as a mining engineer. Instead, from childhood onwards, Ernest wanted to be a doctor, modelling himself on the family doctor who had lived nearby for a while: an elegant, dashing, and handsome man who gave Ernest his first dog and, in his childhood fantasies, had played a more important part than his father in the birth of his younger sister.

The autobiography also tells how Jones had no difficulty in accepting Freud's sexual theory, given his experience of a disenchanted and barely "repressed" childhood, at least compared to Freud's or Ferenczi's. He slept between his parents in the same bed and became aware of sexual intercourse from the age of six or seven. He remembered how the nursemaid, who spoke only Welsh, had taught him "two words to designate the male organ, one for it in a flaccid state, the other in an erect," adding, "It was an opulence of vocabulary I have not encountered since."[10]

But in early adolescence Jones had suffered intensely from a sense of guilt. "He had been preoccupied with religion, philosophy and salvation of his soul, which he came to interpret as atonement with the father."[11]

With his curiosity stimulated early in life, he soon became passionate about foreign languages (he taught himself Italian grammar at the age of ten!), was interested in codes, and quickly learned Pitman shorthand. Despite his Welshness, he developed an accent and manner designed to make him more English than the English. He later learned German with similar ease, perfectly adopting his teacher's Prussian pronunciation: coupled with his military bearing, this never failed to arouse hilarity in his Viennese colleagues.

His adolescence was still more of a trial, divided between the marked cruelty and obscenity of the boys at Llandovery College, a hotbed of Welsh rugby at the time, and the romantic fantasies he cultivated in solitude, skating artistically on the frozen river, dreaming of an enchanted waltz with a Viennese girl.[12]

Being a dedicated student, he gained excellent results at school and enrolled at Cardiff College at the age of only 16 before qualifying as a doctor in London in record time.

In 1898, before his graduation, his father had rewarded him with a journey to Europe – France, Switzerland, and Italy – a sort of *grand tour* together which did not prevent the final clash. Having decided to cut his ties with his father and his homeland, armed with a first-class degree and a doctorate, Jones anticipated rapid progress in his career and an ascent to a higher social class. At this point, his temperament, his difficult relationship with paternal authority, and of course his Welsh origin were serious stumbling blocks for him.

Paradoxically, this ambitious Welshman's worst fault, besides his extreme efforts to defend a precarious primogeniture, was his indisputable talent, his quick Celtic intelligence, his almost ferocious stamina,[13] the conviction of never being in the wrong, that he could always do things better than anyone else:[14] from chess to rose-growing to figure-skating (about which he even wrote a manual [1931]), it was hard to match him. These unquestionable gifts, fed by his arrogance (the sharp tongue his mother had always scolded him for), ended up leaving him empty inside: in London, after hurting the feelings of many senior colleagues, he became a *persona non grata*.

It happened that while at the North-Eastern Hospital for Children in Bethnal Green, where he exasperated the staff with his efficiency and his incessant questions, he impudently challenged the diagnosis of none other than a director of the National Hospital in Queen's Square. He was probably correct, but he was forced to resign, and, despite his gold medal and his doctorate, he failed to be appointed to the National Hospital. With his character, his gaffes, and the incidents he was starting to accumulate, Jones was not a suitable candidate.

Anyone else would have gone back to Cardiff to practise as a neurologist with neither shame nor glory. Not Jones, who wasn't cut out for that kind of worthy, serene existence and who, above all, could not go home defeated to his father.

He was a tireless organiser and, according to Putnam in 1910, "the most energetic, precise and prompt and efficient individual" you could ever meet.[15] Imperturbable and tenacious, his job applications were bringing a growing pile of rejections, and he also had to give up a promising betrothal (his first relationship, with a woman ten years older from an excellent family).

Jones seemed trapped in a vicious circle dominated by what psychoanalysis is accustomed to calling a castration complex. He later tried to interpret this deeply rooted condition:

> The premature weaning and early ill-health had combined with internal factors to induce a deep feeling of insecurity and inferiority, against which the life force (for I must have had somewhere an unusual amount of vitality) had reacted by building up a defence of the opposite extreme, an unwarranted belief in the omnipotence of my wishes.
>
> (1959, p. 115)

His small stature, the fact that he seemed younger than his years, his Welsh sense of inferiority, his aspiration towards a higher social and cultural level, and finally the very common surname "Jones" (which he often thought of changing) combined to provide material for the complex that had begun to manifest itself in this young attention-seeking doctor who was also good-looking, elegant and capable of making women fall in love with him: his lively eyes, his rapid, fluent speech, the animation of his gestures, and the elegance of his appearance made him "irresistible to women."[16]

And so, undaunted, Dr Ernest Jones had opened a private neurology clinic together with an older colleague, the neurosurgeon Wilfred Trotter (who later became his brother-in-law) and, being short of funds, tried to survive by performing every kind of medical activity. Meanwhile, despite having ruled out work in a psychiatric hospital, he began to specialise in psychopathology.

Unlike what was happening on the Continent, psychiatry in Britain was a backward discipline at that time, more aimed at custody than study and care of the mentally ill: what was required of a psychiatrist was the ability to play cricket, as he scornfully claimed in his autobiography.[17]

His interests turned towards psychoanalysis: in 1903, Trotter had shown him the review of *Studies on Hysteria* in the journal *Brain*, and together they discovered that Freud listened to every nuance of his patients' communications, paying them the same attention that medicine gave to bodily symptoms. So they decided to take German lessons in order to begin studying the *Traumdeutung* right away.

Jones also started practising psychoanalysis, the first in Britain to do so, as he never failed to repeat.[18] But psychoanalysis was no help in his professional life: when one of his first patients, the sister of a colleague, was cured of hypnosis and decided on a divorce, the ex-husband, a neurologist in New York, persecuted him for years because of this unwanted consequence of the treatment. He later became responsible for a couple of unpleasant and much more embarrassing incidents while busily applying psychoanalysis to his research on sexuality by interviewing children at the School for the Mentally Handicapped and the West End Hospital for Nervous Diseases. Principally, he failed to respect the rule about visiting children in the presence of a third person.[19] This time he was accused of paedophilia and indecent behaviour, spending a day under arrest, and the London newspapers bore the headline *"Harley Street Man – Sexual Offences."*[20] There were no legal repercussions, and the magistrate closed the case with a letter full of cordiality – so Jones informs us – but many colleagues were inclined to believe the accusation "under the doubtfully flattering pretext that clever people were apt to be queer," and he added that "disagreeable echoes of that episode, in increasingly distorted forms, reached me for very many years after."[21] Money-Kyrle who, before going to Freud in 1922, had 18 months analysis with Jones, was advised against this by his GP with hints at dark secrets in his analyst's private life. And even Jones's own receptionist used to come into the waiting room and beg the patients to "give up such wicked nonsense."[22]

In any case, whatever Jones said on the subject, because of the liberties he had taken with those children, he was compelled to resign his position.[23]

This and subsequent episodes of naiveté, imprudence, or outright abuse, which cast a dark shadow over the man who was to become the world's chief exponent and ambassador of psychoanalysis, caused the decisive collapse of his reputation.

And so it was that at the age of 29, Ernest Jones was forced to emigrate and, having unexpectedly obtained a recommendation from Sir William Osler, who was originally from Canada, he moved to the British Dominion and took a senior post in the new Ontario Clinic for Nervous Diseases at the University of Toronto. Off he went, to seek a restored virginity in the New World.

Freud's first pupil from Britain

Every cloud has a silver lining: before leaving London for his exile in Canada, Jones decided to make a long trip to the Continent for study, visiting the best psychiatric clinics.

He met Janet in Paris and attended Kraepelin's seminars in Munich: here he became a friend and pupil of Kraepelin's assistant, Otto Gross,[24] one of the wildest geniuses in psychoanalysis, who, having embraced Freud's sexual theory, distributed interpretations and maxims from his regular table at the Café Passage in Schwabing.

In the summer of 1907 at the International Congress of Neurology in Amsterdam, Jones met Carl Jung, who was defending Freud's theory against the attacks of Gustav Aschaffenburg, and followed Jung to Zurich, to the *Burghölzli*, where he was put in contact with Freud.

They met in April 1908 at the first Psychoanalytic Congress in Salzburg, which Jones promoted with Jung.

> My first impression of Freud was that of an unaffected and unassuming man. He bowed and said: "Freud, Wien," at which I smiled, for where else did I think he came from? This German custom of announcing one's name and town on being introduced, which by the way is a very sensible one, was still novel to me. His first remark was to say that from my appearance I couldn't be English; was I not Welsh? This greatly surprised me.
>
> (Jones, 1959, p. 156)

As it had done for Jung and Ferenczi, the meeting with Freud changed the whole course of Jones's life. It was nothing like an initiation: rather than falling under his spell, Jones seemed to take possession of Freud as if he had identified something deeply congenial in him. Jones was not yet 30, and he was struck by the intellectual power of the 50-year-old Freud, who seemed such an "unaffected and unassuming man," and to whom Jones listened as he spoke off the cuff in his usual fascinating way for a full four hours. Freud presented the case that later became famous as *The Rat Man* (1909b).

> As is well known, Freud was no orator and all arts of rhetoric were alien to him. He spoke as in a conversation, but then his ordinary conversation was so distinctive as to be worthy of a literary recording. His ease of expression,

his masterly ordering of complex material, his perspicuous lucidity, and his intense earnestness made a lecture by him – and I was to hear many – both an intellectual and an artistic feast.

(Jones, 1959, p. 166)

At the Congress, Jones read a paper entitled *Rationalization in Everyday Life*, the only work presented there in a language other than German.

It should be recalled that he was not the only representative of Britain, since his friend Trotter had also come to Salzburg. Although Trotter was interested in psychological processes (he had proposed the term "rationalisation" and would later make a study of the "herd instinct"),[25] he was principally a surgeon and mistrusted any kind of dazzling intuition. Unlike Jones, he was very ill at ease in this environment, above all because of Freudian ideas about sexuality. During the official lunch, greatly troubled by the bursts of wild interpretation being made by some of the delegates, he confided to his friend, "I console myself with the thought that I can cut a leg off, and no one else here can."[26] Greatly disappointed, he packed his bags and left before the end of the congress.[27]

In a fit of enthusiasm, Ernest Jones, who still had some time to spend in Europe before taking up his post in Canada, decided to follow the professor to Vienna; from there he went on to spend some days in Budapest at the invitation of Sándor Ferenczi, whom he had just met. Later, after he had settled in Canada, Freud's trip to America, with Jung and Ferenczi, gave him the opportunity to meet his colleagues again, when he joined the group to attend the lectures at Clark University in Worcester in the fall of 1909.

As for Freud's impression of Jones, we read in the correspondence with Jung how, despite his sometimes arousing a feeling of mistrust and almost of "racial strangeness," Freud found him interesting, of value to the cause, but too fanatical. "He denies all heredity," wrote Freud. "To his mind even I am a reactionary." Jung was also puzzled by certain enigmatic features of the Welshman, by what he might be hiding under his admiration and opportunism.[28]

I thought you knew more than I about Jones. I saw him as a fanatic who smiles at my faint-heartedness and is affectionately indulgent with you over your vacillations. How true this picture is, I don't know. But I tend to think that he lies to the others, not to us.

(Freud to Jung, 18/07/1908)

Freud and Jung were soon exchanging speculations about his regrettably uninhibited sexual life, so remote from the bourgeois morality to which they both adhered, in spite of everything. In Canada, perhaps having fallen under the influence of Gross and his theory of polygamy, Jones went on to cohabit blissfully with several women, his "harem":[29] there was the fascinating Jewish Dutchwoman, Loe Kann, who had followed him from London and whom he introduced as his "wife," and also his sisters, Elizabeth and Sybil Jones, and last but not least, Lina, Loe's one time governess. They all gathered around him.

> By nature he [Jones] is not a prophet, nor a herald of the truth, but a com-promiser with occasional bendings of conscience that can put off his friends. Whether he is any worse than that I don't know but hardly think so, though the interior of Africa is better known to me than his sexuality.
>
> (Jung to Freud, 7/03/1909)

We can imagine how, as they shared their impressions of the Welshman's sexual-ity, Freud and Jung were scrutinising each other at the same time: after the curios-ity and fantasies stirred up in Jung by the professor living with the two Bernays sisters, the revelations about the Spielrein *affaire* were brewing.

As we have come to understand, sexuality was the first and most heated prob-lem for psychoanalysis: not only because of its fundamental role in patients' psy-chopathology but because of its active presence in the care relationship through the phenomena of transference and countertransference. The question of this pre-cious but also highly risky material was becoming inescapable for analysts.

Dr Jones's stethoscope: rationalisation and censorship of excess countertransference

Freud's journey to America gave Jones an opportunity to meet him again. He came to New York from Canada to join Brill in accompanying the professor to Worces-ter and could see how immersed he was in his analytic adventure with Jung and Ferenczi. He probably suffered greatly at his exclusion from the trio, given how much he had done to prepare the ground and make it ready to accept the new theo-ries, shuttling between Toronto and Boston via his university contacts with Morton Prince and Putnam.[30] Freud noticed Jones's bad mood and towards the end of the festivities took the trouble to accompany the Welshman to the station and see him off on his train back to Canada, saying goodbye with particular warmth and gratitude.

Jones took some time to get over his disappointment and tried to begin a self-analysis: unlike Jung, as he got to grips with clinical work, he quickly recognised the emergence of his own sexual transference, which he considered responsible for his medical vocation.

In June 1910, he confided to Freud about the outcome of his first working-through, provoked by a patient's question.

> In unpacking my furniture and arranging it in this house last year I came across a single, wooden stethoscope and . . . was impelled to place it on the consulting-room desk, between me and the patient's chair. I never use it, and keep all other medical instruments in drawers. A few months ago a patient asked me what it was, and why I kept it there. . . . It started me thinking.
>
> (28/06/1910)

First of all, he had recalled the doctor who had lived with his family until he was three years old, a man who had the habit of keeping a straight stethoscope handy inside his hat: when he went to see patients he only had to take off his hat and "pull it out."

As a little boy, he had greatly admired and loved that "handsome dare-devil fellow" and had developed fantasies about him and the birth of his sister: specifically "the double phantasy that she was the child (a) of the doctor and myself, (b) of myself and my mother." And he confided to Freud that he had been in love with the doctor but also terrified of him:

> He never much liked me, owing chiefly to my disturbing him then by crying etc. One day in a rage he hung me in a high water-butt, which with other traumata formed the later basis for a phobia of heights.
>
> (*ibid.*)

From the transference onto this powerful, idealised, and persecutory image which had been reawakened by meeting Freud, Jones continued his analysis by trying to understand the phobic symptoms from which he had suffered in childhood and confronted himself in the present with the conflicts stirred up by the experience of his erotic investment in his female patients, confiding "I have always been conscious of sexual attractions to patients; my wife was a patient of mine."

After reporting a dream in which a man was attacking him with a sword (which he described as "plainly of a homosexual-masochistic nature") the associations which followed led him to understand the complex function of that old stethoscope which, like the drawn sword in the Volsung-Nibelung legend placed by Sigurd between himself and Brunhild, he had placed between himself and his patients.

Freud included the example in the 1912 edition of *The Psychopathology of Everyday Life*, commenting that it well showed "how close the connection can be between a symbolic action performed through force of habit and the most intimate and important aspects of a healthy person's life":[31]

> The act was a compromise-formation: it satisfied two impulses. It served to satisfy in his imagination the suppressed wish to enter into sexual relations with any attractive woman patient, but at the same time it served to remind him that this wish could not become a reality. It was, so to speak, a charm against yielding to temptation.
>
> (Freud, 1901, p. 196)

Designed to let the doctor's ear communicate with, and keep a distance from, the patient's chest, the wooden instrument may have reminded Jones on the one hand of contact with the lost maternal breast (and his own chronic chest complaint dating from the birth of his sister when he was three) and on the other of the exhibitionist and phallic elements in his adolescence. By being placed on the table in his consulting room, the stethoscope served to symbolise the emergence of desire and its limit, the paternal prohibition: a concrete witness of its owner's difficult journey towards the ability to master his sexuality and construct an analytic listening.

In 1916, Jones would write a fine essay on the unconscious origin of symbols. Such objects include Freud's archaeological finds and Winnicott's spatula, which suggest the importance of the concrete and symbolic elements of the setting, the basis for the construction of analytic listening.

Despite these intuitions, Jones's self-analysis resembled a good exercise in rationalisation, on the model that he had described in Salzburg. His letters commenting on Freud's writings and his own psychoanalytic articles – highly erudite but, as he himself admitted, not always endowed with originality – displayed the fracture between his acute intelligence, curiosity, and scientific ability and the instinctuality that was still prompting him to act in an inconsiderate manner.

No wonder doubts periodically arose about his sincerity. On 14/08/1910, Freud wrote about him to Jung: "He made an excellent personal impression; he seems much more secure. He represented only one of his stories as a personal experience, and then my boys told me it was an old anecdote."

However, Jones showed himself to be so capable of accepting criticism, so fervent, energetic, combative, and devoted to the cause that he regained Freud's trust and gratitude every time. This remained the case even when, at the start of 1911 in Canada, Jones became embroiled in further scandals which once again cost him his career and his professorship: not only had he been so maladroit as to disturb his North American colleagues by including over-explicit sexual details in his publications, but he was still having trouble maintaining the boundaries of the therapeutic relationship and dealing with the excesses of the transference.[32]

A rumour had earlier been spread that he was recommending women to masturbate, encouraging men to use prostitutes, and showing them obscene pictures to stimulate their sexuality.[33] Whether he had been careless or downright perverse, or simply didn't know how to manage his own seductiveness, the fact is that his private work had brought protests from two husbands, along with a blackmail attempt and then a formal complaint from a female patient.

By this time, even Toronto was too small for Jones, and he found it "an unpleasant atmosphere for a free thinker."[34]

Having been able to widen his influence across the border, with seminars and conferences in Washington, Baltimore, Chicago, Boston, and New York – Putnam had given him hopes of an appointment at Harvard – he now could do no better than found an American Psychoanalytic Society with the all too obvious aim of absorbing the New York Psychoanalytic Society newly set up by A A Brill. Misunderstandings and jealousies quickly arose between the two, including a disagreement over the translation of Freud's works into English.

At this time, Jung foresightedly commented, "I think a big success would do Jones good too. The trouble is he is always getting in his own way."[35] He felt especially sorry for Jones over the tie with Loe Kann, having himself been very careful not to compromise himself with Spielrein. He certainly did not suspect that the Welshman would end up in direct competition with him, having quickly identified Jung's mystical weaknesses and the lack of that scientific outlook which represented his own personal credo:

> In private conversations both in Zurich and in Worcester, Jung had revealed himself to me as a man with deep mystical tendencies that prevented a clear vision of a scientific attitude in general or a psychoanalytical one in particular; the superstructure was brilliant and talented, but the foundation was insecure.
>
> (Jones, 1959, p. 215)

By 1912, Jung was increasingly estranged from Freud and the movement, holding a second series of conferences in the United States on Fordham's invitation, at the Jesuit University of New York, without thinking twice about cancelling the planned IPV congress.[36] Jones, noting the desexualised version of psychoanalysis which the Swiss was putting forward in his own area of influence, took the opportunity to try to occupy the space left vacant on this side of the ocean. And while Jung could boast on his return that he had lowered the Americans' resistance with the modifications he had made to the sexual theory, almost repairing the damage caused by the Welshman's scandals and provocations, for his part, Jones intended to leave North America now that he had set up his advance guard, and he spent much of the summer in close contact with the professor, using every spell of leave from teaching to go back to Europe.

A prescribed training analysis in Budapest

Jones's professional difficulties in Canada were compounded by a crisis in his "marriage," especially after his youngest sister had left the couple and gone back to live in London, where she married Trotter.

Louise Dorothea Kann (1882–1944) was a beautiful and fascinating woman, brilliant and wealthy, but very unwell both psychologically and physically. She had originally been sent to Jones by one of the first English psychoanalysts, David Eder (1866–1936), to be treated for a manic-depressive condition complicated by dependence on morphine (used for a painful kidney disease). Instead, they had become lovers, and Loe stayed with him, supporting him financially, even when Jones moved to Canada. Her condition worsened there, and Jones started putting pressure on her to be analysed by Freud, with whom he had reached an agreement about this in September 1911 at the Weimar Congress.

Once he had persuaded Loe, he accompanied her to Europe in 1912 and entrusted her to Freud's care, taking advantage of the opportunity to spend long periods in Vienna in his company. It was clear to Jones that this was also a way of having his own analysis with Freud. He tells us that

> Two or three evenings a week would be spent *tête-à-tête* with Freud. . . . we ranged over all sort of topics in philosophy, sociology, and above all psychology. More than once I had to reproach myself for allowing him to continue

till three in the morning when I knew his first patient was due at eight o'clock. Those were days when I got to know Freud well – his fearlessness of thought, his absolute integrity of mind and character, and his personal lovableness.

(1959, p. 197)

Other pupils had been analysed in this same informal way: by correspondence – Eugen Bleuler among others[37] – or, so to speak, reciprocally, as Jung and Ferenczi were during the American expedition. And on Freud's return from the United States in autumn 1909, Max Eitingon had undergone a "peripatetic analysis,"[38] accompanying Freud for a few weeks on his daily walks, trailing behind him as he strode through the noisy streets of Vienna. Indeed, Freud claimed that walking encouraged the flow of free associations.[39]

Putnam had "six hours of analysis" with Freud in Zurich during September 1911, before going to the Weimar Congress.[40]

The time was ripe for serious consideration to be given to the transmission of psychoanalytic thought and to psychoanalytic training. This first extemporised mode was clearly no longer sufficient, or at least not for Ernest Jones. For all the faith he once again had in his pupil, Freud had begun to be troubled by Jones's persistent difficulty in mastering his sexuality:[41] the Master's unhappiness about the news of the Canadian scandal was not mitigated by reading the monograph on nightmare and superstition which Jones had sent him,[42] where he showed how in theory he could diligently interpret the intensity of his repressed incestuous desires.

We know that in 1912, Freud had begun to address Ferenczi's romantic and countertransferential difficulties, and he had concluded Elma's analysis in the spring, convinced that he had resolved the matter. So he started a treatment with Loe Kann which did not, however, look as if it would be equally swift.

The young woman began to revive, despite many fluctuations, and fell in love with another man, the wealthy American Herbert Jones, who became known as "Jones the Second." Freud kept the Welshman informed about the progress and blockages in Loe's analysis and eventually decided it was time for Jones to give up staying in Vienna when he came to Europe. And in 1913, they had a conversation in which Freud convinced Jones to begin an analysis with Sándor Ferenczi in Budapest.

And so it happened that Jones was not analysed by Freud: even though Ferenczi could be considered the most capable and creative clinical talent of the time, Jones was disappointed at being sent to another man, and a possible rival. He did as he was told, however, and interpreted the professor's advice as a way of helping him cope with the difficult separation from his partner of so many years,[43] which he did not seem at all prepared to face (incapable of being alone, he was already consoling himself with Loe's former governess), but, above all, he expected it to aid the achievement of an imminent qualification. The Committee had already been formed in 1912, and by the summer of 1913 relations with Jung were entirely compromised: Freud and Ferenczi were hoping Jones would be a candidate to succeed Jung as president.[44]

This was when Ernest Jones really changed, in spite of himself. When he returned to London from Canada at the end of 1913 after a five-year absence and resumed his psychoanalytic practice despite the ostracism, he was thoroughly transformed: the neurotic complex that had blocked his emancipation seemed to have dissolved.

It is hard to witness such radical changes, and some doubt is permissible, but there were no more scandals, or at least no grossly obvious ones. From now on, Ernest Jones was accused of excessive respectability and too much attention to the conformist, right-minded susceptibilities of his fellow-countrymen's scientific world. It even reached the point that Freud became both irritated and amused by the air of a League of Nations diplomat that he had acquired.[45]

Jones embarked on his analysis with Ferenczi in a state of restored narcissism with a mandate from Freud who provided that paternal recognition and promise of success that he needed so extremely. Initially he seemed to appreciate Ferenczi's gifts:

> an altogether delightful personality which retained a good deal of simplicity and still greater amount of the imagination of the child; I have never known anyone better able to conjure up, in speech and gesture, the point of view of a young child. . . . He had a very keen and direct intuitive perception, one that went well with the highest possible measure of native honesty. He instantly saw into people, but with a very sympathetic and tolerant gaze. Then he had an exceptionally original and creative mind.
>
> (Jones, 1959, p. 199)

He later expressed quite different opinions about his former analyst and his "wild and fantastical" theory, opinions connected to what had not been worked through in that *prescribed* analysis.

It must be said that Ferenczi's own attitude was not without ambivalence and that, both at the time and later on, alongside manifestations of deep affection and appreciation for his pupil, he expressed severe judgements about his ambition, his tendency to intrigue, his smugness and plagiarism, his lack of creativity, and the sadistic elements present in his theorising.

Jones's analysis with Ferenczi was the first to be officially called a *training analysis*, and the Welshman was always proud of this: "At the time it was a revolutionary idea," he wrote,[46] "but it has since become part of the normal procedure" in the preparation of analysts. The event is also important because Jones was the first of the pupils to formally accept an analysis with this aim.

The analysis was challenging, like every experience undertaken by Jones, because of the energy generated by his temperament. He not only attended sessions two or three times a day, but they were conducted in German: for Ferenczi this was a second language, for Jones a language of adoption with which he was still familiarising himself, as he was with the Jewish character of the group around Freud. The analysis only lasted a few weeks, but it was a *full immersion*. The two colleagues spent almost all their time together: after the sessions, which did not have a fixed timetable, they carried on their discussions together or with friends and colleagues from the Hungarian group, read the advance copies of works that Freud sent, and responded with their comments.

Notes

1 Brome, 1983.
2 From 1920 to '24 and from 1932 to '49; later he was honorary president for life.
3 2004, p. 9.
4 He initially fuelled hopes that he might be induced to do good for psychoanalysis in Germany. Jung himself collaborated with Matthias Göring as president (he resigned only in 1940) of the International General Medical Society of Psychotherapy, which, also representing the German Society, maintained the principles of Nazism; Wilhelm Reich accused Jung of opportunism (Roazen, 2001). Many did not understand what direction Nazism would take. It is astonishing that in Berlin, the Göring Institute, which was immediately able to take over the premises and equipment of the Berlin Psychoanalytic Institute supported at the time by Eitingon, dealt with psycho-therapy, financed by the German Labour Front and, after the start of the war, by the Luftwaffe, the National Socialist Party, the "Hitler Youth," the "League of Germans Girls," the Reich Criminal Police Office and finally by the SS-Lebensborn (Cocks, 1985): see Pierri, 2022.
5 Roudinesco, 2014.
6 Falzeder, 1996, p. 79.
7 Rodrigué, 1996.
8 1988, p. 443–444. Gay does not even mention another text by Freud, *A Child Is Being Beaten* (1919a) which, as we will see, is so significant (see Pierri, 2022).
9 Jones, 1959, p. 27.
10 Ibid., p. 30.
11 Hale, 1971, p. 19.
12 Jones, 1959.
13 Winnicott, 1958b.
14 Gillespie, 1979.
15 Putnam to Jones, 14/10/1910. In Hale, 1971,
16 Maddox, 2006, p. 4.
17 1959.
18 Jones, 1945.
19 Maddox, 2006.
20 Brome, 1983, p. 40.
21 Jones, 1959, p. 148.
22 Money-Kyrle, 1979, p. 266.
23 Kuhn, 2015.
24 Son of H Gross, founder of modern criminology.
25 Trotter, 1916 (quoted by Freud in 1921).
26 Jones, 1959, p. 168.
27 Later he had an influence on J Rickman and especially on W Bion. He assisted Jones in the translation of the *Standard Edition* and was on the committee of the Royal Society, which in 1936 awarded Freud the position of honorary foreign member (signed in 1938): "for his pioneering work in psychoanalysis." Founded on November 28, 1660, the United Kingdom's National Academy of Sciences is one of the oldest. In 1978, Trotter also awarded similar recognition to Freud's daughter Anna.

28 Freud to Jung 3/5/1908 and Jung to Freud 12/07/1908.
29 Freud to Jung, 9/03/09.
30 Clark, 1980.
31 See Pierri, 2012a.
32 Jones to Putnam, 13/01/1911, in Freud to Jung, 2/04/1911, note.
33 Hale, 1971; Gabbard, 1995.
34 Jones to Putnam, 13–23/01/1911 and 07/04/1911.
35 Jones to Putnam, 19/04/1911.
36 The *Internationale Psychoanalytische Vereinigung*, IPV, would change its name to the International Psychoanalytical Association, IPA, at the Marienbad congress in 1936.
37 See Marinelli and Mayer, 2002.
38 Freud to Eitingon, 24/01/1922.
39 Gay, 1988.
40 Hale, 1971, p. 39.
41 Gabbard, 1995; Appignanesi and Forrester, 1992; Paskauskas, 1993.
42 Jones, 1912; Freud to Jones, 14/01/1912.
43 Paskauskas, 1994.
44 Jones, II, p. 454.
45 Ibid., p. 176.
46 Ibid., p. 199.

Chapter 16

The intergenerational transmission of psychoanalysis

Love and death: the three women of the three pupils

We are starting to realise that in a brief period of time, Freud was called on to analyse women who were attached, as lovers and/or patients, to more than one of his pupils: the analysis of Elma Pàlos, begun in January 1912, had ended at Easter, and in June, as if taking over from her, Loe Kann arrived in Berggasse. Before the summer vacation, while Loe's treatment was still in progress (unlike Elma's, it lasted several years), Freud also agreed to Sabina Spielrein's analysis in the autumn.

Dr Spielrein had arrived in Vienna in October 1911 and, to all intents and purposes, had become a member of the local psychoanalytic society after presenting a summary of her doctoral thesis, "The Psychological Content of a Case of Schizophrenia (*Dementia Praecox*)." A few Wednesdays later, she had read a new work: "Destruction as the Cause of Coming into Being" (1912) which, identifying the representation of sexual activity "in the opposite direction" in the deadly hallucinations of her patient Martha N, widened Freud's definition of the antithetical meaning of primal words (1910a) and identified mutually antagonistic polarities in the nature of sexuality itself: destruction and rebirth, death and life. For Spielrein, the images of death were the Ego's reaction to the threat of dissolution in sexuality, while the images of rebirth reflected the transforming element of sexuality.[1]

With these concepts, she created a bridge between the destructive aspect of the drives and the problems of psychosis, inserting herself in a masterly fashion into the emerging conflict between Freud and Jung over the concept of the libido. Jung dwelt on an image of a terrible, destructive, incestuous mother responsible for "mythological descent into hellish nether regions," while Spielrein's first writings concerned "the inevitability of destruction as a necessary complement to love." In her view, sexuality harboured "an implicit threat of dissolution of the self." The two texts of 1912, his and hers, writes Gabbard, quoting Kerr, "adjoin each other like severed halves of a forgotten conversation."[2]

DOI: 10.4324/9781003246473-17

Sabina Spielrein's attempts at mediation were unsuccessful, even those under-taken on a personal level by maintaining contact with her former analyst and at the same time completing her interrupted analysis with Freud.[3] Having travelled to Berlin in the spring of 1912, she had intended to return the same year to resume it. Well aware of the transferential elements that Sabina needed to work through, but also of his own irreconcilable difficulties with the Swiss, the professor wrote to her in June:

> I look forward as October approaches to receiving your decision about com-ing to Vienna in order to break your dependence on Jung. I am most grateful for your clever words to Jung; there is no lack of others who are at pains to widen these chinks into a breach.
>
> (14/06/1912)[4]

That analysis never took place. During the summer, Spielrein suddenly decided to marry Pavel Scheftel, a Jewish physician whom he had met in Germany and, like her, of Russian origin. Congratulating her on her "half" cure, Freud advised her against treatment so that she could devote herself to her husband, with the hope that "someone else" would soon present himself to claim his rights. Since the Russian doctor immediately became pregnant, confident that the new relationship could make her happy, Freud urged her to give up her idealisation of Jung and the recurring fantasy of giving birth to his "Siegfried," so as to become aware of her underlying hatred of him. It is not impossible that Freud was relieved at not having to undertake this analysis since, while inviting Spielrein to end her affec-tive investment in the "Germanic hero," he was coming to terms with his own disappointment: "Let us banish all these will-o'-the-wisps!" he wrote to her. "We are and remain Jews. The others will only exploit us and will never understand or appreciate us."[5]

Unlike Spielrein's, the treatments of Elma Pàlos and Loe Kann were implicit, indirect requests for analysis from Ferenczi and Jones, who had sent their female *halves* on ahead, so to speak: an analysis by proxy and a gift to the master – almost *jus primae noctis* – which Jung had never granted him. But in all three situations, he was under pressure from the need to loosen the incestuous ties of those first analytic experiences in which his pupils were still entangled, with potentially seri-ous consequences for each of them. It became evident that these were transference relationships and not amorous ones, which would have brought a quite different satisfaction and pleasure.

It is legitimate to wonder how Freud experienced the task he found himself tak-ing on in these triangles, which were exacerbating the transference being under-gone in the pupils. Apparently impervious to the charms of the young patients, he convinced himself that he was no longer susceptible to the deceptions of love, and yet the situation could have satisfied deep desires, both present and past.

Dante Alighieri wrote in *Canzone of Exile*:

Three women have come round my heart" –
and sit outside it,
for within sits Love,
who holds sway over my life.[6]

In the relationship with the three pupils and their three women, just as in his own family with children who were now growing up (three sons: Jean Martin, Oliver, and Ernst, and three daughters: Mathilde, Sophie, and Anna), Freud was being presented with the oedipal crossroads, the profound fabric, the beating heart of existence,[7] this time from the parental angle. The fact that his father Jacob had had three wives and was already advanced in years when he had married the young Amalia (of whose seven children Sigmund was the eldest) may suggest what an arduous challenge Freud was facing as he tried to identify with such paternal potency in relation to the upcoming new generation.

Incidentally, the three beautiful and problematic young women approached him at the time when, the ardour of love having long since faded, he was making no secret of his dissatisfaction with his own married life and was about to be separated from his beloved daughter Sophie, who had become engaged in 1912 and would marry and move to Hamburg early in 1913. Straight after concluding Elma's analysis, while he was still trying to distract Ferenczi from his infatuation with Gizella's young daughter and gearing up to divide Loe Kann from Jones and Sabina from Jung, Freud began to regard his third daughter, Anna, 17 at the time and still at home, as his own Cordelia: identifying himself with King Lear, well aware of the distinct bond with his youngest child, he was already envisaging her assisting him in his old age and death.

He confided to Ferenczi (who was still struggling with Elma's analysis):

> the introductory scene in Lear must mean the same as the selection scene in the Merchant of Venice. Three caskets are the same as three women, three sisters. The third is always the correct choice. But this third one is peculiar, she doesn't speak, or she hides (Cinderella!), she is mute. Do you remember the words of the song of Paris in "Beautiful Helena"?

> *And the third – yes, the third –*
> *stood beside them and remained mute.*
> *I must give her the apple.*
> *You, oh Kalchas, you know why.*
> (23/06/1912)

Freud was finding himself once again confronted with the Sphinx's riddle about the three ages of man. Behind the attitude of resignation and apparent renunciation, what the emotional investment in his young daughter was making manifest in him too, no less than in his pupils, was love for his mother and his precious, hidden "sexual megalomania."[8] It was this that sustained his hungry curiosity and incessant desire for research, shot through with profound nostalgia for childhood and the maternal. The scene of the three women had already appeared in a dream from the period of his self-analysis:

> I went into a kitchen in search of some pudding. Three women were standing in it; one of them was the hostess of the inn and was twisting something about in her hand, as though she was making Knödel [dumplings]. She answered that I must wait till she was ready. . . . I felt impatient and went off with a sense of injury.
>
> (1900, p. 204)

Then his associations had immediately led him via the three Fates to the mother, source of life and earliest nourishment – "Love and hunger, I reflected, meet at a woman's breast" (*ibid.*) – and to the latent thought: "One should never neglect an opportunity, since life is short and death inevitable" (*ibid.*, p. 207). The incestuous fantasy – the opportunity not to be missed – was so intense and dangerous, and so strongly repressed that it substantially and ineluctably coincided with death as "giving up on living life deeply."[9]

And this is how he forlornly concluded the interpretation of the dream:

> So they really were Fates that I found in the kitchen when I went into it – as I had so often done in my childhood when I was hungry, while my mother, standing by the fire, had admonished me that I must wait till dinner was ready.
>
> (*ibid.*, p. 205)

Returning to this internal theme in spring 1912 with "The Theme of the Three Caskets," Freud reflected on the unconscious meaning of the choice of the third casket and the "third daughter," who, based on Shakespeare (*King Lear*, *The Merchant of Venice*), myths (*Aphrodite*, *Psyche*) and fairy tales (*Cinderella*), he interpreted as a portrayal of death and its ineluctable non-choice. Behind the Three Graces, the third of whom is the Goddess of Love, he saw the looming shadow of the Moerae, among whom Atropos inexorably cuts the thread of life.

As often symbolised in dreams, it was her silence that revealed the deadly nature of the third woman. In support of this hypothesis, in "The Theme of the Three Caskets," Freud will cite, for the first time and without further comment, the example of a telepathic dream (not taken up again):

> More than ten years ago a highly intelligent man told me a dream which he wanted to use as evidence of the telepathic nature of dreams. In it he saw an absent friend from whom he had received no news for a very long time, and reproached him energetically for his silence. The friend made no reply. It afterwards turned out that he had met his death by suicide at about the time

of the dream. Let us leave the problem of telepathy on one side: there seems, however, not to be any doubt that here the dumbness in the dream represented death.

(1913a, p. 295)

After idealising the mother-son relationship, here for the first time Freud clearly addresses the underlying image of the mother as Goddess of Death, an expression of the split in his relationship with her, characterised by emotional coldness and intense eroticisation. Prioritising the latent destructive aspect over the manifest content of the myth – the amorous dilemma, its capricious and fatal nature – Freud could mimic the incestuous nature of his bond with his daughter whom he was tying more closely to him in an attempt to evade death. In June, he wrote to Ferenczi:

So, the motif of the choice between three sisters, the third of whom is mute. With a few associations I came out with the idea that they are the three – sisters of destiny, the Fates, the third of whom is mute, because she – symbolizes death. The compulsion of fate is transformed into the motif of selection. Cordelia, who loves and is silent, is thus actually death. The situation of Lear with Cordelia's corpse in his arms should be reversed, the old man in the arms of the Fate of death. The three Fates are woman in her three principal manifestations: the one who gives birth, the one who gives pleasure, and the one who spoils; or mother, lover, and Mother Earth = death.

(23/06/1912)

We can only imagine how difficult it would have been for Freud at that time, and what a considerable source of envy and jealousy, to identify with the loving desire of his pupils, Ferenczi in particular. . .

For her part, *Annerl*, in her jealousy of Sophie, with whom she quarrelled endlessly, especially after the latter's engagement, had invested her father with her urgent adolescent expectations, trying to find that particular space in his heart where she was certain to have no competitors, no brothers or sisters, and still less her mother: later on, having reached this place, she would defend it at all costs. In the beginning, her true rival was psychoanalysis itself, that all-engrossing work from which *Papa* would emerge punctually at meal times, calling to her with a private signal, "a funny sound, something between a growl and a grunt" addressed only to her, with which he imitated her adolescent noise of protest.[10] Fascinated by what it was that kept him busy in the consulting room with his patients and on Thursday evenings with his colleagues, Anna had begun to read her father's works and to suggest that she become his pupil.

She was a young woman with a lively and acute intelligence, and as early as May 1910, A A Brill, who witnessed an exchange of witticisms between father and daughter over permission to read the newly published "Leonardo," while struck by their excellent relationship, had wondered how a young girl of 15 would be able to take in the complexity of the problems the work dealt with, given the effort that he himself had had to make.[11]

If this were not enough, during that summer, having been left out of her father's holiday with her brothers in the Netherlands, had consoled herself by reading *Delusions and Dreams in Jensen's Gradiva*.[12]

In contrast to his approach with his sisters, Freud did not concern himself with curbing his youngest daughter's curiosity and precocious reading, and now, in the turbulence of the awkward age, she was presenting herself to her father as a patient in need of cure. Prey to dark moods and "bad habits" which left her weary and irritable, she was often immersed in a world of fantasy in which she tried to express her emerging sexuality and explore her own feminine identity. In fact, Anna looked very like her father and seemed to want to present herself as a boy, to conceal her femininity and her attempts at seduction.

Experiences of inadequacy, shame, and compulsion made her feel inferior to Sophie and not very "reasonable."[13] From her holiday in Merano, where she had been encouraged to stay for more than six months, with instructions to put on weight and restore her good humour, far from her sister's wedding celebrations, Anna wrote all too frequently to her father, confiding her discontents, her conflicts with Sophie or her mother, her dislike of her brother-in-law, her unhappiness at not being able to attend his lectures at the university, and telling him about the books or psychoanalytic journals she was reading: "You shouldn't be upset: I'm big now and it is not therefore surprising that they interest me" (31/01/1913).

And in spring 1913, after Sophie had left for Hamburg and Anna had become the only daughter, she succeeded in making a trip to Italy with her father, to Venice, a trip she had dreamed of since she was 13 and had demanded to go with him to America. Her father was more and more proud of her and during the preparations for the holidays he wrote to Ferenczi:

> My next company will be my little daughter, who is now developing so gratifyingly (you have surely long ago guessed this subjective condition of the "choice of caskets").
>
> (9/07/1913)

In the same context, he confessed that he was always opposed to his friend's relationship with Gisella's daughter:

> I never forcefully advised you against Elma only because I was afraid that you would still have wanted to go through with it, following neurotic patterns. . . . What do you want to do now? For each of us fate assumes the form of one (or several) women, and your fate has some rare, precious qualities.

Though in the autumn, it was still Minna who accompanied him to Rome, he now seemed inclined to replace her and Ferenczi with Anna as his travelling companion. Freud was directing towards his third daughter the unconscious Eros diverted from his affective investment in his children and pupils and dormant or restrained in his analytic relations with Elma, Loe, and Sabina.

A decisive role would later be played as intermediary between father and daughter by another female pupil of special fascination and high cultural attainment, Lou Andreas Salomé, who made her entrance onto the scene in Vienna at the end of 1912, regally concluding the series of women who had gone before her.

"If you go to women, don't forget the whip"

> *You have taken away my heart,*
> *my sister, my bride;*
> *you have taken away my heart,*
> *with one look you have taken it,*
> *with one chain of your neck!*

(Song of Solomon)

At the age of 20, Lou Andreas von Salomé, goddess of love, descending from the stars to inspire Friedrich Nietzsche and Paul Rée, muse and lover of Rainer Maria Rilke, was "as shrewd as an eagle and brave as a lion, and yet still a very girlish child."[14] A guest at Weimar in 1911 – in the photo recording the event she is seated in front of Freud and Jung, splendid in spite of her 50 years – Salomé frequented the most prestigious intellectual circles in Europe: friend of the greatest literary figures of the period from Wedekind to Schnitzler, Kraus, von Hofmannsthal, and Strindberg (she had even met Tolstoy), she was herself a writer.

Born in St Petersburg (1861–1937) to German Lutheran parents, the only daughter after five sons, gifted with uncommon intelligence, sharp-wittedness, and sensitivity, from a very early age she had begun to show an interest in the history of religions, philosophy, and philology, engrossed in a personal, pressing, and entirely secular enquiry into the essence of the divine and the human. She first attended the University of Zurich and then made a tour of Europe: at the age of only 20 she began a scandalous but chaste philosophical *ménage à trois* with Friedrich Nietzsche and Paul Rée. The photograph by J Bonnet has gone down in history, showing her driving a carriage drawn by the two men, with reins and whip in her hand. . .

At this point in our journey, in this image of a Holy Trinity orchestrated by Nietzsche and of virginal power, we can detect the courage in a challenge to the masculine assumption about care and control of the female body which revolutionises the representations of the primal scene of care that we have seen developing thus far: from Fliess and Freud's intervention on Emma Eckstein, to the dream of Irma's injection, the operation on Elma Pàlos, and her subsequent analytic treatment by Ferenczi and Freud.

A woman of luciferian fascination, capable of arousing and resisting men's desire, Lou not only won them over with her vivacity and *joie de vivre* but was able to inspire them with her creative freedom. Paul Bjerre promoted the legend that if she formed a passionate attachment to a man, nine months later he gave birth to a book. This happened with *Thus Spoke Zarathustra*, Wedekind's *Lulu*, and the flow of poetic creativity in Rilke.

She herself remained little known as a writer and entirely unknown as a psychoanalyst.[15] Envy and fear of the potent force of her fascination and intelligence, as well as her transgressive defiance (including the unconsummated marriage she contracted in 1887 with the orientalist F C Andreas) meant that she was overshadowed among the pioneers of psychoanalysis, almost an outsider, despite her intimate communion of ideas with the master.

In her dialogue with Freud, Lou Andreas Salomé would be the first psychoanalyst to contradict him and to open up new perspectives on the theorising that had hitherto been based on the psychic centrality of penis envy; she proposed a different conception of femininity and maternal potency as a generative "casket" and not just a symbol of castration or death.

This "beautiful and joyful"[16] woman would propose a conception of the unconscious as a universal and positive womb of life.

It is only astonishing that the meeting between Salomé and Sigmund Freud took so long to come about. Some have speculated that they had met before – for a consultation? – in 1895.[17] She herself gives this impression, describing the first official meeting with Freud almost as a reacquaintance:

> When, returning home from a stay in Sweden, I stood before Freud at the psychoanalytic conference in Weimar in the fall of 1911, he laughed at my impetuous request to study psychoanalysis with him.
>
> (1968, p. 103)

After Weimar, Salomé went to Berlin for a while to study with Abraham, who, in his enthusiasm for his pupil, strongly supported the publication of a paper by her in the *Jahrbuch* of May 1912. It was only in October 1912, after reading all the works of Freud, that this formidable woman arrived in Vienna and spent a term there attending his university lectures. Also being admitted to the Wednesday gatherings, she was present during the period of the splits with Adler and Stekel and witnessed the steady deterioration of the relationship with Jung: we owe to her the most acute intuitions about the pupils' relations with their teacher: Jung, Adler, Tausk, and Ferenczi in particular.[18]

Lou instantly engaged in a relationship of equals with Freud, gaining his admiration and a friendship that would last all his life. In the meantime, she added the most promising, brilliant, and tormented of the Viennese, Victor Tausk, to her list of victims, seducing and abandoning him, but only after they had collaborated on a paper about narcissism for the Munich Congress.[19]

When, in April 1913, that "woman of dangerous intelligence" left Vienna to visit Ferenczi in Budapest, Freud was careful to put the Hungarian on guard: "all the tracks around her go into the lion's den but none come out."[20] Freud may perhaps have been motivated by jealousy, but Lou Salomé never disappointed him, either as a pupil or as a friend. It was a warm and loyal friendship that years later would also embrace Anna Freud: indeed, in 1921, Sigmund would ask Lou to return to Vienna as a guest in his house, to play a maternal role for Anna alongside him.[21] But in the winter of 1912, Anna was stuck in Merano and had no opportunity to meet her.

At school with Freud: the transmission of psychoanalysis

It cannot be repeated too often that psychoanalysis did not *discover* the unconscious[22] but constitutes an entirely original method *invented* by Freud for exploring it, for understanding neurotic symptoms and the maturational development of the individual: a unique method built on the person, on his or her dreams and symptoms – "by its own nature, so to speak, through its own deficiencies."[23] He tried to explain this to Jung, who was complaining about some reticence and incompletely analysed material in *The Interpretation of Dreams*: when Freud tried to make the book less personal, a text about the symbols of dreams which would have tamed the strangeness and singularity of unconscious experience, he was instantly obliged to give up. Again in order to modify its subjective, linguistically German, status for a period, this untranslatable text welcomed additions by other authors (examples of dreams in English by Brill and some chapters by Rank) until it almost took the form of a collective text, but in vain.[24]

The Interpretation of Dreams founded the psychoanalytic method but at the same time gave a structure to the social formation of the movement.[25] It was clear from the start that learning psychoanalysis could not happen through a simple intellectual apprenticeship or through books, but needed an evident personal experience of the manifestations and functioning of the unconscious processes. For this, a formative transferential, flesh-and-blood contact with the author was indispensable: today we understand how psychoanalysis is not solely the product of Freud's psyche but also of his body, of his whole unique existence.

In the encounter, Freud offered his theory and his mode of interpretation, together with all the rich psychic material produced by his resistances which, repressed or dissociated, were yet to worked through, and constituted the unconscious hinterland, the matrix of his thinking. The pupil's relationship with his teacher facilitated other experiences and constructions. Psychoanalysis soon demonstrated itself to be a theorisation *in itinere*, a discipline endowed with a constitution that can never be said to be fully acquired, nor fixed, but which lives in dynamic equilibrium with the person (the transference and the resistances) who is interpreting it and which is thus susceptible to being renewed but also attacked, confused, or diluted by other theories or constructions sustained by more primitive and destructive defensive motivations: which Freud inevitably experienced as wiping out not only his own paternity but himself too.

From Anna O to the first female psychoanalyst, Emma Eckstein, the (female) patients had heroically collaborated in constructing the method with their symptoms and dreams, and it should be no surprise that Freud's first Viennese pupils had begun by being in treatment with him. Later on, every aspiring analyst would inevitably find himself playing the not always anticipated or comfortable role of patient. It is this particular mode of transmitting knowledge, *conscious and unconscious,* from one generation to another that makes psychoanalysis unique among the sciences.

It came about that anyone interested in learning the new art began a correspondence with Freud or came to Vienna, to the workshop of the master craftsman. When the professor wasted no time in getting the visitor to recount his dreams, interpreting them and their unconscious desires in the light of the associations brought to them – the method of free association being the first and fundamental rule of learning and observing – he was simply responding, generously, to the request of whoever wanted to be introduced to the new technique.

In *Recommendations to Physicians Practising Psycho-Analysis,* Freud writes:

> Some years ago I gave as an answer to the question of how one can become an analyst: "By analysing one's own dreams." This preparation is no doubt enough for many people, but not for everyone who wishes to learn analysis. Nor can everyone succeed in interpreting his own dreams without outside help.
>
> (1912, p. 116)

These first exercises in analysis with Freud, improvised and extemporaneous, by letter, on a walk, over lunch, or on a long journey, which occurred over quite brief spaces of time and were intended to prompt a process of *self-analysis*, rapidly set up a transference relationship within a set of arrangements which certainly allowed Freud to catch the resistances and the transference onto his own person but not always to manage them appropriately and still less to recognise the countertransferential elements in play or to address the negative transference. All these aspects of the treatment – from the actual meaning of the transference manifestations to the realm of the countertransference, to developing the rules of the setting – were at work. The fact that in those early years Freud misleadingly termed the analyst's formative process "self-analysis"[26] indicates how the complex question of the personal and training analysis was yet to be defined, along with the control analysis and the teaching that would later establish the organisation of the transmission of knowledge, and power, in the tradition of the new psychoanalytic movement itself.

Difficulties and resistances soon made themselves evident on both sides, the teacher and his pupils, precisely because the transference was actively participating in the process of learning and filiation in the psychoanalytic Saga that was being constructed: a transference that was hard to govern, all the more so for the pioneers.

The desire to disseminate psychoanalysis and the readiness to teach its technique put Freud in the challenging, asymmetrical position of sustaining a paternal transference, of representing an authority with a primal power of a kind to pro-voke inevitable admiration, envy, dependency, jealousy, and rivalry in his pupils, together with reactions of idealisation, imitation, rebellion, schism, and parricide. This type of transference turned out to be even more demanding to deal with than the erotic investment of his patients.

Within the first Viennese group who gathered around Freud, for all the sup-posed "intellectual communism" which inspired the Wednesday evenings, it was clear that a very thin wall divided the room overlooking Berggasse, where the meetings were held, from the consulting room next door.[27] Although the order of the contributions was regulated by the drawing of lots, the pupils were waiting for Freud's conclusions, anxiously receiving his judgement or a sign of attention, just as they would later look for the quotation they longed for in his writings. That small and intimate band of acolytes soon found itself involved in complex dynamics in relation to their leader and in heated sibling conflicts, punctuated by the rhythmic weekly encounter in that first group setting.

The habits of the *Burghölzli* community were quite different. Ever since the time of Forel, a kind of collective interpretation had been established, a practice of mutual self-observation in a group of peers who avoided noting the evident dis-parities between its members. From the senior doctor to the assistants, the nurses, the doctors' wives, they were all invited to associate freely in a kind of ecumen-ism, except for the fact that, in the context of a community spirit of austere moral-ism, the Freudian method of *free association* could not flourish.

It was the method that gave rise to all the difficulties for the Swiss.

Bleuler had had to acknowledge his own incapacity for free association and also recognised the influence of his authoritative presence on the rest of the group's ability to be genuinely open about their thoughts and achieve fully detailed inter-pretations of their dreams. On the occasions when he tried to facilitate the group's analytic work by finding a pretext for leaving the room, he lamented the fact that the interpretations of his own dreams were adversely affected by his wife's uncon-scious complexes. . . .[28]

Even in the epistolary communication which Bleuler very soon began with Freud – an attempt at analysis by correspondence – his difficulties in freely expressing his thoughts continued to be insurmountable.[29] The use of a typewriter, his indispensable companion that he even kept at his bedside when he was ill, although it still gave rise to slips and symptomatic mistypings – which he tried to analyse – in fact allowed him to entrench his violent instinctual reality even more deeply behind his impenetrable ethical armour. The failure of this analysis would be the basis of his withdrawal from the movement.[30]

At Bleuler's urging, Jung had initially concentrated on a mechanical asso-ciative method suited to the *Burghölzli* style: by studying the automatic responses of his patient groups with the galvanometer test, he identified their repressed representations, their *complexes*, reaching them in an impersonal manner. In this way, he avoided becoming aware of the repressed sexual quality of what he called the *personal* factors of the treatment: bypassing the oedipal conflicts, he suggested that Freud emulate the community of equals he experienced with Bleuler.

In spite of his expectations, mutual and democratic analysis found no place in Freudian psychoanalysis, which was founded on the distinguishing of boundaries and on asymmetries, valued the differences of gender and generation, and relied on the transference to set the emancipatory process in motion.

This was how, in his attempt to deny Freud's paternity, his transferential position as the pioneer and teacher, as well as his own responsibility to his patients, Jung began to accuse Freud of infantilising and pathologising his pupils. And similarly, Jung's colleague Alphonse E Maeder (1882–1971), co-opted to lead the Zurich group when Jung took on the presidency of the IPA, claimed that the dependant relationship of the Viennese to Freud was a local Jewish characteristic which the Christian elements of Swiss culture would be able to mitigate.[31]

Freud's rejection of Jung's ideas was inevitable, since the latter's theory was not developing a recognised legacy but, by avoiding the formative process of analysis, tended to disavow the method and legitimise Jung's own paternity. Adler, Stekel, and in 1925 Rank, too, were runaway children taking a bit of the paternal inheritance with them, and they could always be imagined making a prodigal's return, as in the parable: they had isolated and decontextualised certain strands of the Freudian discourse.[32]

Freud would justly compare Jungian psychoanalysis to the famous "knife of Lichtenberg," one that has "changed the hilt" and inserted "a new blade."[33] Jung represented the most expert and thoughtful attempt to relativise and reabsorb the revolutionary significance of his work and the unconscious into the previous cultural tradition, treating the libido as "a piece to be inserted into a fascinating game of potentially endless conceptual assonances."[34]

These illegitimate filiations and Jung's metamorphoses must be distinguished from the later conflict with Ferenczi, in which we can acknowledge Freud's defensive retrenchment and a certain orthodoxy in relation to the development which the Hungarian was able to imprint onto psychoanalysis in fully authentic fidelity to the tradition. In his case, there was no split: Ferenczi was only suppressed because he was ahead of his time.[35]

Notes

1 Corsa, 2010. Freud will quote her about the death instinct (1920b).
2 1995, p. 1121.
3 Corsa, 2010.
4 Carotenuto, 1980.
5 20/01/1913; 28/08/1913 and 29/09/1913, in Carotenuto, 1980.
6 *Dante's lyric poetry* trans. K. Foster and P. Boyde, O.U.P., 1967.
7 Lopez, 1986.
8 Freud to Abraham, 9/01/1908.
9 Lopez and Zorzi, 1999, p. 46.

10 Sachs, 1945, p. 79. Blanton reported that at times Freud also did it in analysis, indicating that he agreed or sympathized, and Cremerius adds a comment on the regressive danger of this communication based on sounds and noises (1985).
11 Brill, 1940.
12 Freud, 1907.
13 Young-Bruehl, 1988, p. 277.
14 Nietzsche to Gast, 13/07/1882.
15 In 1981 Giuseppe Sinopoli wrote the opera *Lou Salomé* about her.
16 Magris, 1980.
17 Pfeiffer, 1972.
18 1958.
19 Roazen, 1969.
20 Freud to Ferenczi 31/10/1912 and 20/03/1913.
21 See Pierri, 2022.
22 Brabant, 2011.
23 Freud to Jung, 17/02/1911.
24 Marinelli and Mayer, 2002.
25 Freud to Jung, 17/02/1911.
26 1926, p 199.
27 Napolitano, 1999.
28 Marinelli and Mayer, 2002.
29 Marinelli and Mayer, 2002; Falzeder, 2005.
30 Falzeder and Burnham, 2007.
31 Marinelli and Mayer, 2002.
32 Fachinelli, 1974.
33 Freud, 1916, p. 454.
34 Fachinelli, 1974, p. 98.
35 Brabant, 2011.

Chapter 17

The secret committee

The transformations and the desertion of Jung

We had left Jung on his return from the expedition to America in 1909, struggling with the tempestuous work of self-analysis provoked by the crisis with Sabina Spielrein and his contact with Freud. For two months, he had not even been able to write. He disguised his true feelings with Freud and played the part of the disciple, brewing his resentment and a secret feeling of superiority which was confirmed by his having so easily gained the position of president.

While Freud and Ferenczi continued their adventures in the exploration of clairvoyance and prophecies – and Jones did his own reflecting on nightmare and superstition – Jung plunged into fantasies about occultism, mysticism, and alchemy. Questioning the sexual nature of *libido*, he spent all his free time with books of mythology and ancient history, looking for confirmations of the impersonal character of the unconscious and pursuing the phylogenetic roots of dream-symbols and psychopathological productions in the primal experience of humanity.

Besides his destructive attack on Freud, a by no means secondary motive for his withdrawal was the need to deny the intensity of the incestuous element still alive and current in relation to Sabina Spielrein, who had got back in contact with him and was busily preparing her graduation thesis.

In works of archaeology, and especially in the psychology of primitive peoples, Jung was looking for support and an indirect route towards self-understanding. It was during this period that he recalled the mythological character of the medium Hélène Smith's fantasies, studied by Flournoy in *Des Indes à la Planète Mars*, a book he had read with passion as soon as he arrived at the *Burghölzli* and had offered to translate. During her trances, the medium proclaimed that she was reliving past lives, including that of a fifteenth-century Indian queen and a noblewoman of Mars: this book, more than any other, legitimised spiritualist phenomena as a territory for effective psychological research.

Approaching Flournoy again, Jung began a friendship with him that formed a sort of counterweight to his relationship with Freud and avoided generational conflict. A cultivated and distinguished personality with a refined education,

DOI: 10.4324/9781003246473-18

a sound spiritual equilibrium, and a clear sense of proportion, a "revered and fatherly friend,"[1] the professor from Geneva was not a physician and represented a neutral figure with whom to speculate about philosophy and the psychology of religions as well as somnambulism and parapsychology, and Jung started organising séances again, involving Bleuler,[2] too, without having to take into account the reality of the body and the sexual transference or finding himself in a position of dependency. In September 1913, he would ask Flournoy to attend the Munich Congress as his supporter.

The re-reading of Flournoy's book and Hélène Smith's own memoirs went on to form the most significant part of the book that Jung was starting to write, *Transformations and Symbols of the Libido*, an arduously composed work, the first volume of which came out in 1911 and the second a year and a half later, marking the definitive break with Freud. Reversing the Freudian perspective, Jung used the symbols of mythology and astrology, and popular psychology, to explain the manifestations of the individual psyche: regarding its archaic features as valuable, he interpreted the content of psychotic delusions and also the visionary creations of the medium as the fruit of a prophetic presentiment of the tasks faced by humankind. This mystical vision was compatible with the emphasis on the creative faculties he attributed to the Preconscious (whose "lower" half recalled the past, while the "higher" half anticipated the future) and enabled him to take a stand on the question of the Christian religion's primacy and on phenomena such as prophetic dreams.[3]

The Freudian concept of sexual libido was uprooted from the body and transformed into a philosophical spiritual energy, the unconscious lost its connotation of repression, symbols figured in it as the very constituents of the psyche, and the function of dream was directed towards presentiment and preparation for the future. The new theory, shared by a large part of the Swiss group, bore no relationship to psychoanalysis, and its cultural operation revealed itself as "divergent, dismembered, indefinable" like the mythic figure that Jung would construct for himself in European culture.[4]

Even since autumn 1911, Emma Jung had noted with concern the divergences present in what her husband was writing and feared the crisis to come.

Shortly after the Weimar Congress, she approached Freud, alarmed also by the echoes from the recent break with Adler and Bleuler's growing distance.[5] On a private level, jealousy and anxieties about the fate of her marriage were prompting Jung's wife to ask the professor secretly for help, perhaps encouraged by some of his confidences about the unsatisfactory nature of his own family life. "Naturally the women are all in love with him,"[6] she lamented to Freud about her husband, and her fears were not groundless. Jung was not only persisting in his ambiguous contacts with Spielrein but had just taken into analysis the young Antonia Wolff – "a remarkable intellect with an excellent feeling for religion and philosophy"[7] – later bringing her with him to Weimar. "Tony" soon became his close collaborator, assisting – and analysing – him during the long period of psychic disarray that followed his break with Freud, and finally appearing beside his wife as his official companion.[8]

Despite Emma's pleas, and even though he intuited his pupil's persistent difficulties, Freud went on hoping for a workable solution and waited until Jung
had completed the second part of *Transformations and Symbols of the Libido*
before pronouncing judgement. In the meantime, he had set to work on his own
area of research, the origin of culture and the history of religions, convinced of
finding evidence to support an integration with his own theory, or at least some
convergence.

With painful slowness, the irreconcilability of their mutual expectations became
apparent.

In their steadily less frequent correspondence, Jung's rebellious urge for
freedom was becoming ever more obvious: he was attributing a different meaning to the incest prohibition or taboo and declared that he wanted to make
substantial modifications to the concept of libido, convinced that in theorising
the presence of incestuous desire, Freud had run into an error as bad as when,
at the start, he had been led to believe in the factual reality of every recollection of abuse.

An incestuous phase was inconceivable for the Swiss, and in support of his
thesis, he criticised Freud's claims in "The Horror of Incest,"[9] crudely challenging him for thinking that children, once adult, could have the slightest nostalgia
for a mother no longer young, "with her sagging belly and varicose veins."[10]
After reading Rank's book on the same subject (1912), he loftily pressed the
point: "people without parents would have no chance to develop an incest
complex."[11]

It was the very foundation of psychoanalysis that Jung could not share: the
unconscious transference which keeps alive the figure of the mother from childhood and causes her to be rediscovered later in the sister, the wife, the daughter,
and, naturally, in young patients. He set out from the conviction that the incest
prohibition originally had nothing to do with desire and the practical necessity of
a limit as a step in the oedipal crisis of transition to the symbolic but was the product of a more elevated and spiritual teleological demand to strengthen the social
foundation of the family, a foundation which for Freud was instead the laborious
inner outcome of the prohibition.

Jung was running away from maternal carnality and contrasting it with the
mother's archetypical significance as a symbol of nature and origins: from his
viewpoint incestuous desire was a thirst for rebirth, a spiritual initiation that did
not involve the bodily and sexual concreteness of the Freudian interpretation. Trying to reach the symbolic level directly, skipping the passage through the Oedipus
complex and its conflicts, Jung lost all contact with the reality of the body. His
"collective unconscious" was reflected in an "individual structure given from the
outset and the same for everyone."[12]

Freud, who, in attempting to understand the birth of culture and religion, was
applying the same method to the psychology of peoples that he had applied to the
individual psyche, was not yet clear about the basis on which the Swiss intended
to build his modified concept of libido and was awaiting the big news which Jung

had announced that he would officially present in New York, in September 1912. In the meantime Freud wrote:

> we do not share the belief of some investigators that myths were read in the heavens and brought down to earth; we are more inclined to judge . . . that they were projected on to the heavens after having arisen elsewhere under purely human conditions.
>
> (1913a)

Jung's Fordham University conferences, a commitment which had caused the postponement of the IPA Congress that summer, were given the provocative title *Theory of Psychoanalysis*. He used them to transmute the entire Freudian edifice: in his thinking, infantile sexuality, the unconscious, repression, the theory of dream as wish-fulfilment, and of the neuroses had partly vanished, partly been reduced to insignificance: "To me the sexual theory was just as occult, that is to say, just as unproven an hypothesis, as many other speculative views," he wrote in his autobiography.[13]

By approaching infantile sexuality and the unconscious by means of spiritual criteria in his attempt to emancipate them from the reality of the flesh, Jung was reducing psychoanalysis, as he had done in the past with philosophy, to an *ancilla theologiae*.[14]

A missed meeting: the "Kreuzlingen gesture"

In the spring of 1912, while the analyses of Elma Pàlos and later of Loe Kann were still in progress and the events were taking shape that would, in a different but equally intimate and indissoluble way, entwine the destinies of Ferenczi and Jones with Freud's, Jung's estrangement reached the point of no return. And Freud began to resent the fact that his president was showing little interest in the politics of the Association, failed to attend the meetings of the executive (even going so far as to postpone the Congress), and buried himself completely in his astrological research, outwardly manifesting towards Freud the violence being hidden under his mysticism. Freud thought that what lay behind all this was the influence of Jung's relationship with Wolff.[15]

At the height of the epistolary confrontation between the two men, when the letters they were exchanging from their different theoretical standpoints still respected the amenities but contained hammer blows that often resulted in a delayed response as the impact was felt, a misunderstanding occurred which took the form of a missed meeting.

With the "compliance of chance,"[16] Freud's visit to Kreuzlingen in May 1912 was interpreted by Jung as a personal affront.

It had not been a holiday: the professor's brief trip to Switzerland, "one of the most remarkable demonstrations of personal friendship" Binswanger had ever experienced,[17] had been motivated by the pupil's illness, his having had an operation on a testicular sarcoma in March. With only a few days at his disposal, Freud had not planned a stay in Zurich, but, imagining that he would have an opportunity to meet Jung too, he had let him know in advance; when Jung neither appeared nor telephoned, Freud attributed the fact to the cooling of their relationship and took no further action. Jung had spent the whole weekend sailing on the lake and only read the note on the Monday. Thinking it had come in that morning's post and that Freud had deliberately sent it late, he had given free rein to his jealousy and his fantasies of exclusion and conspiracy. He was looking for a good excuse: the misunderstanding facilitated his emotional incomprehension.

It is fascinating that this missed appointment which opened hostilities found a witness in Binswanger, who had also been a spectator at the first, exciting encounter between the two men: then he had been a follower of Jung, now a close friend of Freud – to whom he had confided his health problems, and to no one else – and a possible rival, having begun to take the reins of the Swiss group.[18] In the subsequent correspondence, Jung sarcastically alluded several times to the "Kreuzlingen gesture":

> Your Kreuzlingen gesture has dealt me a lasting wound. I prefer a direct confrontation. With me it is not a question of caprice but of fighting for what I hold to be true. In this matter no personal regard for you can restrain me.
>
> (Jung to Freud, 11/11/1912)

Abandoning his usual courtesy, Jung declared himself to be offended and his correspondent, irritated in turn, did not bother to try to understand what he meant. Freud reacted by ceasing to call him *Lieber Freund*, and instead using the more formal *Lieber Herr Doktor*. The incident was only cleared up after Jung's return from the United States, at a meeting of the presidents of the psychoanalytic societies in Munich that autumn, but the relationship was beyond repair. Jones writes:

> Freud and Jung then took a walk together for the two hours before lunch. This was the opportunity to find out about the mysterious "gesture of Kreuzlingen." Jung explained that he had not been able to overcome his resentment at Freud's notifying him of his visit there in May two days late. . . . Freud agreed that this would have been a low action on his part, but was sure he had posted the two letters, to Binswanger and Jung, at the same time on the Thursday before. Then Jung suddenly remembered that he had been away for two days on that week-end. Freud naturally asked him why he had not looked at the postmark or asked his wife when the letter had arrived before levelling his reproaches; his resentment must evidently come from another source and he had snatched at a thin excuse to justify it.
>
> (*II*, p. 164)

It was shortly after the mutual explanations that Freud had his second fainting fit in Jung's presence. Ferenczi, who for some time had been trying to dissuade

Freud from "the frantic effort to appoint a personal successor," had far-sightedly wondered if the malaise at Bremen "wouldn't perhaps be repeated in Munich."[19]

The decisive emotional situation once again came about at lunch, at the culmination of a discussion. Freud was a bit over-excited: after complaining that the Swiss group no longer cited him in their articles, he launched into a dispute with Jung over the origin of religion, taking as his starting point Abraham's essay on the cult of the god Aten in ancient Egypt.[20]

This was a current topic because the sculpted head of the Pharoah Amenophis IV-Akenaton, discovered at Tell el-Amarna in the digs funded by the Jewish patron James Simon,[21] had recently been exhibited in Berlin. The professor countered Jung's enthusiasm for the heretical Pharoah's revolutionary achievement in founding the first monotheistic religion by stressing that he had erased his father's name from all the monuments.[22]

In this discussion at the Park Hotel in Munich, in the narcissistic dynamic of his twin roles, the warrior son and "Conquistador" that Jung personified brought Freud face to face with his own violent and ambitious attack on his father. After the temporary success of the walk in resolving Jung's persecutory-victimising mistake, the deep psychology of Freud's faint gave a vivid representation of the aggression at work – "defeat of the father/rival and fainting as simulated death" – and expressed its convergent libidinal components: Freud "fainted like a hysterical woman, and fell like a king struck down in battle."[23] The fact that Jung took him in his arms and gently laid him in a chair was the finishing touch to the scene. On waking, the look Freud cast at Jung and his sigh – "How sweet it must be to die" – were justly experienced by the Swiss as those of a child looking for his father.[24]

A few months after the episode, the epistolary relations between the two men would be limited to professional matters. In December, the clash became overt and no holds were barred. Jung reacted to the weakness displayed by Freud and insolently challenged him:

> your technique of treating your pupils like patients is a *blunder*. In that way you produce either slavish sons or impudent puppies. . . . Meanwhile you remain on top as the father, sitting pretty. For sheer obsequiousness nobody dares to pluck the prophet by the beard.
>
> (18/12/1912)

Freud had no option but to leave Jung to his own devices:

> I propose that we abandon our personal relations entirely. . . . you have everything to gain, in view of the remark you recently made in Munich, to the effect that an intimate relationship with a man inhibited your scientific freedom.
>
> (3/01/1913)

This time, however, the repetition of his faint[25] called Freud's attention to his own traumatic features and the significance of early bereavements: "in my case it was a brother who died very young, when I was a little more than a year old," he confided to Ferenczi 9/11/1912.

The Committee: the *Männerbund*, and the defence of the "Cause" (*Die Sache*)

During the summer of 1912, Jones, who had spent part of the spring in Vienna with Loe and stayed there after the professor had gone away on holiday, met Ferenczi to discuss the new turn in the situation, both of them worried by Jung's behaviour.

It was then that the idea came up of building a kind of "Old Guard around Freud."[26] With one of Ferenczi's many intuitions as his starting point, Jones put this plan into shape and wrote about it to Freud on 30/07/1912:

> Ferenczi, Rank and I had a little talk on these general matters in Vienna. . . . One of them, I think it was Ferenczi, expressed the wish that a small group of men could be thoroughly analysed by you, so that they could represent the pure theory unadulterated by personal complexes.

Thinking that circumstances were in his favour and sparing no one, Jones expressed the desire that things at the top of the movement might function more satisfactorily: he regretted Jung's inexplicable conduct – "he has the game in his own hands, and refuses to play it" – the difficulties in keeping Ferenczi on side, given his now open interest in telepathy –"he is running a big risk with the *Gedankenübertragung* – true or false –" and Rank too, because of his financial situation, "unless something can be done for him."

He came to the point and proposed setting up an unofficial group of devoted analysts (presenting himself as a prime example) capable of faithfully transmitting and developing the Freudian theory.[27] Freud replied immediately from Karlsbad, enthusiastic about a secret committee to defend *Die Sache* when he was no more, and suggested adding not only Rank, who had already been co-opted, but also Sachs and Abraham:

> What took hold of my imagination immediately is your idea of a secret council composed of the best and most trustworthy among our men to take care of the further development of ΨA and defend the cause against personalities and accidents when I am no more. . . . This committee had to be strictly secret in his existence and his actions.
>
> (Freud to Jones, 1/08/1912)

In its final form, the plan included Ferenczi's proposal that a few people should be analysed by Freud in order to serve as representatives for the teaching of beginners.

The Committee, as this trusted group was called, met in its complete form for the first time on 25 May the following year. It was then that Freud, to symbolise the closeness of the bond, gave each member an antique stone engraved to resemble a ring of his own. It was an especially significant ring stamped with the head of Jupiter which he brought to his lips when he was thinking or stared at, slowly opening and closing his fingers in the slow, precise gestures that accompanied his explanations.[28] There was also an exchange of photographs, and the portraits of the ultra-faithful made their appearance on the walls of the waiting room in Berggasse.

This small united body of "Paladins"[29] devoted to protecting the Freudian *corpus* from the dangers of further schisms and distortions was the movement's first private institution, an element with the ability to set the rules and create the tradition, marking the first generational transition.[30]

In evaluating the significance of the ongoing crisis with Jung, the Committee also found itself faced with the task of solving the problem of succession, a question that increasingly concerned the way to transmit knowledge and skills to new analysts in place of an investiture by Freud. The IPA would be transformed from a simple scientific organisation into a professional body with the goal, among others, of training new analysts: membership of one of its societies would be the conclusion of a maturational journey and a kind of diploma. The rules for admitting new members would in due course be discussed by the Committee before being introduced into the discussions of the International Assembly: in this way, the first steps were taken towards building the identity which the Association took on in the twenties.

During the war, Anton von Freund joined the Committee, but after only a year illness forced him to give way to Max Eitingon, a Russian pupil who had settled in Berlin and founded the German society with Abraham and who also took over von Freund's role as a generous patron: in the post-war period, Eitingon endowed the movement with part of the immense fortune gained from the importation of furs which his father, an émigré from Belarus to Leipzig, had skilfully built up and which his brother and cousin had successfully extended to New York.[31]

Among the Committee members Sachs, Rank and von Freund were "laymen," while Ferenczi, Jones, and Abraham, like Eitingon, were physicians, all with experience at the *Burghölzli* before coming to Vienna. Some were informally analysed by Freud: Eitingon back in 1909, while Rank and Sachs were effectively in continual analytic contact with him, and von Freund joined the circle as a patient.[32]

The immediate outcome of setting up the Committee was Jones's training analysis with Ferenczi in Budapest during summer 1913. Specifically so that he could analyse Jones, Freud asked Ferenczi to wait for a formal analysis with him, a project which the Hungarian had first considered in December 1912 and had already postponed once that same May.

The plan for summer 1914 was that Ferenczi and Rank should go to London, where Jones would analyse Rank and would take the opportunity for a further period of analysis with Ferenczi in return for the hospitality being offered. This was all cancelled on the outbreak of war: Sachs was the only Committee member who made the journey to London in May and so was able to deepen his friendship with Jones.[33] Ferenczi made use of his enforced immobility to complete his own analysis with Freud, which continued in fits and starts during the course of the conflict. Rank's analysis could not happen and, according to Jones, if it had been possible, it might have prevented the troubled departure of the Committee's youngest member, who suffered from periodic depressions.[34]

By contrast, we do not know what the Welshman thought of missing a second period of analysis with Ferenczi, but very soon after the war, his intense conflict, not only with Rank, but also with his former analyst, came to the fore.

Not forgetting Abraham: during the moments of crisis in the group clustered around Jones and Sachs, he was unanimously considered constitutionally healthy and did not subject himself to analysis. Rare, highly normal individuals like Abraham, claimed Freud, can do without it and yet work successfully as psychoanalysts.[35] The German analyst was extremely intelligent and cultivated, with a distinctive scientific clear-sightedness though with little capacity for empathy, which made him detached or even, as Jung complained, downright unpleasant. He knew how to keep the right distance from Freud,[36] and their alliance remained stable, as was his affective and family life. With a strong background in psychiatry, Abraham devoted himself to therapy and the psychoanalytic understanding of the psychoses and the pregenital stages of libidinal development. Elected president in 1924, he soon began to show symptoms of what is today thought to have been lung cancer. In these circumstances, he bizarrely entrusted himself to the care of Wilhelm Fliess and his theories about biological rhythms: Abraham died on 25 December 1925 and was the second such loss to the Committee, after Anton von Freund.

Totem and taboo: unconscious intelligence and intergenerational transmission of thought

The analytic treatments of Sabina, Elma, and Loe had a great influence on Freud's relations with Jung, Ferenczi, and Jones. The Master found himself separating the women from his pupils and taking them transferentially for himself, like the ancestor-father of the "primal horde" that he was portraying in *Totem and Taboo*: "a violent and jealous father who keeps all the females for himself and drives away his sons as they grow up."[37] This is a precursor of the father for whom incest was not prohibited and who did not have a clearly generative role: for this to be acknowledged, and for the father who is fundamental to the birth of culture and religion to be internalised he must be killed, as represented in Greek tragedy by *Oedipus Rex*. Freud's pupils had to challenge his paternity symbolically in order to construct the psychoanalytic tradition in that first transition between generations.

Between sessions during Ferenczi's analysis of Jones in the summer of 1913, the two of them were reading and discussing the proofs of *Totem and Taboo* which Freud was sending in advance to the Hungarian. Alongside comments and indiscretions about the ongoing analyses (Jones's in Budapest but also Loe Kann's in Vienna), the letters exchanged between the three men were full of reflections about the new theoretical achievements.

Freud was trying to theoretically fine-tune not only the roots of social organisation and the transmission of knowledge but also the parental conflict he was living through, the violence of the clash, the topic of parricide, and the more shadowy one of filicide. After "The Horror of Incest," in the spring of 1912, he had published "Taboo and Emotional Ambivalence" and in early 1913 "Animism, Magic and the Omnipotence of Thoughts." His journey was taking him a long way from the territories dear to Jung. In the spring, as he was ending his epistolary relationship with the Swiss, he was concluding the essay with the fourth and most important chapter, "The Return of Totemism in Childhood," which gave the oedipal conflict, the central hallmark of his theory, its worthy place as the fulcrum of individual and social culture, able to link dream with myth. He proposed the hypothesis, "which has such a monstrous air,"[38] that the tyrannical father, the chief of the horde, had been overwhelmed, killed and devoured by a combination of his exiled sons.

> The violent primal father had doubtless been the feared and envied model of each one of the company of brothers: and in the act of devouring him they accomplished their identification with him, and each one of them acquired a portion of his strength. The totem meal, which is perhaps mankind's earliest festival, would thus be a repetition and a commemoration of this memorable and criminal deed, which was the beginning of so many things – of social organization, of moral restrictions and of religion.
>
> (p. 142)

The working-through in mourning of this violent primal act invests the father with his function and initiates society, community, and culture: after the parricide, the chief of the horde changes from being the one who sexually possessed all the women into the one who has generated.

Before the parricide – Freud had already written to Jung – the father is only "one who possesses a mother sexually (and the children as property). The fact of having been engendered by a father has, after all, no psychological significance for a child" (14/05/1912). With this recognition, the bond of kinship is also established between the members of the family.

What Freud is recounting is the birth of the psyche and its development from the primitive magical world to the construction of law which, with the prohibition of incest, establishes the difference between innocence and guilt, giving rise to the social norm and the internal barrier of repression: here he acknowledges the presence not just of a child but of a savage in the depths of every civilised man.

As Greek tragedy had been able to represent, the criminal barbarian who is Oedipus at the start of his journey – Oedipus, let's remember, who is the subject of a debate over whether he is innocent or guilty of having killed his fleshly father and king over precedence, not an insignificant and bullying old man at a crossroads – had to face the recognition that the old man who was blocking his way was not just his father *in fact* but was representing him *symbolically*.

There is relevance for thought transmission in Freud's claims about the tradition of knowledge, fantasies, prohibitions, and unconscious feelings of guilt that is passed between generations. In order to explain the characteristic evolution of the human species, he does not regard manifest communication via oral or written tradition as sufficient, nor the simple imitative learning of usages and customs, but proposes the hypothesis of an intergenerational continuity based on unconscious transmission. He writes

> psycho-analysis has shown us that everyone possesses in his unconscious mental activity an apparatus which enables him to interpret other people's reactions, that is, to undo the distortions which other people have imposed on the expression of their feelings. An unconscious understanding such as this of all the customs, ceremonies and dogmas left behind by the original relation to the father may have made it possible for later generations to take over their heritage of emotion.
>
> (*ibid.*, p. 159)

The psychic processes of one generation, though repressed, would leave a trace which the next generation would be able to interpret without having to acquire their own attitude to existence *ex novo*.

When Freud presents the hypothesis that the father-complex is at the origin of every civilisation and at the basis of the constitution of emotional ambivalence in the proper sense (i.e., of the coexistence of love and hate towards the same object), he also mentions the fantasies of the parents, specifying, though only in a note, that it is not just parents' father-complex but "more correctly, their parental complex" (*ibid.*, p. 157, n. 1). However, Freud does not concern himself with the parental side – the one that describes a tyrannical and violent leader or tells of Laius and Jocasta abandoning their son to death: the meaning of filicide remains in the background of human origins behind parricide, the threat of which he feels all too soon coming from his pupils.

From the first self-analytical formulation that Freud had presented about "typical" dreams (1900), the Oedipus complex is adopted as the generative material of psychic life, vital and enigmatic, that each generation inherits from the previous one, reinterpreted and reacquired, before being transmitted to the next. Part legacy and part initiation, the passage between generations is recognised as the risky place where the parents' errors and guilts, and their unconscious expectations, whatever has not achieved adequate cultural interpretation, falls to the children, from generation to generation, as a traumatic and precious patrimony.

It is about the transmission of an inheritance of unconscious thought, which is not passed by words, cannot be taught and which must be conquered, not simply received, as in the verses of the poet so loved and often quoted by Freud:

> *Was du ererbt von deinen Vätern hast, Erwirb es, um es zu besitzen.*
> (Goethe, *Faust*, Part I, scene 1)[39]

> *Das Beste, was du wissen kannst, Darfst du den Buben doch nicht sagen.*
> (*Faust*, Part I, scene 4)[40]

It is the point of conversion which Ferenczi compared to the *hypomochlion*, the fulcrum of a lever, where the foetus begins its rotation ready for birth, on the basis of which "one can unravel all the secrets of the soul." And later, in his enthusiastic comment on the manuscript of *Totem and Taboo*, he adds: "I find new and outstanding the idea of transmission by means of unconscious understanding."[41]

Notes

1 Jung, 1961, p. 162.
2 Fodor, 1971, p. 195.
3 Jung, 1912.
4 Fachinelli, 1974, p. 96.
5 30/10/1911 and 30/11/1911.
6 Carotenuto, 1980, p. 154.
7 Jung to Freud, 29/08/1911.
8 Kress-Rosen, 1993.
9 First short chapter of Totem and Taboo, 1912–13, *Imago*, 1912.
10 Jung to Freud, 8/05/1912.
11 Jung to Freud 2/08/1912.
12 Fachinelli, 1974, p. 101.
13 1961, p. 151.
14 Abraham, 1914.
15 Freud to Binswanger, 14/04/1912.
16 See Freud to Jung, 6/04/1909.
17 Binswanger to Martha and Anna Freud, 2/09/1939.
18 Kerr, 1993.
19 Ferenczi to Freud, 6/08/1912 and 28/11/1912.
20 Abraham, 1912.
21 The splendid head of Nefertiti, his wife, had also been found, but it was not exhibited immediately for fear that the Egyptians might claim possession of it (Illies, 1913).
22 Akhenaton, the pharaoh's younger son, who ascended the throne on the death of the eldest, also supported the abolition of magic and spells and the introduction of monogamy. His rebellion against polytheism was doomed to be forgotten, as his successor Tutankhamun abolished the new religion. The theme stayed in Freud's mind, and he returned to it with the hypothesis that the cult of Aten could be at the origin of Judaism (1934–1939).
23 Lopez, 1991, pp. 133–134 e 135.
24 Jones, 1953, p. 348; Jung, 1961, p. 194.
25 There was also a third one, the first in the series, during one of the last meetings with Fliess in that same Park Hotel in Munich (Schur, 1972; Freud to Binswanger, 1/01/1913).

26 Jones *II*, 172 and 1956, p. 639.
27 Jones, 1956, p. 639 and Jones to Freud 30/07/1912.
28 Sachs, 1945; Weiss, 1970a.
29 Jones to Freud, 7/08/1912.
30 Pierri, 2001b.
31 In 1988, shortly before the fall of the Berlin Wall, there were rumours in some publications, including the New York Times, speculating that Max Eitingon had been a
KGB agent, like his namesake, and perhaps distant relative, Leonid Eitingon (Wilmers,
2010).
32 See Pierri, 2022.
33 Grosskurth, 1991.
34 (Ibid.) Otto Rank had begun to read Freud and had introduced himself to him in September 1906. At the next meeting of the Society, he was appointed salaried secretary.
Until then he had not studied the humanities but quickly reached a high degree of
culture and, in the group of Viennese psychoanalysts, he became the most competent
student of the applications of psychoanalysis to art, literature, and mythology. Almost
adopted by Freud, he dined in Berggasse every Thursday and went home with him talking about the Association, the studies in preparation, the commitments of the "movement," and so on.
35 In Cremerius, 1985, p. 188.
36 Lopez, 2001.
37 1913, p. 141.
38 "which has such a monstrous air" replaced "monstrous hypothesis" at the suggestion of
Jones (23/06/1913).
39 *What thou hast inherited from thy fathers, acquire it to make it thine*, in Freud, 1913,
p. 158.
40 *The best thing you know, you may not tell to boys*, in Freud to Fliess, 3/12/1897.
41 Ferenczi to Freud, 23/06/1913.

Chapter 18

1913 – the year before the war

The last congress with Jung

While Freud was completing the theorisation of *Totem and Taboo*, Ferenczi was coming out against Jung, drafting a stern critique of *Transformations and Symbols of the Libido* for the *Jahrbuch*.[1] The stand he took left no room for doubt that psychoanalytic theories differed profoundly from those of the president of the International Association. He began by restating the Freudian concept of the symbol as the product not of a spontaneous psychic fecundity, collective, and/or individual but of complex unconscious operations conveying a significant personal trace (a creation aimed at representing something that the individual has libidinally invested and then subjected to repression).[2] He went on to reiterate the necessity for clinical verification and took the opportunity to set out the distance from Jung on the subject of prophetic dreams, affirming that similar phenomena, for the moment inexplicable, in the future will be able to find a place in the natural sciences (1913b).

In the meantime he was writing to Freud:

> Never in my life has a more unpleasant task befallen me! [. . .] *Dreams tell the future; neurotics are mantically endowed people who foretell the future of the human race.* [. . .] The unconscious knows the present, past, and future; the fate (i.e., the *"task"*) of humanity is revealed in "symbols." All that comes from Jung's *astrological* (?) studies.
>
> (12/05/1913)

In September 1913, at the Munich Congress, the situation came to a head. Lou Andreas Salomé recalls how Jung's behaviour was intended to force the clash and impute responsibility for it to Freud, as if he had provoked it himself by his wretched rigidity:

> At the congress the Zurich members sat at their own table opposite Freud's. Their behavior toward Freud can be characterized in a word: it is not so much that Jung diverges from Freud, as that he does it in such a way as if he had

DOI: 10.4324/9781003246473-19

taken it on himself to rescue Freud and his cause *by* these divergences. If Freud takes up the lance to defend himself, it is misconstrued to mean that cannot show scientific tolerance, is dogmatic, and so forth. One glance at the two of them tells which is the more dogmatic, the more in love with power. Two years ago, Jung's booming laughter gave voice to a kind of robust gaiety and exuberant vitality, but now his earnestness is composed of pure aggression, ambition, and intellectual brutality. . . . Freud was the same as ever, but it was only with difficulty that he restrained his deep emotion; and there was nowhere I would have preferred to sit than right by his side.

(1958, pp. 168–169)

Shortly before the outbreak of war, in the last edition of the *Jahrbuch*, which under his sole editorship had now changed its name to *Jahrbuch der Psychoanalyse*, in an article entitled *On the History of the Psycho-Analytic Movement*, Freud laconically commented:

The fatiguing and unedifying proceedings ended in the re-election of Jung to the Presidency of the International Psycho-Analytical Association, which he accepted, although two-fifths of those present refused him their support. We dispersed without any desire to meet again.

(1914a, p. 45)

This article was an open attack on the dissidents and a manifesto of psychoanalysis as the creation of its author and a part of his own life story. This is how it began:

No one need be surprised at the subjective character of the contribution I propose to make here to the history of the psychoanalytic movement, nor need anyone wonder at the part I play in it. For psycho-analysis is my creation; for ten years I was the only person who concerned himself with it, and all the dissatisfaction which the new phenomenon aroused in my contemporaries has been poured out in the form of criticisms on my head. Although it is a long time now since I was the only psycho-analyst, I consider myself justified in maintaining that even today no one can know better than I do what psychoanalysis is, how it differs from other ways of investigating the life of the mind, and precisely what should be called psychoanalysis and what would better be described by some other name.

(1914a, p. 7)

At the same time, in the first issue of the *Internationale Zeitschrift für Ärztliche Psychoanalyse*, the new journal that had replaced the *Zentralblatt*, after the crisis with Stekel, Karl Abraham doubled down on the condemnation, declaring that the distortions inflicted on Freudian theory at the American conferences meant that the opinions being maintained by the Swiss no longer deserved the name of psychoanalysis. The attacks achieved the aim of dismissing Jung from his post as president and from the Association, and forced him to adopt the new name of *"Analytical Psychology"* for his own doctrine.[3]

In *The Red Book* (1913–1930), Jung will feel the need to write:

> I in no way exclusively stem from Freud. I had my scientific attitude and the theory of complexes before I met Freud. The teachers that influenced me above all are Bleuler, Pierre Janet, and Théodore Flournoy.
>
> (p. 196)

While *Totem and Taboo* (1912–13) had established the bases for the relationship of psychoanalysis with the human sciences, myths, and religion, the works of Freud on theory and technique that immediately followed – *On Narcissism: An Introduction* (1914b), *Further Recommendations on the Technique...* (1913–1915), and *From the History of an Infantile Neurosis* (1914c) – would be written under the developmental impulse of the need to elaborate the separation and differences from his pupil's thinking.

But there remained a thorn in Freud's side.

A *black tide of occultism*

When Ernest Jones had to devote an entire chapter of his Freud biography to occultism, he found himself seriously embarrassed because he had never been able to understand, let alone share, this singular passion of the Master. Jones's text represents the official, and now traditional, position of the International Association. He makes no bones about declaring occultism the residue of a more primitive mentality, and we see him once again puzzling over the Master's naiveté, that tendency to credulity which, in this case, Jones really could not associate with his genius. Unable to understand it – as Anna Freud recognised[4] – he presents it as a profound uncertainty:

> In it we find throughout an exquisite oscillation between scepticism and credulity so striking that it is possible to quote just as many pieces of evidence in support of his doubt concerning occult beliefs as of his adherence to them.
>
> (*III*, p. 402)

Jones was trapped in a misapprehension that made him misinterpret a series of observations and intuitions by Freud about the psyche that he could not locate theoretically and which scandalised him in their originality. He made no distinction between Freud's superstition and his scientific curiosity about the work of fortune-tellers and his hypotheses about the meaning of prophecies, his interest in telepathy and his search for a hidden coherence, a principle of intelligibility in the phenomena of thought transmission, the sole "kernel of truth"[5] that he acknowledged in occultism. Jones repressed these novelties and filed them away but also preserved them diligently for posterity, under the heading of occultism. It is significant that he, a man who mystified success, could never find an explanation for flights of fantasy, for creative illusions: indeed, he felt almost persecuted by them, as if he were afraid of losing himself in them. He was like the perpetrators of occultism in his insistent scepticism, expressing infantile fantasies of omnipotence which had not been fully sublimated and a lack of real faith in knowledge.[6]

The quality of his unease is not surprising, given the candour with which he recalled certain troubling late-night conversations with the professor:

> In the years before the great war I had several talks with Freud on occultism and kindred topics. He was fond, especially after midnight, of regaling me with strange or uncanny experiences with patients, characteristically about misfortunes or deaths supervening many years after a wish or prediction. He had a particular relish for such stories and was evidently impressed by their more mysterious aspects. When I would protest at some of the taller stories Freud was wont to reply with his favourite quotation: "There are more things in heaven and earth than are dreamt of in your philosophy." Some of the incidents sounded like mere coincidences, others like the obscure workings of unconscious motives. When they were concerned with clairvoyant visions of episodes at a distance, or visitations from departed spirits, I ventured to reprove him for his inclination to accept occult beliefs on flimsy evidence. His reply was: "I don't like it at all myself, but there is some truth in it," both sides of his nature coming to expression in a short sentence. I then asked him where such beliefs could halt: if one could believe in mental processes floating in the air, one could go on to a belief in angels. He closed the discussion at this point (about three in the morning!) with the remark: "Quite so, even *der liebe Gott*." This was said in a jocular tone as if agreeing with my *reductio ad absurdum* and with a quizzical look as if he were pleased at shocking me. But there was something searching also in the glance, and I went away not entirely happy lest there be some more serious undertone as well.
>
> (Jones, *III*, p. 408)

Freud was really letting himself go with these striking stories, detecting Jones's weak point and teasing him for his scientific fanaticism. It has to be said that in those years, he frequently addressed the topic of occultism with his Viennese pupils: not during the Wednesday meetings, since the group of members had grown considerably and lost the intimacy of its first period, but after official gatherings, which were now held on the fourth floor of a building in Obere Donaustrasse. A few participants would spend the rest of the evening at the Café Bauer, near the Burgtheater, where Freud would discuss subjects that he did not even mention in the sessions. In this way, he clearly revealed his inclination to believe in telepathy, "though it was held in check by his sobriety and insistence upon scientific evidence."[7] Freud gave some examples and told his colleagues:

> Very strict inquiry indicates the existence of such phenomena, and we must not disregard this because some ignorant or at times some ambitious persons try to fake if they don't succeed in producing the real thing.
>
> (Weiss, 1954, p. 10)

Unlike Jones, who clung to scientific rationality to keep from being terrified by spirits, Lou Andreas Salomé, who had recently joined the group, was able to detect the healthy artistic ingredient in this interest of Freud's. Preferring the poets to the men of science, and having no illusions about the occult but being fully convinced that soul and body were aspects of the same, single reality, on 21/11/912, she noted in her *Freud Journal*:

> Another objection to the occultists, that when they "materialize" psychic stuff, they just for that reason bring nothing psychic to us, and not even as much as matter.
>
> (1958, p. 55)

She had a different version of the creative mixture at the base of Freud's genius. In the open letter that she dedicated to him on the occasion of his 75th birthday (1931), she boldly addressed this scandalous topic from the start with the aim of highlighting Freud's ability to immerse himself in the irrational, even at the cost of sacrificing his own scientific vision.

Referring to his troubles in the face of the "quagmire" that is occultism, Salomé quoted what he had confided to her in Munich in 1913 about those rare instances of thought-transference which tormented him. In the paper he had read at that difficult Congress (centring on the case of Elfriede Hirschfeld, who was also the protagonist of the second unfulfilled prophecy), Freud had hinted at the theme of thought transmission:

> But I have had good reason for asserting that everyone possesses in his own unconscious an instrument with which he can interpret the utterances of the unconscious in other people.
>
> (1913b, p. 320)[8]

He had returned to the subject in a private conversation with the Russian psychoanalyst when they talked about the two cases of unfulfilled prophecy. Salomé was much struck by this and afterwards noted in her *Journal*: "This is a point which he hopes need never again be touched in his lifetime; I hope the contrary."[9] She later rethought the meaning of the prophecy about the patient who was destined to have two children by the age of 32:

> the situation goes like this: one problem involves affects, that the woman had to speak with such emotion after so many years about a fortune-telling which was

not fulfilled as if it had been fulfilled. It was simply because (as was revealed in psychoanalysis) it all had come true in her mother's life, hence as if her mother's life had already modified hers too, while she consciously suffered from her own frustrations. The second problem involved the manner of the transference to the fortune-teller. He reads off to her and expresses in realized form not only her conscious wishes but also those that lie deep behind her consciousness. It is hard to say where there would be any *bottom* to these depths.

Thirdly, there is a question of timeless duration within us. Freud always emphasizes that by "timeless" he means unabreacted and no more. But that does not explain many things; even the fantasies of dementia, which Jung described and which bring to life the mythology of primal antiquity, are themselves like enduring primal wishes and images in the abundance with which they are repeatedly brought to life and in their primitive nature. And in the case at hand the mother had indeed abreacted that which had retained its intensity in the daughter, quite as though it were her own, far beyond her own experience.

Here we approach the psychologistic *boundary*. It is very perilous, for Freud must guard particularly against being confused with the mystagogues. Here a philosophic attitude is no more to be evaded; we do live through more than we are.

(1958, pp. 169–170, my italics)

In the clinical sequence described by Freud, the brilliant and concise intuition which Salomé confided to her *Journal* highlighted an unconscious experience of psychic non-differentiation between mother and daughter and directly posed the problem of the quality of this primitive nucleus and the part it plays in the transition between generations, also comparing it with Jung's theories of psychosis. Salomé remained interested in this primitive psychic condition and often reminded Freud about the value of the primally maternal as a complement to the concept of autoeroticism.

In that September of 1913, Salomé was also curious about the occult and, while staying in Munich with Rilke, had spent an evening with Prof Ludwig Staudenmayer, a Bavarian chemist, investigator of artificially produced states of possession and author of the autobiographical book *Die Magie als experimentelle Naturwissenschaft (Magic as an Experimental Natural Science)*, 1912.[10] On a visit to Freising to meet Staudenmayer the year before,[11] Salomé realised she was in the presence of a madman, which confirmed Ferenczi's opinion of him. Having understood the Hungarian's genius from the start, she noted that "the psychology of mediums who are neither insane nor dishonest remains a problem which I should like to tackle with Ferenczi" (17/09/1920).

The question of telepathy

Ferenczi was still interested in collecting experiences with fortune-tellers, and by now he was making no secret of the fact. In 1912, he had gone back to see Frau Seidler in Berlin, but she was no longer interested in taking part in his experiments. In 1913, he attended a meeting of the Viennese Society with a tele-path whom he brought along with him from Budapest. The performances of this man were very convincing: at least according to Edoardo Weiss[12] (1889–1970), an analysand of Paul Federn, recalling one of the first Wednesday meetings he had been allowed to join:

> Everyone would ask him questions, for example, as to what one held in one's hand, what one was doing while the telepath's back was turned, etc. and the latter could tell immediately. I remember that I asked him what I was hold-ing in my hand, and he replied that it was an Austrian crown; but I also wanted to know the year, and he guessed it correctly.
>
> (Eissler interview, in Accerboni Pavanello, 1990, p. 354)

In October 1913, shortly after the Munich Congress, Ferenczi became enthusi-astic about a certain Prof Alexander Roth who conducted experiments in mind-reading with his wife as the medium: the couple seemed able to communicate with each other telepathically (at a distance of 15 metres) and let Ferenczi study them from the psychoanalytic viewpoint. He intended to travel to London with the Roths that same December to present the results of his observations to the Psycho-medical Society and to the Society for Psychical Research and told Ernest Jones about his plan: in fact, as a way of avoiding the trouble of the journey, he even tried to persuade Jones to read the paper in his place. As if that were not enough, he told Jones that he would give the Roths his address in London so that he could establish their trustworthiness for himself . . . so little was he aware of the Welsh-man's absolute aversion to the subject and how, unlike Jung, Freud, Ferenczi, and many of his colleagues in London, Jones would never dream of joining the SPR!

We do not know how Jones might have responded to this enticing prospect, but the problem never arose because, when Ferenczi presented his first thoughts about it to the Vienna group, his paper on thought transmission was appreciated (though the text is missing from the Minutes of 19/11/1913),[13] but the displays of the Roths had a very different reception. Even our Hungarian had to admit that his experiments had only produced positive results in a very small number of cases. The "telepathic" couple stayed a few more days in Vienna, and the session which took place on 23 November in Berggasse in the presence of Rank, Sachs, Hitschmann, and their wives – as well as the professor, his brother, and their children – went miserably, with not the merest shadow of a success. So Freud wrote irritably to Ferenczi, adding that things had gone the same way at Weiss's house.[14] Many years later, Anna Freud indignantly recalled the dreadful impres-sion made on everyone and how taken aback they were by the rough way in which the poor woman was urged, forced, and hustled to produce results.[15]

In an interview with K Eissler, Weiss recalled being present at a conversation which took place at the Café Bauer in 1914, in the presence of Federn, Rank, Sachs, and Tausk, during which Freud openly proclaimed his belief in the exis-tence of telepathic phenomena.

However, he quickly added that he could not admit this publicly because psychoanalysis was already in the crossfire of many other charges. At this point Tausk commented that those who believed in telepathy could have accused him of communicating telepathically to his patients the complexes attributed to them. This remark triggered Freud's sense of humour and he began to laugh.

(In Accerboni Pavanello, 1990, p. 354)

This is a significant anecdote: besides Fliess's criticism of "mind-reading," what the hypothesis of telepathy brought into play was something capable of dismantling the entire Freudian theory of transference as it had so far been constructed, something that could seriously undermine the scientific applicability of the method. It was no laughing matter.

I should point out that Henri Bergson was beginning his tenure as president of the *Society for Psychical Research* with the conference "Phantoms of Life and Psychic Research" (1913), at which he speculated that telepathy might be a phenomenon to study on the same terms as the facts of physics, chemistry, or biology, originating from the action of one consciousness on another without visible intermediaries, with a mutual interpenetration and exchanges comparable to the phenomena of osmosis:

If telepathy is real, it is consequently a fact that can be repeated indefinitely. I go a little further: if telepathy is real, it is possible that it takes place constantly and around the world, but too feebly to be noticed, or in a way that brain mechanisms can stop its effect, for our sake, as soon as it is about to cross the threshold of consciousness. . . . If this communication exists, nature will certainly have taken precautions to make it harmless and it is likely to believe that certain mechanisms have the specific function of rejecting the images thus introduced into the unconscious, which otherwise could be particularly annoying in everyday life. However, these images may be able to infiltrate contraband, especially when inhibitory mechanisms are malfunctioning.

(1913, pp. 18 and 27)

However, in concluding his astute intuitions on psychic life, Bergson did not rule out the possibility of the soul's survival after death.

It is likely that in those pre-war years, what especially tormented Freud about this subject on the boundaries of occultism, so dear to Jung, was connected to the working-through of the break with him and the need to reappropriate and systematise material which had until then been a common heritage, though delegated to the Swiss rather than Ferenczi.

The question of telepathy had been coming up since the age of magnetism and subsequently in studies of hypnosis and mediums, often simultaneously with the search for a new form of communication to be recovered or invented – the supposed "Sanskrit," the Martian language, the original idiom of mankind, but also

English as the foreign language used by Anna O. This required Freud to investigate the presence of further psychic potentialities that remained unknown and could play a part in, or be activated by, psychoanalytic treatment.

He was always more interested in thought-induction but not entirely ready to consider the activity of unconscious communication between analyst and patient which, in the absence of a theory of relationships, risked bringing suggestion back into play or, through the action of the countertransference, slipping into a mystical or paranoid interpretation.[16]

Freud never gave any of his works the title "Psychoanalysis *and* Telepathy."[17] Up until the abandonment of the seduction theory, in order to construct the psychoanalytic method it had been necessary to negate the presence of the analyst – along with parents, teachers, and so on – keeping the reality of relationships in the background.

The dialogues of the unconscious

Jones attributed Freud's interest in telepathy to the bad influence of Jung and Ferenczi, but he failed to appreciate how different Jung's mysticism was from the deep coherence of the professor's scientific attention to thought transmission and from Ferenczi's integrity in widening the range of his hypotheses to include clinical practice, the analyst's own disposition, and the parent-child relationship.

In 1913, while Freud was writing the third essay of *Totem and Taboo*, "Animism, Magic and the Omnipotence of Thoughts," his Hungarian pupil was simultaneously and in complete harmony with his teacher starting to look for the roots of "telepathy" and superstition in the child's original magical relationship with its mother, in the period of what he called the "illusion" of omnipotence. When Ferenczi wrote, "The mimic expressions that continually accompany thinking (peculiarly so with children) make this kind of thought-reading especially easy for the adults,"[18] he seems to be laying the ground for what Freud would shortly be writing in the fourth essay – "The Return of Totemism in Childhood" – on the unconscious transmission of psychic processes between generations.

They were advancing on two parallel paths: while for Freud, the function of language would remain central as a foundation and limit of the analysis of the subject, from now on, Ferenczi would be interested in the pre-oedipal relationship, the transition from preverbal communication to the symbolism of language and vice versa, concentrating on the person of the object – the analyst and the parent – and going on to develop what Freud was leaving in suspense.

Although the Hungarian would repeatedly get involved in experiences with clairvoyants – this is the period just before the war when he had been persuaded to take part in a telepathic experiment, only to find a friend entirely unexpectedly standing in for the fortune-teller as the receiver of his thoughts by anxiously telephoning him about a dream that matched the subject of the experiment[19] – in his theory, he began to reflect seriously on psychogenic somatic problems and the way they arise in and from relationships.

And so his many brief works on this topic in 1914 – "Sensations of Giddiness at the End of the Psycho-Analytic Session," "On Falling Asleep during Analysis," "Embarrassed Hands," "Rubbing the Eyes as a Substitute for Onanism," "Restlessness towards the End of the Hour of Analysis," and so on – seem to prepare the way for the splendid, concise paper of 1915, "Psychogenic Anomalies of Voice Production," in which he interprets a patient's uncertainties over the change of his voice during puberty as an incestuous fixation, maintained in his relationship with his mother by her attuned unconscious resistance. On this topic, he wrote:

> In my opinion we have to do here with one of those numerous cases that I am in the habit of calling *Dialogues of the Unconscious*, where, namely, the unconscious of two people completely understand themselves and each other, without the remotest conception of this on the part of the consciousness of either.
>
> (1915, p. 145)

It would be the war that sent Freud and Ferenczi their different ways, or rather perhaps Ferenczi's two-part formal analysis with Freud in 1914 and 1916. Or maybe it was the new and important presence of Anna beside her father and the analysis she began in autumn 1918. What we know is that, immediately after the war, Freud resumed his work on thought transmission by himself.

And so, while Ferenczi advanced into the turbulent world of the countertransference and unconscious communication between analyst and patient, clothing the transference in the garb of the mother ready to listen to "the new-born grizzling on the couch,"[20] Freud would keep a firm grip on the Ariadne thread of telepathy and venture into the infant's occult link with the mother and their secret primal language, staying securely attached so that he could re-emerge unscathed.

Regarding telepathy, Freud felt close to a discovery comparable to that of the meaning of the dream.[21] If there was a taste for the occult, for the irrational, if there were even traces of superstition in him, Roussillon states that it was a hidden, masked rationality that he was trying to reconstruct.[22]

One of the reasons for Freud's irritation with Jung was his indiscriminate admixture of science and mysticism, while Freud "restrained himself and adhered only to points for which there was scientific verification."[23] On the other hand, because of the way Ferenczi took into account the fascinating and "occult" aspects of the preverbal area,[24] Freud would by no means have approved of the fact that he was increasing the patient's regression and dependence on the analyst.[25]

So, after the war, Freud started reflecting on his own again about the work of fortune-tellers and unfulfilled prophecies that he had only apparently yielded to Ferenczi. And in autumn 1919, he found himself faced with a strange experience of thought transmission that occurred during an analytic session.

Freud would mention this case, the Forsyth case,[26] more than ten years later in *Lecture XXX*, "Dreams and occultism" (1933) a substantial and systematic essay which only came to be published after a long gestation and which Freud intended as an exhaustive account of psychoanalytic observations on the topic of thought transmission. What had hitherto been private, occult research finally intersected with the public theory and could be included in the *New Introductory Lectures*.

Meanwhile, Ferenczi was in the last months of his life. He was no longer playing with fortune-tellers but had just written "Confusion of Tongues between Adults and the Child: The Language of Tenderness and of Passion" (1933), in which he addressed unconscious parent-child communication and analyst-patient communication from a different perspective. He wrote that he had gradually come to the conclusion that patients have "exceedingly refined sensitivity for the wishes, tendencies, whims, sympathies and antipathies of their analyst" and that "they show a remarkable, almost clairvoyant knowledge about the thoughts and emotions that go on in their analyst's mind."[27]

In the *Clinical Diary* (January–October 1932) which started life as a long letter to Freud,[28] Ferenczi wrote that cases of thought-transference during the analysis of suffering people were extraordinarily frequent but that the reality of such processes encountered strong emotional resistance, and any insights into them had the tendency to come undone like the tissue of dreams, and added:

> It is possible that here we are facing *a fourth "narcissistic wound,"* namely that even the intelligence of which we are so proud, though analysts, is not our property but must be replaced or regenerated through the rhythmic outpouring of the ego into the universe, which alone is all knowing and therefore intelligent.
>
> (14/02/1932, my italics)

A few days later, appreciating the thousands of signs that allow the patient to guess a great deal about analyst's mood and feelings – subtle, barely discernible differences in the handshake, the absence of colour or interest in the voice, the quality of his alertness or inertia in following and responding to what the patient brings up – he stated:

> Some maintain with great certainty that they can also perceive our thoughts and feelings quite independently of any outward sign, and even at a distance.
>
> (16/02/1932)

Ferenczi's pioneering paper (he never published a single article on "telepathy": his folders stayed in his private correspondence) would be dedicated to extending theory and practice in the field of early mother-child communication and the transference-countertransference relationship, those prophecies constituted by the promises for the future which parents invest in their children. This is a legacy of expectations which seduce the child into desiring and growing – but can also be an obstruction – acting in the area of uncertain boundaries where language is yet to appear.

In his experiments on trauma, Ferenczi ventured into this telepathic and occult sphere, the hypnotic dimensions of development and the analytic situation, opening the way to what is now considered the third working model of the mind in psychoanalysis, the area of interpsychic relations.[29]

Today we can certainly claim that all of Ferenczi's clinical and theoretical work may be considered prophetic, a real promise of the future. Lou Andreas Salomé wrote in 1913:

Perhaps publication of Ferenczi's ideas is premature with respect to Freud's present and next endeavours, but they really *are* complementary. So Ferenczi's time *must* come.

(1958, p. 137, 7-9/03/1913)

And it is also the secret "dialogues of the unconscious" between Freud and Ferenczi[30] which Jones censored, together with Freud's interest in telepathy: however, he acknowledged that Freud and Ferenczi must have spent "enjoyable times together when there was no criticizing audience" (*II*, p. 179).

Lecture XXX was the last word that Sigmund Freud would utter in his life about "telepathy": he quickly dismissed and lost interest in problems which he had no hope of solving – recalls Weiss[31] – but at the same time, he encouraged everyone who seriously wanted to study such phenomena to go ahead.

He had not shared the drafting of this work with Ferenczi, nor did it cite his old fellow-researcher with whom he had formulated the first hypotheses on thought transmission. But we like to imagine that perhaps, after Ferenczi's death, he had lost the desire to venture into this field. The sultan had lost his clairvoyant.

"Dreams and Occultism" leaves the reader with the idea that research on the transmission of thought could have precious discoveries in store and widen the scientific view of psychoanalysis but, above all, it suggested the hypothesis that the "occult" treasure concerns precisely the "golden coin" of the original and generative mother-child communication.[32]

Maybe this was the hidden treasure of Spiritualism that Ferenczi had hoped for from the start: this was his true great discovery.

What is certain in the end is that Ferenczi's thinking about the countertransference eventually resurfaced, though not without difficulty, beginning with Heimann's 1950 article and thanks to the work of his pupil Balint, whereas Freud's research into unconscious thought transmission and the "secret language" of analysis (1933) was left isolated in the "crypt"[33] of telepathy, in that chapter of the story to which Jones gave the misleading title "Occultism."

Jones was ironically humorous about it:

Freud was right in his prediction that analysts would be found who would adopt a belief in telepathy; it has even been suggested that use might be made of the process in the course of psycho-analytic treatment! I may recall an incident in this connection that happened to me years ago. A lady who consulted me explained that she could not leave her home a hundred miles away for long, and suggested that I devote an hour a day to analysing her in her absence. When I expressed my regret that her plan was not feasible she gave a sigh and said, "No, I suppose you haven't yet got so far in your work."

(III, p. 435)

Freud's writings on thought-transference waited years to be translated, and they were even more censored and cut than Ferenczi's or forgotten in the Archives like the *Nachtrag* manuscript – the first draft of the Forsyth case – long thought forgotten and then rediscovered in a binder of Sigmund Freud papers, in The Library of Congress in Washington, fascinatingly alongside his obituary of Ferenczi.[34]

But that is another story.[35]

Like Vienna and Budapest, the capital cities of the Empire symbolised by the two-headed eagle and united by the waters of the Danube, Freud and Ferenczi are the twin poles of an extraordinary and passionate journey, present in the conscious and unconscious legacy transmitted to every psychoanalyst of the present generation.[36]

Notes

1 Ferenczi, 1913b.
2 Jones would later explore this concept to differentiate it from occultist conceptions of the symbol (1916).
3 In turn, Adler had called his school "Individual Psychology."
4 See *Prologue*.
5 Freud, 1925c, p. 136.
6 Devereux, 1953.
7 Weiss, 1970b, p. 13.
8 Freud had used Elfriede's case to understand the choice of neurosis, since she was a rare example of the transformation of an hysterical neurosis into an obsessional one, like a "bilingual document" (1913, p. 319). He wrote about her in the short contemporary essays *Two Lies Told by Children* and *An Evidential Dream* (1913c, 1913d) and later in *A Child Is Being Beaten* (1919a).
9 1958, p. 169.
10 The paper had been discussed at the Viennese Society on 11/30/1910 and 1/18/1911 (Nunberg and Federn, 1962).
11 Ferenczi to Freud 3/07/1912 and 13/06/1917. Staudenmayer was one of the cases considered by Tausk in his famous article on the "influencing machine" (1919).
12 Edoardo Weiss was born in Trieste where his father, of Jewish origin, had moved from Bohemia. In 1908, he enrolled in the Faculty of Medicine in Vienna and consulted Freud about devoting himself to psychoanalysis. On his advice, he began analysis with Federn, and in 1913, even before graduating, he joined the Viennese Society. Returning to Trieste after the end of the war, he began to practise psychoanalysis, arousing great interest in the city's cultural circles (at the end of the 1920s, he had the poet Umberto Saba in analysis). In 1931, he moved to Rome, where in 1932, he was one of the founders of the Italian Psychoanalytic Society and of the Rivista Italiana di Psicoanalisi. In 1938, after the racial laws and the dissolution of the Italian Society, Weiss emigrated to America and joined the Chicago Institute.
13 There remains a summary which he himself prepared and sent to Freud (29/11/1913).
14 Freud to Ferenczi, 23/11/1913.

15 Jones, *III*, p. 416.
16 Conrotto, 2009.
17 Pierri 1016 e 2021.
18 1913a, p. 230.
19 Fodor, 1979, p. 109.
20 Granoff, 2000, pp. 22–23.
21 Roazen, 1969.
22 1995, p. 73.
23 Weiss, 1954, p. 2.
24 Roussillon, 1995.
25 Fromm, 1959.
26 On the Forsyth case see Pierri, 2022.
27 1933, pp. 226–227.
28 Martin Cabré, 2016.
29 Bolognini 2004 and 2016. The first model is defined in *The Interpretation of Dreams* (1899) as a distinction of the psychic apparatus into the three areas: Conscious, Unconscious, and Preconscious; the second model will be the structural distinction provided in *The Ego and the Id* (1923) into ego, id, and superego.
30 Pierri, 2002.
31 1954, p. 2.
32 Pierri, 2016.
33 Abraham and Torok, 1987.
34 Pierri, 2010.
35 Pierri, 2022.
36 Martín Cabré, 2005.

Epilogue
A fortune-teller visits Freud in Berggasse

Nevertheless, Freud must still have been worried about the possible confusion between psychoanalysis and occultism. His son Martin recounts an anecdote from the early thirties, probably connected to the publication of *Lecture XXX.*

"Family jokes were welcome," begins Freud's eldest son, who managed the psychoanalytic publishing company from 1931, and he goes on:

> A man in Sigmund Freud's position is bombarded by much correspondence, and it was often heavy work to sort father's mail, which was sometimes addressed to the press and sometimes to Berggasse 19. The cranks and other queer correspondents generally chose the press, and I would carry their letters as well as the more important communications with me on my twice-daily calls on father to discuss business. Father dealt with everything speedily and efficiently. For some weeks there had arrived most regularly letters from a German whose note-paper boldly proclaimed him to be as "Astrologer and Psychoanalyst." The letters asked for a meeting to discuss matters of mutual scientific interest, and their phraseology clearly proclaimed a crank. Father's verdict before destroying each of this man's letters was, "No reply!"
>
> Nevertheless, the letters from the "Astrologer and Psychoanalyst" continued to arrive and grow so much in urgency that I felt bound to advise father to send some kind of reply, signed by me as his secretary, regretting firmly that a meeting could not be arranged. But father was adamant: there should be no reply.
>
> The "Astrologer and Psychoanalyst" became something of a joke, but as it turned out, not quite so much of a joke as I had imagined when I got our printers to design a visiting-card boldly printed with the gentleman's name, address and profession. A theatrical hairdresser having turned me into a grizzly old gentleman with abundant grey hair and long beard. I placed horn-rimmed spectacles on my nose and walked to Berggasse 19, no passer-by taking he slightest notice of me and thus allowing me to know that my disguise was perfect. Paula was in the plot and admitted me and, although we had not confided in the dogs, they accepted me without question and nearly ruined the joke forthwith by their normal display of friendliness.

Paula went before me with the card, and I arrived in time to hear father shout, "By all means keep that man out!"

"Herr Professor," I began in a voice well disguised by my thick beard, "between scientists there are certain rules of behavior, even if they differ in their theories. . ."

This was as far as I got. As father leant back in his chair, giving the "Astrologer and Psychoanalyst" so furious a glare that I must have paled under my disguise, Anna, whom it is very difficult to deceive and who was probably helped by the friendly behavior of the dogs, cried, "It is Martin, Papa!" and the tension eased in general laughter. Nevertheless, I must admit that when I had removed my disguise, I endured an anticlimax that was not pleasant. Father's furious glare, although not really meant for me, had shocked me.[1]

Leaving aside self-proclaimed psychoanalyst-astrologers with whom we know he had unfinished business from the time of Fliess, there is no question of Freud having lost interest in authentic fortune-tellers' ability to read unconscious thought or in comparing their work with that of the analyst.

And his experiences with Ferenczi must have remained dear to his heart.

The following account, also from the thirties, concerns a visit to Berggasse by Francesco Waldner, a more respectable astrologer from Italy, consulted, as astrologers of Freud's time were, by important personalities from politics and the performing arts: Marlene Dietrich, S Niarchos, von Karajan, the Krupp and Savoy families, the Shah of Persia, Farah Diba, G Pompidou, and so on.

He especially claimed to have the gift of clairvoyance, which he had later consolidated by studying astrological calculations. He said he relied on what he considered a "super-intuition," an ability to "identify" in the sense of "seeing," penetrating the other's personality, even when only presented with objects belonging to them.[2] But what mattered most to him was what happened in the other person's presence, which he describes in terms that very much recall the processes known in current psychoanalysis as empathy, reverie, and projective trans-identifications:

Seeing a person stirs a certain sensation in me and, in turn, this sensation mixes with senses hidden in the depth of my subconscious which create a sort of picture in me and this is the reason why I say to myself: this will happen to this person.

(Waldner, 1962, p. 15, [originally translated from French by the author])

In a booklet entitled *Mes Aventures surnaturelles* (1962), he recalls his first childhood experience of clairvoyance and connects it with a traumatic situation, a violent clash with his father when he was at primary school. When, on his teacher's suggestion, he asked if he could be the standard-bearer at a parade in Bolzano, his father's brutal prohibition caused a nervous crisis that left him in a state of semi-consciousness. After a sleepless night, he arrived at school early and had his first vision – a black coffin covered in flowers in the classroom – which antici-pated the real scene he had to attend the following Sunday at his teacher's funeral. There had been a bloody political battle during the procession and the teacher had been killed trying to rescue the standard-bearer (pp. 22–24).

Waldner tells us that he lived in Vienna for several years as a young man and had a meeting with Freud, who had been interested in examining his singular abilities. The account of this visit is particularly touching.[3]

> He was expecting me at four o'clock precisely at IX. Berggasse 19, where I found a plate saying Dozent Dr Sigmund Freud. It was Sunday and he was not seeing patients. I rang the doorbell and found myself in the presence of a distinguished old gentleman with a piercing gaze who scrutinised me closely as he asked me to come in; I felt immediately at ease speaking to him in German. He invited me to sit down in front of his little desk that was adorned with flowers in a glass vase and a row of statuettes as paperweights. He kindly asked me if I could examine three or four objects and a photo enclosed in various envelopes, to give him an idea of what associations or, rather, images they evoked for me.
>
> I was very moved by the reassuring tone of his voice and the kindly expres-sion on his face. He was holding a pen and a small notebook; as he handed me the first packet he smiled with a trusting expectancy. I started talking and he never interrupted me until the end of each experiment, when he asked some specific questions, without commenting or suggesting the answers. Some-times he encouraged me as he handed me an envelope. I understood that he did not have much time as he noted down what I was suggesting. After the fifth object he said that would be enough, he didn't want to wear me out. He was very interested in what I had said. Some of my answers matched what he knew, while others left him rather puzzled; he promised to reflect on this.
>
> He thanked me, insisting that I had been a great help, but he added, "I want to do something for you too, naturally if you were prepared to take time for yourself and have some sessions." I told him I was well. "On the contrary," he replied. He said I had unfinished business with my father, to which I responded by saying he had been dead for some years. At this point he came over to me and hit the centre of my chest with his fist, saying "He's still here, you must free yourself from him if you want to give full rein to these extraordinary gifts of yours!" Could he have deduced this from some of my answers during the experiment? It was only at that moment that I under-stood as never before, the truth of what he had just told me, how I was being blocked by a sense of guilt. As I was saying goodbye, I was tempted to tell him that I wouldn't be seeing him again and yet at the same time I had an intense desire to do so.
>
> He had already undergone surgery for a cancer of the pharynx, but he didn't ask me about his life and health, which I found amazing. Quietly and

calmly he said, "You must take advantage of this, my boy. I don't expect to live much longer and you really need my help if you want to avoid interference and keep your problems out of your visions." In the end I did go back to see him several times and at each session I felt worse because of what I was finding in myself. I who could see clearly into other people's minds, I thought, had denied myself this precious knowledge.

The years passed quickly in a climate that was anything but serene in Europe until, in 1938 Freud was forced into exile in London, and on 23 September '39 I learned from mutual friends that this great man of absolute humility was dead. I felt abandoned by him after he given a helping hand to my troubled adolescence, which had been far from easy to live through, showing me how self-knowledge is the way to serenity. Thank you again, Professor Sigmund.

We do not know what Freud had noted down in the course of the consultation, but we know that during that single meeting he had "seen" the complex at the heart of Waldner's visionary ability.

Notes

1 1957, pp. 381–382.
2 See Pierri, 2021.
3 Document read by Alessandro Bruni (following a personal communication from Massimo Fornicoli) at the *Giornata dell'editoria*, Centro Psicoanalitico di Roma, 14 September 2019: the launch of M. Pierri (2018) *Un enigma per il dottor Freud. La sfida della telepatia.*

Correspondence

Andreas-Salomé L., and Freud A. (2001), *À l'ombre du père. Correspondance 1919–1937*, Paris: Hachette.

Boehlich W., ed. (1989), *The Letters of Sigmund Freud to Eduard Silberstein 1871–1881*, Cambridge, MA: Harvard UP, 1990.

Bonaparte M., Freud A., and Kris E., eds. (1950), *The Origins of Psychoanalysis: Sigmund Freud's Letters to Wilhelm Fliess, Drafts and Notes from the Years 1887–1902*, New York: Basic Books.

Brabant E., Falzeder E., and Giampieri-Deutsch P. (1993), *The Correspondence of Sigmund Freud and Sándor Ferenczi 1908–1914*, Cambridge, MA: Belknap.

Eros F., Szekacs-Weisz J., and Robinson K. (2013), *Sándor Ferenczi–Ernest Jones Letters 1911–1933*, London: Karnac.

Falzeder E. (2002), *The Complete Correspondence of Sigmund Freud and Karl Abraham 1907–1925*, London and New York: Karnak.

Falzeder E., and Brabant E. (1996), *The Correspondence of Sigmund Freud and Sándor Ferenczi 1914–1919*. Cambridge, MA: Belknap.

Falzeder E., Brabant E., and Dupont J. (2000), *The Correspondence of Sigmund Freud and Sándor Ferenczi 1920–1933*. Cambridge, MA: Belknap.

Fichtner G. (2003), *The Sigmund Freud-Ludwig Binswanger Correspondence 1908–1938*, London: Open Gate Press, Incorporating Centaur Press.

Fortune C. (2002), *The Sándor Ferenczi-Georg Groddeck Correspondence, 1921–1933*, London: Open Gate Press.

Freud E. L. (1961), *Letters of Sigmund Freud 1873–1939*, London: The Hogarth Press.

Freud E. L. (1970), *The Letters of Sigmund Freud and Arnold Zweig*. New York: University Press, 1987.

Freud S. (2010), *Lettres à ses enfants*, Paris: Aubier, 2012.

Freud S., and Bernays M. (2011), *Die Brautbriefe, Bd 1: 'Sei mein, wie ich mir's denke' (Juni 1882-Juli 1883)*, eds. G. Fichtner, I. Grubrich-Simitis, and A. Hirschmüller, Frankfurt a. M.: Fischer.

Freud S., and Bernays M. (2013), *Die Brautbriefe, Bd. 2: Unser Roman in Fortsetzungen (Juli 1883–Dezember 1883)*, eds. G. Fichtner, I. Grubrich-Simitis, and A. Hirschmüller. Frankfurt a. M.: Fischer.

Freud S., and Bernays M. (2015), *Die Brautbriefe, Bd. 3: Warten in Ruhe und Ergebung, Warten in Kampf und Erregung (Januar 1884–September 1884)*, eds. G. Fichtner, I. Grubrich-Simitis, and A. Hirschmüller. Frankfurt a. M.: Fischer.

Freud S., and Bernays M. (2019), *Die Brautbriefe, Band 4. Spuren von unserer komplizierten Existenz. (September 1884 – 1885)*, eds. G. Fichtner, I. Grubrich-Simitis, and A. Hirschmüller. Frankfurt a. M.: Fischer.

Freud S., and Eitingon M. (2004), *Correspondance 1906–1939*. Paris: Hachette, 2009.

Hale N. G. (1971), *James Jackson Putnam and Psychoanalysis: Letters between Putnam and Sigmund Freud, Ernest Jones, William James, Sándor Ferenczi, and Morton Prince, 1877–1917*. Cambridge, MA: Harvard University Press.

Masson J. M. (1985), *The Complete Letters of Sigmund Freud to Wilhelm Fliess, 1887–1904*. Cambridge, MA: Belknap.

McGuire W. (1974), *The Freud/Jung Letters*. Princeton, NJ: Princeton University Press.

Meyer-Palmedo I. (ed., 2014), *Sigmund Freud, Anna Freud Correspondence 1904–1938*. Cambridge: Polity Press.

Paskauskas R. A. (1993), *The Complete Correspondence of Sigmund Freud and Ernest Jones 1908–1939*. Cambridge, MA: Harvard University Press.

Pfeiffer E. (1963), *Sigmud Freud and Lou Andreas-Salomé Letters*. London: Hogarth Press.

Vincent C. (1996), *Lettres de Famille de Sigmund Freud et des Freuds de Manchester 1911– 1938*, Paris: PUF.

For the Committee's circular letters, unless otherwise specified:

a: Grosskurth P. (1991), *The Secret Ring*. London: Cape.
b: Freud S., and Ferenczi S. (2000), *Correspondence 1920–1933*. Paris: Calmann-Levy.

The quotes from Jones E. *Sigmund Freud Life and Work* Vol *I* 1953, *II* 1955, *III* 1957, are here abbreviated to: Jones *I, II, III*.

Bibliography

Abraham K. (1912[1935]), Amenhotep IV (Ikhnaton) – A Psychoanalytic Contribution to the Understanding of His Personality and the Monotheistic Cult of Aton. *Psychoanal. Q.*, 4:537–569.

Abraham K. (1914), C. G. Jung, Versuch Einer Darstellung der Psychoanalytischen Theorie. Neun Vorlesungen, Gehalten in New-York im September 1912. Jahrb. f. Psychoanalyt. Forsch., Bd. V. Buchausgabe: Wien, F. Deuticke 1913. *Int. Zeit. für Psychoanal.*, 2:72–82.

Abraham N., and Torok M. (1987), *The Shell and the Kernel*, Chicago: The University of Chicago Press, 1994.

Accerboni Pavanello A. M. (1990), Sigmund Freud as Remembered by Edoardo Weiss, the Italian Pioneer of Psychoanalysis. *Int. Rev. Psycho-Anal.*, 17:351–359.

Adorno T. W. (1957), *The Stars Down to Earth: And Other Essays on the Irrational in Culture*, Abingdon: Routledge, 2001.

Albrecht A. (1909), The Eminent Vienna Psychotherapeutist Now in America, 1968. *Psychoanal. Rev.*, 55:334–339.

Andreas Salomé L. (1931), *My Thanks to Freud*, Open Letter to Professor Freud for his 75th birthday, Int. Psych. Verlag Vienna. https://archive.org/details/MeinDank

Andreas Salomé L. (1958), *The Freud Journal*. London and New York: Quartet Books.

Andreas Salomé L. (1968), *Looking Back: Memoirs*, Boston: Da Capo Press.

Anzieu D. (1959), *Freud's Self-Analysis*, London: Hogarth Press, 1986.

Appignanesi L., and Forrester J. (1992), *Freud's Women*. New York: Basic Book.

Archer-Hind R. D. (1888), *Plato The Timaeus*. New York: Macmillan and Co, https://www.google.it/books/edition.

Arendt H. (1968), Walter Benjamin 1892–1940. In *Men in Dark Times*. New York: Harcourt, Brace World.

Balint M. (1968), Préface. In Ferenczi S. (ed.), *Psychanalyse Oeuvres Complètes*, Vol. I. Lausanne: Payot, 1990.

Balsamo M. (2000), *Freud et le destin*. Paris: PUF.

Balsamo M., and Napolitano F. (1998), *Freud, lei e l'altro. Sulla genesi della teoria psicoanalitica* (Freud, She and the Other. On the Genesis of Psychoanalytic Theory). Milano: Angeli.

Barron J. W., Beaumont R., Goldsmith G. N., Good M. I., Pyles R. L., Rizzuto A. M., and Smith H. F. (1991), Sigmund Freud: The Secrets of Nature and the Nature of Secrets. *Int. R. Psycho-Anal.*, 18:143–163.

Basile G. (2001), *Le parole nella mente: relazioni semantiche e strutture della mente* (Words in the Mind: Semantic Relationships and Structures of the Mind). Milano: Angeli.

Behling K. (2002), *Martha Freud, A Biography*. Cambridge: Polity Press, 2005.

Benjamin W. (1938), Berlin Childhood Around 1900. In *The Work of Art in the Age of Its Technological Reproducibility, and Other Writings on Media*. Cambridge, MA: Harvard University Press, 2008.

Bergson H. (1886), De la simulation inconsciente dans l'etat d'hypnotisme. *Revue Philosophique*, 22:525–531.

Bergson H. (1913), Phantoms of the Living and Psychical Research. *Proc. Soc. Psy. Res.*, 27:157–175.

Berman E. (2004), Sándor, Gizella, Elma: A Biographical Journey. *Int. J. Psycho-Anal.*, 85(2):489–520.

Bernays Freud A. (1940), My Brother, Sigmund Freud. In Ruitenbeek H. M. (ed.) (1973), *Freud as We Knew Him*. Detroit: Wayne State University Press.

Binswanger L. (1956), *Sigmund Freud – Reminiscences of a Friendship*. New York: Grune & Stratton, 1957.

Binswanger L. (1957), My First Three Visits with Freud in Vienna. In Ruitenbeek H. M. (ed.) (1973), *Freud as We Knew Him*. Detroit: Wayne State University Press.

Blanton S. (1971), *Diary of My Analysis with Sigmund Freud*. New York: Hawthorn Books.

Bleuler E. (1896), Review Studien Über Hysterie, Sigmund Freud with Josef Breuer. *Münchener medizinische Wochenschrift*, 43:524–525.

Bleuler E. (1905), Consciousness and Association. In Jung C. G. (ed.), *Studies in Word-Association*. London: William Heinemann, 1918.

Bleuler E. (1911), *Dementia Praecox or the Group of Schizophrenias*, New York: Intl Universities Pr Inc, 1966.

Blum H. P. (1996), The Irma Dream, Self-Analysis, and Self-Supervision. *J. Amer. Psychoanal. Assn.*, 44:511–532.

Bolognini S. (2004), Intrapsychic-Interpsychic. *Int. J. Psycho-Anal.*, 85:337–358.

Bolognini S. (2016), The Interpsychic Dimension in the Psychoanalytic Interpretation. *Psychoanal. Inq.*, 36:102–111.

Bonomi C. (1994), Sexuality and Death in Freud's Discovery of Sexual Aetiology. *Int. Forum Psychoanal.*, 3:63–86.

Bonomi C. (1999), Il giudizio di Jones sul deterioramento mentale di Ferenczi: un riesame. In Borgogno F. (ed.), *La partecipazione affettiva dell'analista*. Milano: Angeli.

Bonomi C., ed. (2006), *Sándor Ferenczi e la psicoanalisi contemporanea*, Roma: Borla.

Bonomi C. (2007), *Sulle soglie della psicoanalisi. Freud e la follia infantile* (On the Threshold of Psychoanalysis: Freud and the Insanity of the Child). Torino: Bollati Boringhieri.

Borch-Jacobsen M., and Shamdasani S. (2012), *The Freud Files: An Inquiry into the History of Psychoanalysis*. Cambridge: Cambridge University Press.

Borgogno F. (1999), Sándor Ferenczi's First Paper Considered as a Calling Card. *Int. Forum Psychoanal.*, 8:249–256.

Borgogno F. (2010), Ferenczi, «l'analista introiettivo». *Rivista Psicoanal.*, 56:561–576. Eng. trans. Borgogno F. (2011), Sándor Ferenczi, the "Introjective Psychoanalyst". *American Imago*, 68:155–172.

Botella C., and Botella S. (2001), *The Work of Psychic Figurability: Mental States without Representation*. Brunner: Routledge, 2005.

Brabant E. (2003), Les voies de la passion. Les rapports entre Freud et Ferenczi. *Le Coq-Héron*, 174:100–113.

Brabant E. (2011), Éditorial À quoi sert l'historien de la psychanalyse? *Le Coq-Héron*, 207:7–10.

Breuer J. (1893a), Fräulein Anna O, Case Histories from *Studies on Hysteria*. (1893–1895). *SE*, 2, II, 19–47.

Breuer J. (1893b), Theoretical, from *Studies on Hysteria* (1893–1895). *SE*, 2, 183–251.

Breuer J., and Freud S. (1893), On the Psychical Mechanism of Hysterical Phenomena: Preliminary Communication, from *Studies on Hysteria* (1893–1895). *SE*, 2, 1–17.

Breuer J., and Freud S. (1893–95), *Studies on Hysteria*. In *SE*, 2.

Brill A. A. (1940), Reminiscences of Freud. *Psychoanal. Q.*, 9:177–183.

Britton R. (2003), *Sex, Death, and the Superego: Experiences in Psychoanalysis*, London: Routledge.

Brome V. (1978), *Jung: Man and Myth*. Looe Cornwall: House of Stratus, 2010.

Brome V. (1983), *Ernest Jones: A Biography*. London: Norton Company.

Bruni A. (2019), presentazione di: M. Pierri (2018) Un enigma per il dottor Freud. La sfida della telepatia *Giornata dell'editoria*, Centro Psicoanalitico di Roma, 14 Settembre.

Cacciari M. (1980), *Dallo Steinhof. Prospettive viennesi del primo Novecento*, Milano: Adelphi, 2005.

Canestri J. (1983), Nota su due saggi di G. C. Lepschy. *Riv. Psicoanal.*, 29:395–404.

Carloni G. (1989), Prefazione. In Ferenczi S. (ed.), *Opere*, vol. I. Milano: Cortina.

Carotenuto A. (1980), *A Secret Symmetry: Sabina Spielrein between Jung and Freud*, New York: Pantheon books, 1982.

Carpenter D. W. (1854), On the influence of suggestion in modifying and directing muscular movement independently of volition, *Royal I. G. Brit.*, March 12th.

Carrington H. (1909), *Eusapia Palladino and her phenomena*, New York: Dodge &Company. https://archive.org/details/eusapiapalladino1909carr

Casement P. (1985), *Learning from the Patient*, New York: The Guilford press.

Cavarero A. (2003), *For More Than One Voice: Toward a Philosophy of Vocal Expression*, Stanford: Stanford University Press, 2005.

Chevreul M. E. (1833), Lettre a M. Ampère sur une classe particulière de mouvements mousculaires. *La Revue des Deux Mondes*, April–June, pp. 258–266.

Chevreul M. E. (1854), *De la baguette divinatoire, du pendule dit explorateur et des tables tournantes, au point de vue de l'histoire de la critique et de la méthode expérimentale*, Mallet-Bachelier, Paris: Book digitiz. from the collections of New York Public Library.

Cimatti F. (1997), Linguaggio tempo e rappresentazioni mentali. Gli stati mentali come eventi spazio temporali. In Ferretti F., and Gola E. (eds.), Scienze cognitive e filosofia della mente, *Il Cannocchiale*, 2, 125.

Clark R. W. (1980), *Freud the Man and the Cause*, New York: Random House.

Cocks G. (1985), *Psychotherapy in the Third Reich: The Goring Institute*, New York: Oxford University Press.

Coleridge S. T. (1817), *Biographia Literaria*, CreateSpace Independent Publishing Platform, 2015.

Conan Doyle A. (1926), *The History of Spiritualism*, London: Cassell.

Conci M. (1996), Why Did Freud Choose Medical School? *Int. Forum Psychoanal.*, 5:123–132.

Conci M. (2000), *Sullivan Revisited, Life and Work*, Trento: Tangram ed, scient, 2010.

Conci M. (2016), Le lettere del giovane Freud a Emil Fluss (1872–1874). *Rivista Psicoanal.*, 62:1057–1084.

Conrotto F. (2009), Ricezione del transfert e processo della cura. *Rivista Psicoanal.*, 55:49–66.

Corsa R. (2010), Il contributo di Sabina Spielrein alla comprensione della schizofrenia – Notazioni storiche sull'istinto di morte. *Rivista Psicoanal.*, 1:73–94.

Cranefield P. F. (1958), Josef Breuer's Evaluation of His Contribution to Psychoanalysis. *Int. J. Psycho-Anal.*, XXXIX:319–322.

Cremerius J. (1985), *Il mestiere dell'analista*, Torino: Bollati Boringhieri.

De Sanctis S. (1899), *I sogni studi psicologici e clinici di un alienista*, Torino: Bocca.

De Waal E. (2010), *The Hare with Amber Eyes: A Hidden Inheritance*, London: Chatto & Windus.

Deutsch F. (1956), Reflection on Freud's One Hundredth Birthday. In Ruitenbeek H. M. (ed.) (1973), *Freud as We Knew Him*. Detroit: Wayne State University Press.

Devereux G., ed. (1953), *Psychoanalysis and the Occult*, New York: International Universities Press.

Drummond H. (1894), *The Ascent Man*, New York: The Lowell lectures James Pott. https://archiv.org

Einstein A. (1945), *Mozart, His Character, His Work*, Oxford: Oxford University Press.

Eissler K. R. (1993), *Three Instances of Injustice*, New York: International Universities Press, Inc.

Ellenberger H. (1970), *The Discovery of the Unconscious; the History and Evolution of Dynamic Psychiatry*, New York: Basic Books.

Elms A. C. (2001), Apocryphal Freud, *Ann. Psychoanal.*, 29:83–104.

Evrard R., Massicotte C., and Rabeyron T. (2017), Freud as a Psychical Researcher: The Impossible. *IMÁGÓ Budapest*, 6(4):9–32.

Fachinelli E. (1974), *Il bambino dalle uova d'oro*, Milano: Feltrinelli.

Falzeder E. (1994), My Grand-Patient, My Chief Tormentor: A Hitherto Unnoticed Case of Freud's and the Consequences. *Psychoanal Q.*, 63:297–331.

Falzeder E. (1996), Whose Freud Is It? In Haynal A., Falzeder E., and Roazen P. (eds., 2005), *Dans les secrets de la psychanalyse et de son histoire*. Paris: PUF.

Falzeder E. (2005), Sigmund Freud et Eugen Bleuler: L'histoire d'une relation ambivalente. In Haynal A., Falzeder E., and Roazen P. (eds.), *Dans les secrets de la psychanalyse et de son histoire*. Paris: PUF.

Falzeder E. (2007), Is There Still an Unknown Freud? *Psychoanal. Hist.*, 9(2).

Falzeder E., and Burnham J. C. (2007), A Perfectly Staged "Concerted Action" Against Psychoanalysis: The 1913 Congress of German Psichiatrists. *Int. J. Psycho-anal.*, 88:1223–1244.

Faraday M. (1854), The Table Turning Delusion. *New Hampshire J. Med.*, IV–V.

Ferenczi S. (1899[1963]), Spiritism. *Psychoanal. Rev.*, 50:139–144.

Ferenczi S. (1908a), A neurozisok Freud tananak megvilagitasaban 6s a pszichoanalizis (The Neuroses in the Light of Freud's Ideas and Psychoanalysis). *Gyogyaszat*, 48:232–235, and 252–255.

Ferenczi S. (1908b), The effect on women of premature ejaculation in men. In *Final Contributions to the Problems and Methods of Psycho-Analysis*. London: Karnac, 291–294.

Ferenczi S. (1908c), Psycho-Analysis and Education. In *Final Contributions*, 280–290.

Ferenczi S. (1911a), On the Organization of the Psycho-Analytic Movement. In *Final Contributions*, 299–307.

Ferenczi S. (1911b), On Obscene Words. In *First Contributions to Psycho-Analysis*. London: Karnak, 132–153.

Ferenczi S. (1911c), On the Part Played by Homosexuality in Pathogenesis of Paranoia. In *First Contributions*, 225–241.

Ferenczi S. (1913a), Stages in the Development of the Sense of Reality. In *First Contributions*, 213–239.

Ferenczi S. (1913b[2005]), Criticisms and Review: C.G. Jung, Transformations and Symbols of the Libido. *Psychoanal. Hist.*, 7:63–79.

Ferenczi S. (1914a), Sensations of Giddiness at the End of the Psycho-Analytic Session. In *Further Contributions to the Theory and Technique of Psycho-Analysis*. London: Hogarth, 239–241.

Ferenczi S. (1914b), On Falling Asleep During Analysis. In *Further Contributions*, 249–250.

Ferenczi S. (1914c), Embarrassed Hands. In *Further Contributions*, 315–316.

Ferenczi S. (1914d), Rubbing the Eyes as a Substitute for Onanism. In *Further Contributions*, 379.

Ferenczi S. (1915a), Restlessness Towards the End of the Hour of Analysis. In *Further Contributions*, 238–239.

Ferenczi S. (1915b), Psychogenic Anomalies of Voice Production. In *Further Contributions to the Theory and Technique of Psycho-Analysis*. London: Karnac Books, 105–109.

Ferenczi S. (1917), My Friendship with Miksa Schachter. *British J. Psychoth.*, 1993, 9:430–433.

Ferenczi S. (1932), *The Clinical Diary*, J. Dupont ed., Cambridge, MA: Harvard University Press, 1995.

Ferenczi S. (1933), Confusion of Tongues between Adults and the Child. In *Final Contributions*, 156–167.

Ferenczi S., and Rank O. (1923), *The Development of Psychoanalysis*, Madison, CT: IUP, 1986.

Flammarion C. (1900), *L'Inconnu et les problèmes psychiques*. http://gallica.bnf.fr/ark:/

Fliess E. (1982), Robert Fliess: A Personality Profile. *Am. Imago*, 39:195–218.

Fliess R. (1942), The Metapsychology of the Analyst. *Psychoanal. Q.*, 11:211–227.

Fliess R. (1953), Countertransference and Counteridentification. *J. Amer. Psychoanal. Assn.*, 1:268–284.

Fliess W. (1893), *Neue Beiträge zur Klinik und Therapie der nasale Reflexneurosen* Deuticke, Wien: Leipzig.

Flournoy T. (1899), *Des Indes à la Planète Mars, étude sur un cas de somnambulisme avec glossolalie*. http://gallica.bnf.fr/ark:/

Flournoy T. (1911), *Esprits et Mediums, mélanges de metapsichique et de Psychologie*. http://gallica.bnf.fr/ark:/

Fodor N. (1971), *Freud, Jung and Occultism*, New York: University Books.

Fodor N. (1979), Les aventures psychiques de Sándor Ferenczi. *Le Coq-Héron*, 213:105–112.

Forrester J. (1990), *The Seductions of Psychoanalysis. Freud, Lacan, Derrida*, Cambridge: Cambridge University Press.

Forrester J., and Cameron L. (1999), A Cure with a Defect' A Previously Unpublished Letter by Freud Concerning *Anna O. Int. J. Psycho-Anal.*, 80:929–942.

Freud A. (1936), *The Ego and the Mechanisms of Defence*, Hove: Routledge, 1992.

Freud M. (1958a), *Sigmund Freud – Man and Father*, New York: J. Aronson.

Freud M. (1958b), Freud: My Father. In Ruitenbeek H. M. (ed., 1973), *Freud as We Knew Him*. Detroit: Wayne State University Press.

Freud S. (1884), On Coca. In *The Cocaine Papers*. New York: Plume, 1975.

Freud S. (1888a), Hypnotism and Suggestion. In Untranslated Freud (1946) (10). *Int. J. Psychoanal*. 27:59–64.

Freud S. (1888b), Hysteria. *SE*, 1:37–59.

Freud S. (1888d). Referat über Obersteiner, *Der Hypnotismus mit besonderer Berücksichtigung seiner klinischen und forensischen Bedeutung*, Wien 1887 (1888). *Gesamm. W.: Texte aus den Jahren 1885 bis 1938*, 105–106.

Freud S. (1889), Review of August Forel's Hypnotism. *SE*, 1:89–102.

Freud S. (1890), Psychical (or Mental) Treatment. *SE*, 7:281–302.

Freud S. (1891), *On Aphasia*, London and New York: Int. UN. Press, 1953.

Freud S. (1892), Preface and Footnotes to the Translation of Charcot's Tuesday Lectures. *SE*, 1:131–143.

Freud S. (1893a), Charcot. *SE*, 3:7–23.

Freud S. (1893b), Frälein Elisabeth von R, Case Histories from *Studies on Hysteria*. *SE*, 2:135–181.

Freud S. (1893a), Frau Emmy von N, Case Histories from *Studies on Hysteria*. *SE*, 2:48–105.

Freud S. (1893b), Miss Lucy R, Case Histories from *Studies on Hysteria*. *SE*, 2:106–124.

Freud S. (1893c), Katharina, Case Histories from *Studies on Hysteria*. *SE*, 2:125–134.

Freud S. (1893d), The Psychotherapy of Hysteria from *Studies on Hysteria*. *SE*, 2:253–305.

Freud S. (1894a), On The Grounds for Detaching a Particular Syndrome from Neurasthenia under The Description 'Anxiety Neurosis'. *SE*, 3:85–115.

Freud S. (1894b), Draft H. Paranoia. *The Complete Letters of Sigmund Freud to Wilhelm Fliess, 1887–1904* 42:107–112.

Freud S. (1894c), The Neuro-Psychoses of Defence. *SE*, 3:41–61.

Freud S. (1895[1950]), Project for a Scientific Psychology. *SE*, 1:281–391.

Freud S. (1896a), Heredity and the Aetiology of the Neuroses. *SE*, 3:141–156.

Freud S. (1896b), Further Remarks on the Neuro-Psychoses of Defence,. *SE*, 3 157–185.

Freud S. (1897), *Infantile Cerebral Paralysis*, University of Miami Press 1968.

Freud S. (1899[1941]), Appendix A Premonitory Dream Fulfilled in *The Interpretation of Dreams*. *SE*, 4:623–625.

Freud S. (1900), The Interpretation of Dreams. *SE, 4:ix-627*.

Freud S. (1901a), The Psychopathology of Everyday Life. *SE*, 6:vii–296.

Freud S. (1901b), On Dreams. *SE*, 5:629–686.

Freud S. (1905a), Fragment of an Analysis of a Case of Hysteria (1905 [1901]). *SE*, 7:1–122.

Freud S. (1905b), Three Essays on the Theory of Sexuality (1905). *SE*, 7:123–246.

Freud S. (1905c), Jokes and their Relation to the Unconscious. *SE*, 8. 1–247.

Freud S. (1907a), Delusions and Dreams in Jensen's Gradiva. *SE*, 9:1–96.

Freud S. (1907b), The Sexual Enlightenment of Children (An Open Letter to Dr. M. Fürst). *SE*, 9:129–140.

Freud S. (1909a), Analysis of a Phobia in a Five-Year-Old Boy. *SE*, 10:1–150.

Freud S. (1909b), Notes upon a Case of Obsessional Neurosis. *SE*, 10:151–318.

Freud S. (1910a), The Antithetical Meaning of Primal Words. *SE*, 11:153–162.

Freud S. (1910b), The Future Prospects of Psycho-Analytic Therapy. *SE*, 11:139–152.

Freud S. (1910c), Five Lectures on Psycho-analysis. *SE*, 11:1–56.

Freud S. (1910d), Leonardo Da Vinci and a Memory of his Childhood. *SE*, 11:57–138.

Freud S. (1911a), Psycho-Analytic Notes on an Autobiographical Account of a Case of Paranoia (Dementia Paranoides). *SE*, 12:1–82.

Freud S. (1911b), Formulations on the Two Principles of Mental Functioning. *SE*, 12:213–226.

Freud S. (1912a), Recommendations to Physicians Practising Psycho-Analysis *SE*, 12:109–120.

Freud S. (1912b), A Note on the Unconscious in Psycho-Analysis. *SE*, 12:255–266.

Freud S. (1913–1915), Further Recommendations on the Technique of Psycho-Analysis (I, II, III). *SE*, 12, 121–171.

Freud S. (1913a), The Theme of the Three Caskets. *SE*, 12:289–302.

Freud S. (1913b), The Disposition to Obsessional Neurosis, a Contribution to the Problem of the Choice of Neurosis. *SE*, 12:311–326.

Freud S. (1913c), Two Lies Told by Children. *SE*, 12:303–310.

Freud S. (1913d), Preface to Bourke's Scatalogic Rites of all Nations. *SE*, 12:333–338.

Freud S. (1913e), Totem and Taboo: Some Points of Agreement between the Mental Lives of Savages and Neurotics (1913 [1912–13]). *SE*, 13: vii-162.

Freud S. (1914a), On the History of the Psycho-Analytic Movement. *SE*, 14:1–66.

Freud S. (1914b), On Narcissism: An Introduction. *SE*, 14:67–102.

Freud S. (1914c[1918]), From the History of an Infantile Neurosis. *SE*, 17:1–124.

Freud S. (1915), The Unconscious. *SE*, 14:159–215.

Freud S. (1916), Some Character-Types Met with in Psycho-Analytic Work. *SE*, 14:309–333.

Freud S. (1916–17), *Introductory Lectures on Psycho-analysis*. *SE*, 15:1–463.

Freud S. (1919a), The 'Uncanny'. *SE*, 17:217–256.

Freud S. (1919b), 'A Child Is Being Beaten'. *SE*, 17:175–204.

Freud S. (1920a), A Note on the Prehistory of the Technique of Analysis. *SE*, 18:261–265.

Freud S. (1920b), Beyond the Pleasure Principle. *SE*, 18:1–64.

Freud S. (1921), *Group Psychology and the Analysis of the Ego*. *SE*, 18:65–144.

Freud S. (1923a), Dr. Sándor Ferenczi (on his 50th Birthday). *SE*, 19:265–270.

Freud S. (1923b), The Ego and the Id. *SE*, 19:1–66.

Freud S. (1925a), An Autobiographical Study. *SE*, 20:1–74.

Freud S. (1925b), Josef Breuer. *SE*, 19:277–280.

Freud S. (1925c), Some Additional Notes on Dream-Interpretation as a Whole, *SE*, 19:123–138.

Freud S. (1926), The Question of Lay Analysis. *SE*, Freud 20:177–258.

Freud S. (1933a), Lecture XXX Dreams and Occultism, *New Introductory Lectures on Psycho-Analysis*. *SE*, 22:1–182.

Freud S. (1933b), Sándor Ferenczi. *Int. J. Psycho-Anal*. *SE*, 14:297–299.

Freud S. (1936), A Disturbance of Memory on the Acropolis. *SE*, 22:237–248.

Freud S. (1937), Analysis Terminable and Interminable. *SE*, 23:209–254.

Freud S. (1939), Moses and Monotheism: Three Essays. *SE*, 23:1–138.

Freud S. (1888), Some Points For a Comparative Study of Organic and Hysterical Motor Paralyses. *The Standard Edition of the Complete Psychological Works of Sigmund Freud SE* 1.

Freud S. (1938), *BBC Interview*. http://www.archive.org/details/Bbcnterview

Fromm E. (1959), *Sigmund Freud's Mission*, New York: Peter Smith.

Gabbard G. O. (1995), The Early History of Boundary Violations in Psychoanalysis. *J.A.P.A.*, 43:1115–1136.

Gallini C. (1983), *La sonnambula meravigliosa. Magnetismo e ipnotismo nell'Ottocento italiano*, Roma: L'asino d'oro, 2013.

Gay P. (1988), *Freud A Life for Our Time*, New York, London: Anchor book.

Geller J. (2000), Some More Additional 'Day Residues': The First Review of "Studien Uber Hysterie", Ilona Weiss and the *Dream of Irma's Injection*. *Psychoanal. Hist.*, 2:61–75.

Giacomelli R. (2008), Pseudo-glossolalia e affioramenti linguistici inconsci nella personalità profonda della celebre medium Helene Smith. *ACME LXI*, II:311–321. http://www.ledonline.it/acme/allegati/Acme-08-II-14-Giacomelli.pdf.

Gilhooley D. (2002), Misrepresentation and Misreading in the Case of Anna O. *Modern Psychoanal.*, 27(1):75–100.

Gillespie W. (1979), Ernest Jones: The Bonny Fighter. *Int. J. Psycho-Anal.*, 60:273–279.

Goleman D. (1985), Freud's Mind: New Details Revealed in Documents. *New York Times*, 12 November, pp. C1, C3.

Gori E. C. (1985), Freud e la costruzione del linguaggio. In AA.VV. (ed., 1987), *La Cultura Psicoanalitica*. Pordenone: Studio Tesi.

Graf M. (1942), Reminiscences of Professor Sigmund Freud. *Psychoanal. Q.*, 11:465–476.

Granoff W. (2000), *Lacan, Ferenczi et Freud*, Paris: Gallimard.

Grosskurth P. (1991), *The Secret Ring: Freud's Inner Circle and the Politics of Psychoanalysis*, London: Cape.

Grubrich Simitis I. (1981), Siegfried Bernfeld: Historiker der Psychoanalyse und Freud-Biograph. *Psyche – Z Psychoanal.*, 35(5):397–434.

Grubrich Simitis I., and Lortholary B. (2012), Germes de concepts psychanalytiques fondamentaux. À propos des lettres de fiancés de Sigmund Freud et Martha Bernays. *Revue franç. Psychanal.*, 3(76):779–795.

Gurney E., Myers F. W. H., and Podmore F. (1886), *Phantasms of the Living*. https://archive.org/

Gyimesi J. (2018), Spiritualism, Telepathy and the Budapest School of Psychoanalysis. In Pócs E. (ed.), *Body, Soul, Spirits and Supernatural Communication*. Necastle: Cambridge scholars pub.

Hale N. G. (1971), *Freud and the Americans*, New York: Oxford University Press.

Hale N. G. (1971), *James. Jackson Putnam and Psychoanalysis Letters between Putnam and Freud, Jones, James, Ferenczi, and Morton Prince, 1877–1917*, Ann Arbor: University of Michigan.

Harris A., and Kuchck S., eds. (1993), *The Legacy of Sándor Ferenczi*. Hillsdale, NJ, London: The Analytic Press.

Hartman F. R. (1983), A Reappraisal of the Emma Episode and the Specimen Dream. *J. Amer. Psychoanal. Assn.*, 31:555–585.

Haynal A. (2002), *Disappearing and Reviving: Sándor Ferenczi in the History of Psychoanalysis*, Abingdon: Routledge.

Heimann P. (1950), On Counter-Transference. *Intern. J. Psychoanal.*, 31:81–84.

Hirschmüller A. (1978), *The Life and Work of Josef Breuer*, New York: University press, 1989.

Hirschmüller A. (1986), Briefe Josef Breuers an Wilhelm Fließ 1894–1898. *Jahrb. Psychoanal.*, 18:239–261.

Hirschmüller A. (1994), The Genesis of the *Preliminary Communication* of Breuer and Freud. *Cahiers Psychiatriques Genevois*, Special Issue:17–30.

Illies F. (2013), *1913: The Year before the Storm*, London: Profile Books, 2014.

James W. (1889), Notes on Automatic Writing. In James W. (1869–1909), *Essays in Psychical Research*. Cambridge, MA: Harvard University Press, 1986.

Janet P. (1889), *L'Automatisme Psychologique: essai de psychologie expérimentale sur les formes inférieures de l'activité humaine.* https://archive.org

Jastrow J. (1900), *Fact and fable in psychology*, Boston, New York: Houghton Mifflin. https://www.questia.com/library/702816/fact-and-fable-in-psychology

Jones E. (1908), *Rationalisation in everyday life in Papers on Psychoanalysis*. London: Karnak Books, 1977.

Jones E. (1910), On the Nightmare. *Am. J. Insanity*, 66:383–417.

Jones E. (1912), *Der Alptraum in seiner Beziehung zu gewissen Formen des mittelalterlichen Aberglaubens*, Lipsia e Vienna: Hanse.

Jones E. (1916), *The Theory of Symbolism in Papers on Psychoanalysis*, London: Karnak Books, 1977.

Jones E. (1931), *The Elements of Figure Skating*, London: Methuen.

Jones E. (1945), Reminiscent Notes on the Early History of Psycho-Analysis in English-Speaking Countries. *Int. J. Psycho-Anal.*, 26:8–10.

Jones E. (1953), *Sigmund Freud Life and Work I: The Young Freud 1856–1900*, London: Hogarth Press.

Jones E. (1955), *Sigmund Freud Life and Work II: Years of Maturity 1901–1919*, London: Hogarth Press.

Jones E. (1956), Our Attitude Toward Greatness. *J. Amer. Psychoanal. Assn.*, 4:626–643.

Jones E. (1957), *Sigmund Freud Life and Work III: The Last Phase 1919–1939*, London: Hogarth Press.

Jones E. (1959), *Free Associations Memories of a Psycho-Analyst*. New York: Basic books.

Jung C. G. (1902), *On the Psychology and Pathology of So-Called Occult Phenomena Collected Works*, 1:1–88, Princeton: Princeton University Press.

Jung C. G. (1906), Psychoanalysis and Association Experiments, *CW*, 2:288–317.

Jung C. G. (1907a), The Psychology of Dementia Praecox, *CW*, 3:1–151.

Jung C. G. (1907b), On psychophysical relations of the associative experiment, *CW*, 2:483–491.

Jung C. G. (1908), The Freudian Theory of Hysteria, *CW*, 4:10–18.

Jung C. G. (1909), Psychic Conflicts in a Child, *CW*, 7:1–35.

Jung C. G. (1912), The Concept of Libido: Symbols of Transformation, *CW*, 5:132–170.

Jung C. G. (1913–1930), *The Red Book*, Shamdasani S. ed. New York: Norton e Comp, 2009. https://archive.org/details/jungtheredbook.

Jung C. G. (1961), *Memories, Dreams, Reflections*, Jaffé A. ed. New York: Random House, 2011.

Kamieniak J. P. (2000), *Freud: un enfant de l'humour?* Lausanne: Delachaux.

Kandel E. R. (2012), *The Age of Insight: The Quest to Understand the Unconscious in Art, Mind, and Brain, from Vienna 1900 to the Present*, New York: Random House.

Kandinsky W. (1912), *On the Question of Form. The Blue Rider*. In Herschel B. Chipp (ed.), *Theories of Modern Art: A Source Book by Artists and Critics*. Berkeley and Los Angeles: University of California Press, 1968.

Kerr J. (1993), *A Most Dangerous Method: The Story of Jung, Freud, and Sabina Spielrein*, New York: Vintage, 1994.

Kress-Rosen N., (1993), *Trois figures de la passion*, Paris: Springer/Arcanes.

Kris E. (1950), Introduction, to Freud S. (1950). In Bonaparte M., Freud A., and Kris E (eds.), *The Origins of Psychoanalysis: Letters to Wilhelm Fliess, Drafts and Notes From the Years 1887–1902*. New York: Basic Books, London: Imago Publishing Co., 1950.

Kuhn P. (2015), In 'The Dark Regions of the Mind' A Reading for the Indecent Assault in Ernest Jones's 1908 Dismissal from the West End Hospital for Nervous Diseases. *Psychoanal. His.*, 17:7–57.

La Cecla F. (2006), *Surrogati di presenza. Media e vita quotidiana*, Milano: Bruno Mondadori.

Laplanche J. (1997), The Theory of Seduction and the Problem of the Other. *Int. J. Psycho-Anal.*, 78:643–666.

Lavaggetto M. (1985), *Freud, la letteratura e altro*, Torino: Einaudi, 2001.

Lipps T. (1903), *Ästhetik. Psychologie des Schönen und der Kunst. I. Grundlegungen der Ästhetik*, Hamburg: Voss.

Lomas R. (1999), *The Man Who Invented the Twentieth Century: Nikola Tesla, Forgotten Genius of Electricity*. London: Headline.

Lopez (2001), Andare avanti: l'emancipazione. In Pierri M., and Racalbuto A. (eds.), *Maestri e allievi*. Milano: Angeli.

Lopez D. (1979), Un sogno di Jung. *Gli Argonauti*, I(3):235–237.

Lopez D. (1986), *La trama profonda*, Milano: Coliseum.

Lopez D. (1991), *Il mondo della persona*, Milano: Cortina.

Lopez D. (2004), Problemi di identità e differenza in Freud e nei suoi allievi. *Gli Argonauti*, 101:143–158.

Lopez D., and Zorzi L. (1999), *La sapienza del sogno*, Milano: Elsevier Dunod.

Luckhurst R. (2002), *The invention of telepathy: 1870–1901*, Oxford: Oxford University Press.

Maddox B. (2006), *Freud's Wizard: Ernest Jones and the Transformation of Psychoanalysis*, Cambridge: Da Capo press.

Magris C. (1980), postfazione a. In Peters H. F. (ed.), *Mia sorella, mia sposa. La vita di Lou Andreas-Salomé*. Milano: Mondadori, 1962.

Mahony P. (2002), Freud's Writing. *J. Amer. Psychoanal. Assn.*, 50:885–907.

Malcolm J. (1983), *In the Freud Archives*, New York: Review Books.

Mannoni O. (1968), *Freud, The Theory of Unconscious*. London, New York: Verso, 1971.

Marinelli L., and Mayer A. (2002), *Dreaming by the Book*, New York: Other Press, 2003.

Markel H. (2011), *An Anatomy of Addiction*: *Sigmund Freud, William Halsted, and the Miracle Drug Cocaine*, New York: Pantheon Books.

Martín-Cabré L. (1997), Freud-Ferenczi: Controversy Terminable and Interminable. *Int. J. Psychoanal.*, 78:105–114.

Martín-Cabré L. (2005), Il pensiero clinico di Ferenczi: una strana e silenziosa trasmissione. In Bonomi C. (ed., 2006), *Sándor Ferenczi e la psicoanalisi contemporanea*. Roma: Borla.

Martín-Cabré L. ed. (2016), *Autenticità e reciprocità. Un dialogo con Ferenczi*, Milan: Franco Angeli.

Masson J. M. (1984), *The Assault on Truth. Freud's Suppression of Seduction Theory*. New York: Farrar, Strauss & Giroux.

Masson J. M. (1985a), Introduction to *The Complete Letters of Sigmund Freud to Wilhelm Fliess*, 1887–1904. Cambridge, MA: Belknap, 1–13.

Masson J. M. ed. (1985b), *The Complete Letters of Sigmund Freud to Wilhelm Fliess, 1887–1904*, Cambridge, MA: Harvard University Press.

May U. (2019), Willy Haas: Ein junger Münchner Philosoph in Analyse bei Freud. *Luzifer-Amor: Zeitschrift zur Geschichte der Psychoanalyse*, 32:35–57.

Meghnagi D. (2000), Introduzione to Freud S., Zweig A. (2000), *Lettere sullo sfondo di una tragedia (1927–1939)*, Venezia: Marsilio.

Mèszàros J. (1993), Il periodo preanalitico di Ferenczi nel contesto delle correnti cuturali *fin de siècle*. In Aron L., and Harris H. (eds.), *L'eredità di Sándor Ferenczi*. Roma: Borla, 1998.

Mèszàros J. (2008), *Ferenczi and Beyond: Exile of the Budapest School and Solidarity in the Psychoanalytic Movement During the Nazi Years*, Abingdon: Routledge, 2018.

Migone P. (1984), Cronache psicoanalitiche: il caso Masson. *Psicoterapia e Scienze Umane*, XVIII(4):32–62.

Migone P. (2002), Storia dello scandalo Masson. *Il Ruolo Terapeutico*, 89:58–69; 90:47–58, 91:67–77.

Miller A. I. (2009), *Jung, Pauli, and the Pursuit of a Scientific Obsession*, New York: W. W. Norton & Company.

Milutis J. E. (2006), *The Nothing That Connects Everything*, Minneapolis: NED-Univ. Minnesota Press.

Money-Kyrle R. E. (1979), Looking Backwards and Forwards. *Intern. Review of Psycho-Ana.*, 6:265–272.

Moreau C. (1976), *Freud et l'occultisme: l'approche freudienne du spiritisme, de la divination, de la magie et de la telepathie*, Toulouse: Edouard Privat.

Morselli E. (1908), *Psicologia e spiritismo. Impressioni e note critiche sui fenomeni medianici di Eusapia Palladino*. https://archive.org/

Munthe A. (1929), *The Story of San Michele*, New York: Dutton.

Myers F. W. H. (1897), Report of the Literary Committee. *The Proceedings of the Society of Psychical Research*, XIII.

Myers F. W. H. (1903), *Human Personality and Its Survival of Bodily Death*, London: Longmans Green.

Napolitano F. (1999), *La filiazione e la trasmissione nella psicoanalisi*, Milano: Angeli.

Natale S. (2012), Fantasie Mediali. Storia dei media e la sfida dell'immaginario. *Studi Culturali*, IX(2):269–284. https://dspace.lboro.ac.uk/

Natale S. (2016), *Supernatural Entertainments. Victorian Spiritualism and the Rise of Modern Media Culture*, University Park: Penn. State University Press.

Nietzsche F. (1882), *The Gay Science*, trans. T. Common, New York: Dover Publications, 2012.

Nunberg H., Federn E., eds. (1962–1975), *Minutes of Vienna Psychoanalytic Society*, 4 vols, Madison CT: International Universities Press.

Oeri A. (1977), Some Youthful Memories. In McGuire W. R., and Hull F. C. (eds), *C.G. Jung Speaking: Interviews and Encounters*. London: Macmillan.

Papini G. (1934), A Visit to Freud. In Ruitenbeek H. M., (ed., 1973), *Freud as We Knew Him*. Detroit: Wayne State University Press.

Paskauskas R. A. (1993), Notes to Letter from Ernest Jones to Sigmund Freud, October 12, 1919. In *The Complete Correspondence of Sigmund Freud and Ernest Jones 1908–1939*. Cambridge, MA: Harvard University Press.

Paskauskas R. A. (1994), Ferenczi's Analysis of Jones in Relation to Jones' Self-Analysis. *Cahiers Psychiatriques Genevois*, Special Issue:225–255.

Peters H. F. (1962), *My Sister, My Spouse: A Biography of Lou Andreas-Salome*. New York: Norton.

Petrella F. (1995), La "materia inquieta" e le sue trasformazioni. In Mattioli Rossi I. (ed.), *Boccioni 1912. Materia*. Milano: Mazzotta.

Pfeiffer E. (1972), Notes to Sigmund Freud and Lou Andreas-Salomé: Letters. *Int. Psycho-Anal. Lib.*, 89:211–240.

Phillips A. (2001), *Houdini's Box. The Art of Escape*, New York: Pantheon Books.

Pierri M. (2001b), Eredità e creatività nella tradizione. In Pierri M. and Racalbuto A. (eds.), *Maestri e allievi. Trasmissione del sapere in psicoanalisi*. Milano: Angeli.

Pierri M. (2001c), Errori dei padri e dei figli. In Pierri M. (ed.), *Qui e ora . . . con me. Aperture psicoanalitiche all'esperienza contemporanea*. Torino: Bollati Boringhieri.

Pierri M. (2002), Countertransference Phenomena and Telepathy: The Great Discovery of Ferenczi, International Conference *Clinical Sándor Ferenczi*, Torino.

Pierri M. (2010), Coincidences in Analysis: Sigmund Freud and the Strange Case of Dr. Forsyth and Herr von Vorsicht. *Int. J. Psychoanal.*, 91:745–772.

Pierri M. (2012a), Lo stetoscopio del dr. Ernest Jones. *Riv. Psicoanal.*, 4:903–911.

Pierri M. (2012b), Freud e le due profezie non avverate. *Gli Argonauti*, 135:305–320.

Pierri M. (2015), Il lavoro sull'argine: esplorazioni psicoanalitiche in psichiatria fra genitori e figli. *Riv. Sper. Freniatria*, 139(1):75–90.

Pierri M. (2016), Una lontana vicinanza, ovvero la moneta d'oro della telepatia, introduzione a S. Freud (2016), *Telepatia*, BUR Rizzoli, Milano.

Pierri M. (2018), *Un enigma per il dottor Freud. La sfida della telepatia*, Franco Angeli Milano.

Pierri M. (2021), 1921, un anniversario freudiano poco conosciuto: *Vorbericht-Nachtrag*, il "saggio segreto" dello Harz sul transfert del pensiero. *Riv Psicoanalisi*, LXVII(3):597–616.

Pierri M. (2022), *Freud and the Forsyth Case. Thought Transference and Coincidences in Vienna*. London: Routledge.

Pollock G. H. (1968), The Possible Significance of Childhood Object Loss in the Josef Breuer-Bertha Pappenheim (*Anna O.*)-Sigmund Freud Relationship-I. Josef Breuer. *J. Am. Psychoanal. Ass.*, 16:711–739.

Prochnik G. (2006), *Putnam Camp: Sigmund Freud, James Jackson Putnam and the Purpose of American Psychology*. New York: Other Press, *Kindle*.

Proust M. (1920a), *Guermantes Way*, Chatto e Windus London internet archiv/org.

Proust M. (1920b), *In Search of Lost Time, Volume III: The Guermantes Way*, New York: Random House Publishing Group 2000.

Putnam J. J. (1909), Personal impressions of Sigmund Freud and his Work. In Ruitenbeek H. M. (ed., 1973), *Freud as We Knew Him*. Detroit: Wayne State University Press.

Quinodoz J. M. (2004), *Reading Freud a Chronological Exploration of Freud's Writings*, Abingdon: Routledge, 2005.

Rank O., (1912), *The Incest Theme in Literature and Legend: Fundamentals of a Psychology of Literary Creation*, Baltimore: Johns Hopkins University Press, 1992.

Rhine J. B. (1949), Storia delle prime ricerche. In Guarino S., (ed.), *Telepatia di ieri, di oggi, di domani*. Napoli: IEM, 1972.

Richet C. (1922), *Traité di metapsychique*. http://Gallica.bnf.fr/ark. Thirty Years of Psychical Research – A treatise on Metapsychics 1923 The Macmillan Company, New York. https://archive.org/details/30YearsRichet

Roazen P. (1969), *Brother Animal: The Story of Freud and Tausk*. Knopf: NewYork.

Roazen P. (1975), *Freud and His Followers*. Knopf: NewYork.

Roazen P. (1993), *Meeting Freud's Family*, Amherst: University of Massachusetts Press.

Roazen P. (1995), *How Freud Worked: First-Hand Accounts of Patients*, New York: Jason Aronson.

Roazen P. (1998), Elma Laurvik, Ferenczi's Stepdaughter. *Am. J. Psychoanal.*, 5:271–286.

Roazen P. (2001), *The historiography of Psychoanalysis*, London: Transaction Pub. New Brunswick.

Roazen P. (2003), *On the Freud Watch: Public Memories*, London: F.A.b.

Roazen P. (2004), Presentation. *Le Coq-Héron*, 2(177): 9–10.

Robert M. (1964), *The Psychoanalytic Revolution: Sigmund Freud's Life and Achievement*. New York: Harcourt Brace and World, 1966.

Rodrigué E. (1996), *Sigmund Freud: el siglo del psicoanalisis*, Buenos Aires: Editorial Sudamericana, 2 vol.

Rossi P. G. (2006), Diario di viaggio intorno all'Uomo dei lupi. *Rivista Psicoanal.*, 52:1035–1055.

Roudinesco E. (2014), *Freud in His Time and Ours*, Cambridge, MA: Harvard University Press, 2016.

Roussillon R. (1995), *Logiques et archéologiques du cadre psychanalytique*, Paris: PUF.

Roussillon R. (1995[1997]), *Il setting psicoanalitico*, Roma: Borla.

Ruitenbeek H. M., ed. (1973), *Freud as We Knew Him*, Detroit: Wayne State University Press.

Saban M. (2016), Jung, Winnicott and the Divided Psyche. *J. Anal. Psychology*, 61:329–349.

Sabourin P. (1985), *Ferenczi paladin et grand vizir secret*, Paris: Editions Universitaires.

Sachs H. (1945), *Freud: Master and Friend*, London: Imago pub.

Salyard A. (1992), Freud's Narrow Escape and the Discovery of Transference. *Pychoanal. Psychol.*, 9(3):347–367.

Sartre J. P. (1958), *The Freud Scenario*, London, New York: Verso, 2013.

Schatzman M. (1973), *Soul Murder: Persecution in the Family*, New York: Random House.

Schnitzler A. (1968), *My Youth in Vienna*, trans. C. Hutter. New York: Holt, Rinehart and Winston.

Schreber D. P. (1903), *Memoirs of My Nervous Illness 1988*, Cambridge, MA: Harvard University Press.

Schröter M. (2003), Fliess Versus Weininger, Swoboda and Freud: The Plagiarism Conflict of 1906 Assessed in the Light of the Documents. *Psychoanal. Hist.*, 5(2):147–173.

Schur M. (1972), *Freud: Living and Dying*, London: Hogarth Press and the Institute of Psycho-Analysis.

Schwartz J. (1999), *Cassandra's Daughter: A History of Psychoanalysis*, London, New York: Karnak, 2003.

Sconce J. (2000), *Haunted Media: Electronic Presence from Telegraphy to Television*, Durham: Duke University Press.

Searles H. F. (1958), The Schizophrenic's Vulnerability to the Therapist's Unconscious Processes in (1986) *Collected Papers on Schizophrenia and Related Subjects*, Abingdon: Routledge, 2018.

Shamdasani S. (1998), *Cult Fictions: C. G. Jung and the Founding of Analytical Psychology*, London: Routledge.

Silver A. L. (2006), L'influenza di Ferenczi negli stati Uniti. In Bonomi C. (ed.), *Sándor Ferenczi e la psicoanalisi contemporanea*. Roma: Borla.

Skues R. (2019), Elfriede Hirschfeld: Freuds Großpatientin und Hauptplage neu betrachtet. *Luzifer-Amor: Zeitschrift zur Geschichte der Psychoanalyse*, 32:58–69.

Skues R. (2019), Freuds Entzauberung der Telepathie. *Analyse der Gedankenübertragung und die Geschichte eines unveröffentlichten Aufsatzes Luzifer-Amor*, 63:7–34.

Skues R. A. (2006), *Sigmund Freud and the History of Anna O. Reopening a Closed Case*, New York: Palgrave Macmillan.

Spadolini B. (2004),*Educazione e società. I processi storico-sociali in Occidente*, Roma: Armando.

Spengler O. (1919), *The Decline of the West, Vol. I: Form and Actuality*, Must Have Books.

Spielrein S. (1911), Über den psychologischen Inhalt eines Falles von Schizophrenie – Dementia Praecox [On the psychological content of a case of schizophrenia – Dementia praecox]. *Jahrbuch für psychoanalytische und psychopathologische Forschungen*, 3(1):329–400.

Spielrein S. (1912), Destruction as Cause of Becoming. *Psychoanalysis and Contemporary Thought*. 1995(18):85–118.

Steiner R. (1991), To Explain Our Point of View to English Readers in English Words. *Int. Rev. Psycho-Anal.*, 18:351–392.

Steiner R. (2013), *Die Brautbriefe*: The Freud and Martha Correspondence. *Int. J. Psycho-Anal.*, 94(5):863–936.

Stewart W. A. (1967), *Psychoanalysis: The First Ten Years, 1888–1898*, Abingdon: Routledge, 2015.

Swales P. (1982), Freud, Fliess, and Fratricide: The Role of Fliess in Freud's Conception of Paranoia. In L. Spurling (ed.), *Sigmund Freud. Critical Assessments, vol. 1: Freud and the Origins of Psychoanalysis*. London/New York: Routledge, 1989.

Swales P. (1986), Freud, His Teacher, and the Birth of Psychoanalysis. In P. Stepansky (ed.), *Freud Appraisals and Reappraisals*, V. 1. Abingdon: Routledge, 2020.

Szymborska W. (2002), "Receiver," *Moment* in *Map Collected and Last Poems*, Clare Cavanagh, Stanislaw Baranczak translators, Boston: Houghton Mifflin Harcourt, 2015.

Tanner A. (1910), *Studies in Spiritism*, New York: D. Appleton and Company.

Thompson C. (1964), Interpersonal Psychoanalysis. In *The Selected Papers of Clara M. Thompson*. New York: Basic Books.

Tögel C. (2009), Sigmund Freud's Practice: Visits and Consultation, Psychoanalyses, Remuneration. *Psychoanal. Q.*, 78(4):1033–1058.

Traversa C. (1984), Non facciamone una tragedia. *Gli Argonauti*, 6(22):239–247.

Trotter W. (1916), *Instincts of the Herd in Peace and War*, London: T. F. Unwin ltd. https://archive.org/details/instinctsofherdi00trot

Twain M. (1884), Letter to William Barrett. *J. Soc. Psychic. Res.*:166–167.

Twain M. (1891), Mental Telegraphy. A Manuscript with a History, *Harper's Monthly Magazine*, December 1891; in *The American Claimant, and Others Stories and sketches*. https://archive.org/

Twain M. (1895), Mental Telegraphy Again. *Harper's Monthly Magazine*, September.

Twain M. (1906), *Autobiography*, 2 vol. Project Gutenberg of Australia eBook.

Varèze C. (1948), *Allan Kardec*, Roma: Ed. Mediterranee, 2001.

Vigneri M. (2001), L'incidente di Palermo. *Sicilia*, 2:91, pièce, 26/03/2010, https://www.spiweb.it/category/chi-siamo/riferimenti-storici/freud-in-italia/.

Vischer R. (1873), On the Optical Sense of Form: A Contribution to Aesthetics. In Mallgrave H. F., and Ikonomou E. (eds., 1994), *Empathy, Form, and Space: Problems in German Aesthetics, 1873–1893*. Santa Monica: Chicago University Press.

von Hofmannsthal H. (1922), Vienna Letter. *The Dial*, 73:426–433.

Waldeyer W. (1891), Uber einige Newere Forschungenim Gebiete der Anatomie des Centralnerven-systems. *Deu. Med. Wochenschrif*, 17:1213–1356.

Waldner F. (1962), *Mes Aventures Surnaturelles*, Paris: Fayard.

Weiss E. (1954), Impression of S. F., *Sigmund Freud Papers*: Interviews and Recollections, 1914-1998; Set A, Recollections; Weiss, Edoardo, undated; Folder 2.

Weiss E. (1970a), An Impression of Freud. In Ruitenbeek H. M. (ed.), *Freud as We Knew Him*. Detroit: Wayne State University Press, 1973.

Weiss E. (1970b), *Sigmund Freud as a Consultant: Recollections of a Pioneer in Psycho-analysis*, Hove: Routledge, 1991.

Wilmers M. K. (2010), *The Eitingons: A Twentieth-Century Story*, London: faber faber.

Winnicott D. W. (1958a), Ernest Jones: Funeral Addresses – Spoken at Golders Green Crematorium on Friday, 14 February, 1958. *Int. J. Psycho-Anal.*, 39:304–307.

Winnicott D. W. (1958b), Ernest Jones. *Int. J. Psycho-Anal.*, 39:298–304.

Winnicott D. W. (1962), Recensione a *Lettere di S. Freud 1873–1939*. In (1989) *Esplorazioni Psicoanalitiche*. Milano: Cortina, 1995.

Winnicott D. W. (1964), Memories, Dreams, Reflections: By C. G. Jung. (London: Collins and Routledge, 1963). *Int. J. Psycho-Anal.*, 45:450–455.

Wiseman R. (2011), *Paranormality: Why We See What Isn't There*, London: Macmillan Publishers Ltd.

Young-Bruehl E. (1988), *Anna Freud. A Biography*, New Haven: Yale University Press.

Zaretsky E. (2004), *Secrets of the Soul*, New York: Alfred A. Knopf.

Zweig S. (1931), *Mental Healers: Franz Mesmer, Mary Baker Eddy, Sigmund Freud*, 1932 New York: Ungar, 1962. http://archiv.org

Zweig S. (1942), *The World of Yesterday: Memories of a European*, Hallam ed., London, 1953. http://archiv.org

Index

Abraham, Karl 55, 112, 158; death of
225; Eitingon and 224; essay on cult of
god Aten 222; Fliess' influence on 63;
Salomé as pupil of 211; secret council
and 223; Stekel and 231
abstinence: alcohol 137; sex 19; smoking
68, 163
Accerboni Pavanello, A 236, 237
Achensee (Tyrol) 65, 83, 86
Adler, Alfred 84, 159, 176; resignation
from International Association 178
Adlerian movement 170
Adorno, T 96
alchemy 4, 132, 217
Albrecht A 140
Allgemeines Krankenhaus, Vienna 14
Alzheimer, Alois 159
amnesia 9
anaesthesia or anaesthetic 1, 2, 7; cocaine
as 20; as sign of hysteria 25, 28, 30;
somatic 21
Andreas, F C 211
Andreas (von) Salomé, Lou 57, 79, 185;
brief biography of 210; in dialogue with
Freud 211; occultism and 234, 241;
recollections of Jung 230
animal magnetism 5–6, 102
animals, languages of 15
"Animism, Magic, and the Omnipotence
of Thoughts" (Freud) 226
Annales des Sciences Psychiques 103
"Anna O" 10, 21, 23, 27, 28, 35, 39–44,
45–47, 51–53, 69
antisemitism 65, 70n18, 188
Anzieu, D 51, 58, 71
aphrodisiac 2
Aphrodite 207
arc en ciel (symptom of hysteria) 26

Ariman (lower God) 165, 174
Aschaffenburg, Gustav 39, 112, 194
astrologer 67, 168; Ferenczi as "court
astrologer" 169–170; Waldner 245
"Astrologer and Psychoanalyst," letters to
Freud from 244–245
astrology 105; Jung's immersion in 190,
220, 230; symbols of 218
automatogramaphone 96
automatic talking 32
automatic writing 94, 98, 104, 110, 131,
143; telepathic 105
automatism 94
autosuggestion 8, 110
Avril, Jane 26

Babinski, Joseph 27
Baker, Mary 93
Balint, Michael 129, 183, 184, 241
Balsamo, M 176
Barnum & Bailey's Circus 94
Bell, Alexander Graham 101; *see also*
telephone
Benjamin, Walter 100
Bergson, Henri 97, 98, 237
Berlin Psychoanalytic Institute 202n4
Berlin Society of Physics 65; purging
of 189
Bernays family 77
Bernays girls 20, 196
Bernays, Martha *see* Freud, Martha
Bernays
Bernays, Minna 18, 79, 115
Bernhardt, Sarah 26
Bernheim, Hippolyte 8, 9, 14, 29, 30, 35
Binswanger, Ludwig 113–115, 220–221
bisexuality 66, 85, 86; *Human Bisexuality*
(Freud) 83; psychic 11

Bishop, Washington Irving 99
bisexuality 11, 66, 83, 85, 86
Blanton, S 216n10
Bleuler, Eugen: Adler and 218; coldness
  towards Freud 176; as director of
  Burghölzli hospital 10, 39, 111;
  epistolary relationship with Freud 112,
  200, 214; first psychoanalytic Congress
  in Salzburg established by 112; Jung's
  relationship with 214–215, 218, 232;
  review of Freud's *Interpretation of
  Dreams* 84; Schreber case (Freud) and
  165; séances with 218; Spielrein case
  (Jung) and 122, 124
Bonaparte, Marie 57, 63–64, 79, 86
Bonnet, J 210
Bolognini, Stefano xi–xii
Bondy, Ida 55, 65
Bondy, Melanie 59n10
Börne, L 71–72
Botella, C and S 76
Braid, James 7
Brentano, Franz 13
Breuer, Josef: cathartic method of 140;
  charisma of 25, 49; Ferenczi and
  186; Freud and 28, 29, 30, 34–47,
  51–52, 54, 58, 61–62, 65–68, 76, 77,
  104, 105, 120; hysteria and 34–47;
  Myers' consideration of works of 105;
  *Psychical Mechanisms of Hysterical
  Phenomena* 130; transferential
  experience and *defect* of 49, 50;
  treatment of "Anna O" by 10–11, 20–22,
  27, 35, 47, 51–52, 54
Breuer, Mathilde 55
Brill, A A 158, 208, 212; at *Burghölzli* 111,
  112; Freud, Jones, and 196; in New York
  City 138; New York Psychoanalytic
  Society and 198
Britton, R 47n22
Brouardel, Paul 27, 32n5
Brouillet, Pierre André 27
Brücke, Ernest W 13, 14, 65
Bruni, Alessandro 247n3
Buffalo Bill's Wild West Show 94
Bunker, Chang and Eng 94
Burghölzli psychiatric hospital: "ardent
  and enthusiastic" circle of psychiatrists
  at 111–112; Bleuler as director 10, 39,
  111; community style at 214; Forel as
  director of 9, 10, 39; Jones at 194, 224;
  Jung's arrival and work at 111–113, 217;

Jung's resignation and departure
  from 122–123; Spielrein as patient at
  113, 120
Burghölzli rule, Jung's breaking of 117

Cäcilie M (Anna Von Lieben) 35
"capercaillie" patient 35
Cardiff Giant 94
Carotenuto, Aldo 119
Carpenter, W 97
Carrol, Lewis 98
Casement, P 91n13
castration anxiety 58, 68, 75
castration complex 139, 192
castration symbol 211
castration theory of femininity 16
catalepsy 7, 9, 108, 110
Charcot, Jean Martin 14; Brouillet's
  painting of 27; Cäcilie M (Anna Von
  Lieben) consultation 35; as director of
  Salpêtriere 25–28; Fliess' visit in Paris
  65; Freud and 22, 28, 29, 163; hypnosis
  used to treat hysteria by 8; Janet as
  successor of 32, 111; lesson of 25–33,
  58; studies in hysterics 141
Carrington, Hereward 97
Chevreul, M E 97
Christian Science 93
clairvoyance 7, 92, 103, 153, 217;
  Waldner's claim to gift of 245
clairvoyant: Ferenczi as 127–129; Ferenczi
  and Freud's study of 149, 151; Seidler
  151, 153–158; Tesla as son of 102;
  writer's imaginations inspired by stories
  of 7; *see also* mind reader; thought
  transference; thought transmission
Clark University, Worcester 121, 127, 135,
  139, 142, 195
Coca Cola 2
cocaine: as anesthetic 1, 20; Fliess'
  recommendations for nasal use of 65,
  68; Freud's dirty syringe dream and
  56; Freud's use of 2–3, 19–20, 57, 86;
  Freud's monograph on 2, 22; Gross'
  addition to 113
Coleridge, S 3
Columbus, Christopher, bones of 94
Conan Doyle, Arthur (Sir) 96, 98
concatenations 28, 36, 72
Coney Island 94
Cook, Florence 104
countertransference, theory of 129

Cremerius, J 216n10
Crookes, William 96, 97, 98, 101, 104, 132
cryptomnesia 104, 177
Cumberland, Stuart 99
Curies, Marie and Pierre 97, 98

Darwin, Charles 65, 97, 188
daydreams *see* dreams
dead, the: communicating with 101, 104,
    105, 141, 150; executed (criminal)
    27; Jung's familial deaths, impact of
    108–111; manifestations (apparitions
    or ghosts) of 103; return of 106; spirits
    of 93; soul murder of 165; telepathic
    automatic writing and 105; world of
    108–111; *see also* mediums
dementia 235
dementia praecox 9, 112–113; Freud's
    patient diagnosed with 175; Jung's
    competition with Freud on subject of
    159; *Psychology of Dementia Praecox*
    (Jung) 121; Spielrein's doctoral thesis
    on subject of 204
demonic possession 5, 93
De Sanctis, Sante 133, 136
Deutsch, Felix 68
Dickens, Charles 96
diorama 94
double consciousness 8, 40
double personality 7
double, the 76, 80n18, 167
dowsing 97
dream-life 8
dreams: Binswanger 115; Bleuler's
    interpretation of own 112; daydreams
    21, 52, 77, 89, 144; as desire (Freud)
    56–60; Frau B 89; Freud's intuiting
    of significance of 69; Freud's own 69,
    71–76, 139; *Interpretation of Dreams*
    (Freud) 53, 56, 61–63, 73, 84, 86–87,
    103, 104–105, 111, 118–119, 132, 150;
    Jung's own 114, 144–148; meanings
    of typical 73; oedipal 15; *On Dreams*
    (Fliess) 82; poetic intuition and 101;
    premonitory 116; prophetic 116, 124,
    133; spirits and 109; telepathic 99; *see
    also* Irma/Irma dream; nightmare
Dreams and Hysteria (Fliess) 83
dreams, slips, and free associations
    138–140
dream-work 55, 88, 116
Dr. Jekyll and Mr. Hyde, story of 7

Drummond, Henry 99
dual personalities 7

Eckstein, Emma 50–51, 55, 64, 68–69,
    76–79
Eckstein, Fritz 50
ecstasies 7, 110, 125, 153
ectoplasm 94, 96
Edison, Thomas 101
Eibenschütz, Ròza 130
Einstein, Albert 63, 102
Eissler, K R 64, 115, 145, 236
Eitingon, Max: Abraham and 224;
    Berlin Psychiatric Institute 202n4; at
    Burghölzli 112; peripatetic analysis of
    200; as suspected KGB agent 229n31
Eliot, George 97
Elisabeth von R (patient) 35
Emmanuel Movement 139
Emmy von N (patient) 30, 35, 51
Episcopal Church, Boston 139
Eros 11, 27, 209
exorcism 5, 105

fairy tales 21, 72, 73, 207
Faraday, Michael 97
Farah Diba 246
Faria (Abbé) 7
Federn, Paul 84, 236, 242n12
Felletàr, E 131
Ferenczi, Layos 160
Ferenczi, Sándor: accompanying Freud
    to America 125, 137–151; aversion to
    "occultist" views 106; bilingualism
    of 130; biographical overview of
    129–130; "brother complex" of 130,
    157; at *Burghölzli* 112; clairvoyance and
    153–158, 168–173; Clara Thompson
    as pupil of 127; closeness with Freud
    63; conversations and confidences with
    Freud 76, 81, 83, 90, 139, 206; as "court
    astrologer" 169; Elma Pàlos and 160–
    162, 175–186, 200, 205–206, 209–210,
    220, 225; "equinoctial storms" of 176,
    177; fear of masturbation and prostitutes
    133; Felletàr, friendship with 131; Frau
    G and 149, 157, 179, 180; Frau Isolde
    and 148–149, 155; Frau Jelinek and
    156–157; Frau Seidler, consultation
    with 151, 157; as Freud's *alter id* 188;
    Freudianism of 128; Freud's eulogy of
    127, 242; Freud's first meeting with 115,

133–135; Gizella Pàlos and 133, 154–155, 157, 160–161, 175–185; Groddeck and 163–164; incestuous impulses of 175–179; interest in spiritualism and the psyche 106, 131–133; jealousies of 158; Jung and 133, 135, 137, 158; Palermo incident and 162–168; physical appearance of 133; on repression 134; Schachter and 133; Schächter-Freud as teacher of 162; "Spiritism" 132; Stein and 133; theory of counter-transference 129; views of Freud 16; *see also* Frau G

Ferenczi, Zsigmond 130, 153, 154, 156
Fiji Mermaid 94
filial passions 163
filial transference 160
filicide 226, 227
First International Conference on Hypnotism 30
Flammarion, Camille 97, 98, 103
Fleishl, E 2
Fliess, Jacob 84
Fliess, Robert 57, 69
Fliess, Wilhelm 34, 49–51, 58; Abraham and 225; accident with Eckstein 82, 210; accusations of plagiarism from 85–90; affection for Freud 76; criticism of "mind reading" 237; disease restricting growth of 65; Freud's break with 167; Freud's correspondence with 71–74; Freud's dependency on 77; Freud's friendship with 169, 176; hostilities with Freud 83; Ida Bondy 55; influence of 119; Melanie Bondy 59n10; invention of psychoanalysis and 61–69; reflex nasal neurosis and 68

Fluss, Emil 17
Fluss, Gisela 16
folie à deux 27
Forel, Auguste 9, 29, 39; Breuer's letter to 47; as president of *International Order for Ethics and Culture* 10, 159
Forsythe case 239, 242
fortune tellers: Adorno on 96; Ferenczi on 235, 238–239, 240; Freud's reflections on the work of 172, 232, 239–240; Salomé on 234–235; visit to Freud by 244–247
Fox, John 93
Fox sisters (Maggie and Kate) 93–96
Fraenkel, Baruch 129
Franklin Commission 6

Franz Joseph (Emperor) 15
Frau Arold 170
Frau B 89
fraud 141, 142
fraudsters 96
Frau Emmy von N 30, 35, 51
Frauen(zimmer) 148
Frau Fluss 16
Frau G (Gizella) 145, 157, 175, 176, 178–180, 186; elder daughter of (Elma) 162, 169, 175, 178
Frau Isolde 148–149, 155
Frau Jelinek 156–157
Fräulein Elma *see* Pàlos, Elma
Frau Mathilde 20
Frau Seidler *see* Seidler (Frau)
Freiburg, Moravia 15, 16, 74, 76, 78
Freudianism 128
Freud, Alexander 84, 87, 130, 134
Freud, Amalia Nathanson 14–15, 57, 206
Freud, Anna (Sigmund's daughter) xi; at 16 years of age 176; at 17 years of age 206; on Alexander Roth and his telepathic wife 236; boyish presentation of 209; decision to study psychoanalysis by 51; dreams by or with 57; on Ego defences 186; on Emma Eckstein 79; on Ferenczi's "mutual analysis" 128; as Ferenczi's pupil 129; Fliess unknown to 63; as Freud's Cordelia 206; as Freud's pupil 208; Freud's "superstitious" remembered by 88; Jung's patient pseudonyms named after 140; permission granted to publish Freud-Fliess correspondence by 64; Salomé's friendship with 212; Salomé's letters to 185
Freud, Anna (Sigmund's sister) 17, 18, 19, 21, 74, 149; birth of 75, 78
Freud, Ernst 206
Freud, Jacob Kolloman 14–15, 80n14; three wives of 206
Freud, Jean Martin 134, 206; *see also* Freud, Martin
Freud, Julius 15; death of 75
Freud, Mathilde 20, 56, 115, 134, 206
Freud, Martha Bernays (Sigmund's fiancée, then wife) 2, 18–22, 26, 29, 89; 34th birthday of 54; "accidental" conception of 6th child (Anna) 51; correspondence between Sigmund and

16, 22, 43, 71; correspondence with mother 45; engagement to Sigmund 44, 89; hallucinatory states of 45; possibility of emigrating 150; pregnancy of 55; Sigmund's first meeting with 14, 20; Sigmund's affectivity towards 63

Freud, Martin 63, 100, 244, 245

Freud, Oliver 206

Freud, Philipp 74

Freud, Sigismund Scholomo (Sigmund): death of brother Julius 167; early life of 14–16; "equinox letter" of 73; end of adolescence of 17; first love of 16–18; Ferenczi's transferential encounter with 133–135; Fliess' accusations of plagiarism against 85; heart trouble (disease) of 67–69; nicotine poisoning/psychosomatic symptoms of 68; *Psychopathology of Everyday Life* 85; *see also* Anna O; dreams; Freud, Martha Bernays; hysteria; Irma; libido; mediums; Pappenheim, Bertha; telepathy; thought transference

Freud, Sophie 55, 57, 206

Freund, Anton von 139, 224, 225

Fröhlich, Peter 189–190; *see also* Gay, Peter

Gabbard, G 204

Galton, Francis (Sir) 32, 59, 97, 98, 99; dream interpretation by 111; reaction test devised by 112

galvanometer test 214

Gambetta, Léon 25

Gassner, Johann J (Father) 5

Gay, Peter 28, 45, 166; life of 189–190; *Life for Our Time* 189

George, Lloyd 188

ghosts 60n13, 106, 125, 190

Goethe, J W 13, 72, 108

Goethe Prize 71

Göring Institute 189, 202n4

Göring, Matthias 202n4

Graf, Max 84, 90

Groddeck, Georg 128, 163–164, 182–184

Gross, H 202n24

Gross, Otto 113, 194, 195

Halley's comet 160

Hall, Stanley 98, 121, 135, 139, 141–142

hallucination 21, 41, 42, 45; deadly 204; Emmy von N 35; as Freud's experience

with 89; neuro-psychosis of defence 36; paranoid 165

Halsted, W S 1

Hauff, Friedericke 7

Heller, Hugo 84, 153

Helmholtz School 13

Herman von Helmholtz Association 70n17

Hirschfeld, Elfriede 138, 174n54, 234

Hitler, Adolf 12

Hitler Youth 202n4

Hitschmann, Eduard 84, 236

Hofmannsthal, Hugo von 11, 210

Home, Daniel D 96

homosexual desire 165

homosexuality: masochistic 197; paranoia and 169; unconscious 166, 180

homosexuals 131, 160–161

homosexual tendencies 163

homosexual transference 76

Houdini, Harry 94

Houston, John 64

Hugo, Victor 97

Hungarian Medical Society 132

Huxley, Thomas 97

Hydesville incident 93–95

hyperaesthesia 7, 98, 153

hypnosis 7–10, 20; Anna O's self-hypnosis 52; Breuer's patient as subject 44; collective 95; Ferenczi's interest in 130, 132, 186; Freud's abandonment of (Elisabeth von R) 35; Freud's return to use of 51; Freud's use of 27–31; hysteria and 3; mind reading and 98; SPR's research on 104; telepathy and 98, 237; three stages of 26; Wittman as subject of 26

hypnotherapy 35

hypnotic sleep 9, 27

hypnotism: hysteria and 26; International Conference on Hypnotism 30; Nancy school of 8; neuro-hypnotism 7; Quimby's public experiments in 92; SPR and 143; telepathic 98

hypnotist: Jung as 109, 111

hypochondria 36, 68

hypocrisy 71, 80

hysteria 8; aetiology of 39, 54; Anna O 39, 41; anxiety 68; arc en ciel of 26; Charcot's study of 8, 25–27; as degenerative organic process 27; dispositional 34; *Dreams and Hysteria* (Freud) 83; Freud's work on 28, 34–37,

58, 59, 140; 'grand' 25; hypnosis and
3; hypnotic state, resemblance to 98;
male 8, 12n16; Mesmer's cure for 5;
psychically acquired 34; retention 40;
Studies on Hysteria (Breuer and Freud)
10, 35–37, 39, 51, 77, 84, 105, 134,
193; sexual origins of 41, 120; 'Therapy
of Hysteria' (Freud) 77; traumatic 8;
treatment of 46, 58
hysteria of defence 36
hysterical symptoms 21–22
hysterics: Blanche Wittmann 26; Jelinek
156; Spielrein 120

ideomotor phenomenon 97, 99
impotence 77, 165, 174n49
incest 11, 35; "Horror of Incest" (Freud)
219, 226; occultism and 124; parricide
and 73; perversions and 116; prohibition
of 59
incest complex 219
incest prohibition 219, 225
incestuous: bond (Freud with daughter)
208; desire 172, 200, 219; element
(Jung with Spielrein) 217; fantasy
207; Ferenczi's countertransferential
incestuous storms 175–183; fixation
239; maternal image 147, 204; ties 205
Infantile Cerebral Paralysis (Freud) 28
infantile complexes 135, 148, 158; *From
the History of Infantile Neurosis*
(Freud) 232
infantile desire 58; Anna O and projection
of 51; fathers as object of 160; revenge-
driven 186
infantile drives 166
infantilism 181, 215
infantile sexuality 11, 140, 220; discovery
of 71–79; Forel's reservations
regarding 39
infantile sexual fantasy 56, 147
International Congress of Neurology,
Amsterdam 194
International General Medical Society of
Psychotherapy 202n4
International Order for Ethics and Culture
10, 159
*Internationale Psychoanalytische
Vereinigung* (IPV) 113, 125n29, 174n47,
199, 203n36
International Psychoanalytical Association
(IPA) 69n7, 125n29, 203n36; Bleuler's

exit from 125n28; Congress of 1912
220; Congress in Berlin in 1922 126n45;
president of 188; transformation into
professional body 224
interpsychic relations 241
IPA *see* International Psychoanalytical
Association
IPV *see* Internationale Psychoanalytische
Vereinigung
Irma/Irma dream (Freud's subject)
51–59, 62–63, 72, 75, 79, 80, 144, 148,
162, 210

James, Henry 32
James, William 32, 98, 104, 105, 140–143,
150
Janet, Pierre 9, 36; as director of
Salpetrière 32; 107; experiments with
automatic writing 111; experiments in
telepathic hypnotism 98; hypnosis of
Léonie by 97; Jones and 194; Jung and
232; interest in occult phenomenon 131
Jastrow, Joseph 97
Jelinek (Frau) 156, 156
Jensen: *Gradiva* 106, 158, 209
Jones, Ernest 2, 3, 17, 112, 220;
biography of Freud 19, 62, 64, 67, 68,
128, 134, 232; courtship of Anna 115;
on "Anna O." 43; on Emma Eckstein
79; life of 188–202; on Freud's interest
in telepathy 238, 241; Loe Kann and
79, 185, 195, 198–200; on nightmare
and superstition 217; *Occultism* 90,
233, 234; planned analysis of Rank
224; Roazen's admiration for 188–189;
Society for Psychical Research and
236; Third Reich and 189; on William
James 143
Jung, Augusta 108
Jung, Carl Gustav 39, 45; association
experiment 140; Easter 1909 and
116–119; break-up with Freud 135,
147; concept of libido 204; decision
to lie to Freud 145; "Divided Self" of
146; dreams (i.e. own dreams recorded
by) 144–146; epistolary confrontation
with Freud 220; Ferenczi and 153–158;
Freud's rivalry with 167; incompatibility
between Freud and 147; infantile
psyche 140; infantile sexual fantasy
and 147–148; James' interest in psychic
experience and 141; Laboratory for

Experimental Pyschology established by 112; libido, questions related to nature of 217; mummies of Bremen Cathedral, visit to see 137; mysticism of 177, 199; in New York City with Freud 138; occultism, mysticism, alchemy explored by 217; physical appearance of 133; Piper as fraudulent medium and 142; as president of IPA 159, 215; *Psychic Conflicts in a Child* 140; Society of Pyschical Research and 143; Spielrein affair 79, 176, 206, 217; spiritualism and 106, 108–125, 131; stable bigamy practiced by 115; theory of complexes 140; Wolff affair 217, 220; Word Association Test 112
Jung, Emma 112, 121, 144, 145, 218–219
Jungians 146

Kabbalah 87
Kandinsky, W 11
Kann, Loe: Eder and 199; Jones and 79, 185, 195, 198–200; Freud and 185, 204, 205, 206, 220
Kann, Louise Dorothea *see* Kann, Loe
Kassowitz, Max 54
Katharina (patient) 35
Kennedy, Richard 93
Kerner, Justinus 7
Kerr, J 111, 126n43, 204
Klein- Hüningen presbytery 111
Klein, Melanie 129
Klimt, Gustav 10, 11
Kluski, Franek 96
knife of Lichtenberg 215
Kohn (Herr) 177
Kokoschka, O 11
Koller, K 1
Königstein, Leopold 1, 18
Kraepelin, Emil 9, 159
Kraepelin school 112
Krafft-Ebing, R 29
Kris, Ernst 64, 69n10
Kronich, Aurelia 35

Laplanche, J 80n13
Laurvik, John N 184
levitation 94, 96
Levy, Lajos 182
Lichtheim, Anna Hammerschlag 53, 54, 55
libido: conflict between Freud and Jung over concept of 204, 215; Eckstein case

and 76; fixation of 121; Freud's concept of 11, 29; Jung's questioning of sexual nature of 217; Jung's reworking of meaning of term 115, 156; Schreber's withdrawing of 165; *Transformations and Symbols of the Libido* (Jung) 177, 218–219, 230
libido theory 171
Liébault, Ambroise-Auguste 8, 9, 30
Lodge, Oliver 98, 101, 102
Lombroso, Cesare 97, 132
Louis XVI 5
lucid sleep 7
Lucy R (patient) 35
Lueger, Karl 65, 70n18

Maeder, Alphonse E 215
magnetism 5, 7, 237; animal magnetism 5–6, 102
Marconi, E 102
Martha N (patient) 204
"Masson case, the" 69n12
Masson, J M 63, 64, 79, 166
masturbation 36, 41; childhood 78; Emma Eckstein 77–78; Ferenczi's fear of 133; Freud 17; Freud's conclusions regarding 36
matriarch 155
matrix (womb) 8
McComb, Samuel (Rev) 140
McDougall, W 98
medium (spiritualism) 94; Bishop's debunking of 99; Ferenczi's encounters with and explanation for 131–132, 178; Flournoy's study of 103–104; Freud's thoughts and research on 31, 150, 237; Janet and 97–98; Jung's explanation for powers of 110–111, 218; language of 104; menopause and 142; psychology of 235; Schnitzler's enthusiasm for 29; Society for Psychic Research and 143; Walter Benjamin on 99; *see also* automatic writing; ectoplasm; Fox sisters; Frau Jelinek; Home, Daniel; levitation; mentalism; Muller; Palladino; Piper; Roth; Smith; Williams
mediumism 143
mediumistic trance 7, 104; *see also* second condition
megalomania 58, 157; sexual 55, 207
Meghnagi, D 15
mentalism 99

mental telegraphy 97–103
Mesmer, Franz A 5–7
mesmerism 98
Meucci, Antonio 99
Meynert, Theodor 9, 14
Miller, A 125n7
Mill, John Stuart 13
mind cure 92
mind-reader: pseudo 99; as readers of
   thoughts 82–84, 99, 169
mind reading 98, 236, 237
mise en abîme 27
Modrzejewski, Teofil 96
Money-Kyrle, R 193
Morse code 102
Morselli, Enrico 97
Moser, Fanny 30, 35
Mozart, Wolfgang Amadeus 5, 70n32, 119
Muller, Catherine-Elise 103, 104
Munthe, Axel 26, 104
Murray, Gilbert 98
Myers, Frederic 97, 98, 104, 105

Napolitano, F 176
Nazi Germany 43, 166, 202n4; books
   burned by 63
Nazism 189; Jung and 202n4
neuroanatomy 13, 20, 22; lectures given by
   Freud 65
neuro-hypnotism 7
neurologists 142; "clinicians' conspiracy"
   of 159; *see also* Ferenczi; Jones; Gross;
   Putnam
neurology 14, 28, 72; International
   Congress of Neurology, Amsterdam 194
neurone theory 1, 9, 10
neuropathology 25
neuro-psychoses of defence 36, 40;
   Freud's book on 91n13
neuroscience 72
neuroses 3, 19; actual 36; adult 158;
   dreams and 230; dream structure and
   52, 57; Fliess' theory of 82; Freud's
   new treatment of 61; Freud's own
   62; *History of Infantile Neurosis*
   (Freud) 232; hysteria as 83; hysterical
   242n8; importance of sexuality in 39;
   infantile roots of 158; "infantilism"
   as 277; mixed 35; in mythology 156;
   obsessional 39, 90, 171; prayer and
   Christian Science as response to 140;
   pre-conscious dream-work tied to 88;

"reflex nasal neurosis" 68; thought
   transference and 168
neurosurgeon *see* Trotter
neurotica (theory of neuroses): Freud's
   disavowal of 73
neurotic patients 2
neurotic patterns 209
neurotic symptoms 212
Niarchos, S 245
Nietzsche, Friedrich 11, 69, 126n30
nightmares 151, 190, 200, 217

Oberholzer, Emil 112
Obersteiner, H 29
occult 125, 142, 149–151; infant and
   mother link and 239; telepathy and 239
occult phenomena 110, 131, 172
occultism 137; American 92–106, 140;
   animal magnetism and 6; Andreas
   von Salomé and 234, 241; black
   tide of 232–234; boundaries of 237;
   clairvoyance and 153; "Dreams and
   Occultism" 241; Ferenczi and 131,
   178, 185; Fliess' mysticism and 66,
   82; Freud's linking of dreams to 86;
   hypnosis and 9; Jones' aversion to 190,
   232–234; Jung and 109–110, 177, 217;
   Jung's conflict with Freud over 123;
   incest and 124; psychoanalysis and 244;
   question of 171; Salomé's curiosity
   regarding 235; of sexual theory 116,
   220; spiritualism and 9; the Uncanny
   effect of and 124; *see also* spiritualism
Occultism (Jones) 90
oedipal: cathexis 122; coincidences
   179–183; conflict 11, 75, 138, 183, 215;
   crisis 68, 219; crossroads 206; images of
   parents 190; transference 176
Oedipus 89–90, 180, 226
Oedipus complex 14, 15, 46, 87, 158, 186,
   219, 227
Oedipus Rex (Sophocles) 73, 75, 225
opium addiction 113
Ossipov, Nikolaj 112
Ouija board 94

paediatrician *see* Kassowitz; Rie
paediatric neurology 28; *see also*
   neurology
paediatrics 56
paedophilia 193
Palladino, Eusapia 96, 97

Pàlos, Elma 160–162, 175–186, 200, 205–206, 209–210, 220, 225
Pàlos, Gizella 133, 154–155, 157, 160–161, 175–185
Pankeyev, Sergei Konstantinov 158
Papini, Giovanni 71
Pappenheim, Bertha 11, 21, 126n30; see also "Anna O"
paranoia 9, 159; homosexuality and 169; hysteria contrasted to 177; Palermo incident and 162–168; psychosis and 165; symptoms of 166; of Schreber 163–167, 169
paranormal, the 96, 98, 101, 104
parapsychology 116, 117, 218
parricide 11, 73, 138, 145, 214, 226, 227
Phantasms of the Living (Gurney, Podmore, Myers) 99
phantasy pregnancy 47n25
Piaget, Jean 119, 126n45
Pinel, Philippe 25
Piper, Leonora E 98, 141, 142, 150
planchette 94
Planck, Max 102
Plato: *Timaeus* 8
platonic relationships 6, 182
Pompidou, G 245
precognition 7, 109, 117
Preiswerk, Helena 110
Prince, Morton 196, 249, 256
Prochnik, G 140, 150
prophecy: Delphic Oracle 89; Ferenczi's discovery of significance of 187, 190, 217; Freud's exploration of 190, 217, 232; Freud's weakness for 86; gypsy 177; the unconscious and 90; unfulfilled 168–173, 234, 239
prophetic dreams 103, 116, 133, 230
propyls 53
Proust, M 100
psychic: associations 36; bisexuality 11; causes of disease 58; centrality of penis envy 211; concatenations 28; cure 5; development 148; dynamism 82; excitement 47n10; fecundity 230; form 75; illness 66; induction 153; life 181, 183; material 37, 212; materialization of psychic stuff 234; mechanism 145; non-differentiation between mother and daughter 235; pathology 45; potentialities 238; process 34, 227; rays 170; reality 14; role 8; sessions 150;

states 31; strength 97; work 90, 153, 168–173
Psychic Conflicts in a Child (Jung) 140
psychic, the (psychology): nucleus of 59; repression of 73
psychical: chain 52; definition of attribute 105; foreign bodies 40; need 41
Psychical or Mental Treatment (Freud) 30
psychical, the 33n9
psychics *see* mediums; Society for Psychical Research
psychoneuroses 50, 73
psychoanalysis: as new term/practice 105
psychopathology 22, 38, 67; Jones' specialization in 193; Laboratory for Experimental Psychopathology (Jung) 112; primal experiences of humanity and 217
Psychopathology of Everyday Life (Freud) 61, 82, 88–90, 105; 1907 edition of 89; 1912 edition of 197; Freud's cabin steward discovered to be reading 138; last chapter of 88, 90; new edition of 116
psychotherapeutics 9
psychotherapy: Breuer and Freud's writing on 36; cathartic 11; church (Christian religion/piety) and 140; Freud's experimentation with 30, 31; International General Medical Society of Psychotherapy 202n4; memories reawakened via 34; as new term/practice 105
Putnam Camp 142
Putnam, James Jackson 139, 141, 192, 196, 198
Puységur (Marquis de) 6

Quimby, Phineas 92–93
Quinodoz, J 166

radiation 98
radiation poisoning 26
radio broadcasting 100, 102
radiology 25, 26
radiotelegraphy 102
railway spine 8
Rank, Otto 84, 128, 212, 215, 223–225; at Café Bauer 236; on incest/incest complex 219; Jones' proposed analysis of 224

readers of thoughts *see* mind readers; thought readers
Rée, Paul 210
Rembrandt van Rijn 27
Rhine Falls 108
Richet, Charles 97–99, 102–103
Rie, Oskar 53
Riklin, Franz 112
Rilke, Rainer Maria 210, 211, 235
Roazen, Paul 119, 136n30, 183, 185; admiration for Jones 188–189
Rodrigué, E 68, 190
Rolland, Romain 87
Rosanes, H 77
Roth, Alexander 236
Roussillon, R 239

Sabourin, P 187
Sache, Die (the cause) 223–225
Sachs, Hanns 90n7, 223, 224, 225, 236
Sadger, Isidor 84
Salem witch hunt 93, 141
Salomé *see* Andreas (von) Salomé, Lou
Salpêtriere, La: brief history of 25; Charcot as director 8, 25–26, 32; Freud's observation of demonstrations at 27; Janet as director 32; Wittman as patient 26
Salyard, A 50
Salzburg, Vienna 65; Psychoanalytic Congress 112, 125, 134, 194; Trotter in 195
Sartre, Jean Paul 64
Saussure, Ferdinand de 104
Schächter 162
Schächter-Freud 162
Schachter, Miksa 132, 133
Schaffhausen 112
Schatzman, Morton 166
Scheftel, Pavel 205
Schiele, E 11
schizophrenia 112, 204; childhood 146; *see also* dementia praecox
Schnitzler, Arthur 10, 29, 30, 210
Schreber, Daniel Paul 158, 163, 165–166, 168–169
Schur, Max 68, 79
Schuschnigg 100
Schwabing 194
Schwab Paneth, Sophie 65
science fiction, invention of genre of 103
séance 97, 104, 105, 140; Bleuler and 218; Ferenczi and 131; Jung and 108–111,

124, 218; table-turning 96; trance atmosphere of 69n3
Searles, H 166
"secondary superstition" 96
second condition 7–9; *condition seconde* 34; dream as mother of all second conditions 51
second industrial revolution 10, 102
second personality 6; sexual repression and 111
"secret arts" 177
secret Committee 217–228; *Die Sache* (the cause) and 223–225; "Paladins" of 224
secretions 65, 77
secret language 16
secret names 17
"Seidler dossier" 160
Seidler (Frau) 151, 153–158, 235–236
Semmelweis, Ignaz Philipp 130
Siamese twins 94
Sidgwick, Henry 98
Silberstein, Eduard 16, 17, 71, 76
Simon, James 222
Skues, R 45
Smith, Hélène 103, 104, 217, 218
Society for Psychical Research 104–105, 141; Bergson as president of 237; Ferenczi's presentation to 236; Freud as corresponding member of 143, 178; founding of 98; Jung as member of 143
somnambulism: Ferenczi's interest in 156; Flournoy on 103, 218; Jung's interest in 110, 218; symptoms of 156
somnambulists 5–12; *see also* Hauffe; Mesmer
Spielrein, Sabina 79, 185; in *Burghölzli* as patient 113–114, 120; Freud and Jung's exchange of letters regarding 137–138, 145, 147; Freud's apology to 124; Freud's planned analysis of 204–205; Jung's affair with 119–125, 137, 196, 198, 217–218; marriage of 205
spiritualism: American origins of 7; Cumberland's hostility to 99; Crooke's discovery of 104; Ferenczi's interest in 131–133, 241; hypnosis and 9; hypotheses of the unconscious and 103–106; Jung's interest in 106, 108–111; Jung's mother's family's immersion in 108; as new sphere of serious scientific research 97; number of adherents in 1870 96; psychological research into 217; rise in America of 92–97, 140;

sociological explanation for popularity of 96; spread to Old World of 92; *Yorkshire Spiritual Telegraph* 101
spiritism 7; Jones' thoughts on 143; popularity in US of 150; *Studies on Spiritism* (Tanner) 141
"Spiritism" (Ferenczi) 131, 142
Stead, W T 105
Stevenson, R L 7
Strachey, Alix 64
Strachey, James 3
Staudenmeyer, Ludwig 235, 242n11
Steiner, Maximilian 90n7
Stein, Philippe 133
Strindberg, August 26, 210
St Vincent de Paul 25
suicide 43, 84, 174n38, 179, 183; Dora Breuer's death erroneously identified as 48n31; Jung's unconscious impulse towards 108; "three caskets" dream and 207–208
suicide attempt: Bertha Pappenheim 46
Szymborska, W 101

talking cure 11, 21–22, 27, 58; *see also* automatic talking
Tanner, Amy 141, 142
tarot card games 18, 54, 158
Tausk, Viktor 211, 236, 237; influencing machine of 242n11
telegraph 98; Ministry of Posts and Telegraph 102; wireless 102; Yorkshire Spiritual Telegraph 101
telegraphy: mental 97–103; radiotelegraphy 102; Twain's thoughts regarding 101
telekinesis 96
"telectrophone" 99
telepathy 98–99; Bergson on 237; "cosmic ether" and 102; dog's ability 136n19; *Dreams and Telepathy* 190; Ferenczi's interest in 132, 155, 158, 241; Flammarion's belief in 103; Flournoy's skepticism regarding 104; Freud's interest in 151, 208, 232, 234, 238, 239, 241; hypnosis and 9, 98; Jung's capacity for 109; Jung's interest in 111; Mark Twain's support of 101; Myers' coining of term 98; question of 235–238; SPR and 143, 237
telephone 98, 99; Freud's dislike of 100; Twain's thoughts regarding 101
telephone number 14362, Freud's interpretation of dream of 118–119

telescope 98
Tennyson, Alfred 98
Tesla, Nikola 102
Thanatos 27
Third Reich 189; *see also* Nazi Germany
three caskets, theme of (Freud) 207
third sphere of psychoanalysis 241
Three Graces 207
Titanic 102, 105
Tobler, G C 13
Tolstoy, Leo 210
totem and taboo 225–228; "Return of Totemism in Childhood" (Freud) 226, 238
"Taboo and Emotional Ambivalence" (Freud) 226
Totem and Taboo (Freud) 238, 230, 232, 238
Toulouse-Lautrec, Henri 26
training analysis 199–202, 224
transmission of knowledge: Freud and 226; unconscious 227, 228
transmission of psychic processes 238
transmission of psychoanalysis: Freud and 212–215; intergenerational 204–215
transmission of thought: intergenerational 225–228
Thompson, Clara 127, 129, 135n1
thought transference 150, 168, 170, 177, 179, 187; Ferenczi's writings on 240, 242; *Few Recent Observations (On the Theme of Thought Transference)* 160; Freud's writings on 242; Goethe/Tobler on 13; Salomé on 234
thought transmission 6–7, 30, 97; Jones on 232, 236, 238; Ferenczi's and Freud's interest in 149, 154, 158, 187, 190, 241; Ferenczi's legacy of 129; Ferenczi's paper on 160–161, 181; Flammarion's collection of examples of 103; Freud's experience with (Forsythe case) 239; Freud's return to theme of 168, 170, 234; hypnosis and 98; Jung's interest in 110; SPR's study of 98–99; *see also* telepathy
thought induction 161, 169, 177, 238
thought reader 102, 161; reader of thoughts 82–84, 169
thought reading 156, 161, 169, 186, 238; reading of thoughts 124
Tourette, Gilles de la 27
trance: Faria's ability to induce 7; hypnotically induced 26, 98; séance

and 69n3; spirit mediums' entry into 94, 96, 104, 141, 150, 217
trance crisis 110
trance states 110, 111
Trotter, Wilfred 193, 195, 199
Tulp, Nicolaes 27
Turkish War 72
Twain, Mark 93, 98, 101

Uncanny (*Unheimlich*) 18, 124, 233; *Heimlich-Unheimlich* 23n16
unconscious, the: first hypotheses about 103–106; Freud as interpreter of 15; Freudian hypothesis of 104; Freud's analysis of own 72; Freud's clarification of term 105; Freud's reading of Jung 114; Freud's "solving" of 89; jokes and 61; Jung's psychology of 110; occultism and 103; prophetic and telepathic properties of 86; individual and group xii; repressed 11; transferential infantile sexuality and 104; Vienna's role in explorations of 5–7, 10
unconscious: desire 52; fantasy 58; homosexual transference 76; hostility 62; identification 78; sexuality 38, 39, 42; suicidal impulse 108
underworld, the 150
University of Zurich 210

vampires 190
Van Emden 112
Verne, Jules 103
Vorsicht 66

Waldeyer, W 1
Waldner, Francesco 245, 246, 247
water divining 97
Wedekind 210, 211
Wednesday Society 84, 134, 153, 154, 211, 214, 233, 236
Weimar, Germany 176, 178, 210
Weimar Congress 199, 200, 211, 218
Weiss, Edoardo 232m 241, 242n7
Weiss, Ilona 35
Weisz, Ehrich 94
Wells, H G 103
werewolves 190
Williams, Charles 97
willing game 99
Winnicott, Donald W 14, 108, 129, 145–146; spatula 198
Wittmann, Blanche 26
witches and witchcraft 5, 8, 93, 96, 141, 190
witch trials 5; *see also* Salem witch hunt
Wolff, Antonia 217, 220
Wolf Man case 158
Word Association Test (Jung) 112
writing cure 71–76
Wundt, Wilhelm 9, 135

Yeats, William Butler 98

Zurich Clinic 111–113, 122, 133, 159, 194; *see also Burghölzli* psychiatric hospital
Zurich University 9; *see also* University of Zurich
Zweig, Stefan 44, 45, 92

For Product Safety Concerns and Information please contact our EU
representative GPSR@taylorandfrancis.com
Taylor & Francis Verlag GmbH, Kaufingerstraße 24, 80331 München, Germany

www.ingramcontent.com/pod-product-compliance
Lightning Source LLC
Chambersburg PA
CBHW050632280326
41932CB00015B/2618